英汉烟草农业词汇

叶协锋 主编

黄河水利出版社
·郑州·

图书在版编目(CIP)数据

英汉烟草农业词汇/叶协锋主编. —郑州:黄河水利出版社,2016.9
ISBN 978-7-5509-1556-5

Ⅰ.①英… Ⅱ.①叶… Ⅲ.①烟草-词汇-英、汉 Ⅳ.①S572-61

中国版本图书馆 CIP 数据核字(2016)第 228356 号

出 版 社:黄河水利出版社
　　　　地址:河南省郑州市顺河路黄委会综合楼 14 层　邮政编码:450003
发行单位:黄河水利出版社
　　　　发行部电话:0371-66026940、66020550、66028024、66022620(传真)
　　　　E-mail:hhslcbs@126.com
承印单位:河南省瑞光印务股份有限公司
开本:850 mm×1 168 mm　1/32
印张:12.5
字数:395 千字　　　　　　　　　印数:1—1 000
版次:2016 年 9 月第 1 版　　　　　印次:2016 年 9 月第 1 次印刷

定价:38.00 元

《英汉烟草农业词汇》编委会

主　编　叶协锋

编写人员　(按姓氏笔画顺序排列)

于晓娜　付仲毅　李志鹏

张晓帆　郑　好　周涵君

宗胜杰　凌天孝　管赛赛

前言

近年来，随着科学技术和社会的迅速发展，在科研和教学领域中，相邻学科之间的交叉渗透愈来愈多。因此，一本词汇量多、学科跨度大、涉及范围广的英汉烟草农业词汇，较便于读者应用。为此，我们大量查阅了新近出版的有关词汇、著作、期刊等，大量收集了有关基本词、常用词和新词共计 12 000 多条。收录的词汇主要围绕烟草农业，部分涉及烟草工业。其中烟草农业主要围绕烟草类型、烟草育种、烟草栽培、烟草采收与调制、烟草植保、烟叶分级、烟叶复烤、烟草植物学、烟草水分与灌溉、烟草产量与品质、土壤质量、土壤物理、土壤化学、土壤生物与生物化学、土壤养分、植物营养与肥料、土壤生态、土壤侵蚀与水土保持、土壤环境污染与修复、土壤与全球变化、土壤研究的技术与方法、农业生物技术、种植制度、土壤耕作等。

本书由叶协锋任主编，主要编写人员有于晓娜、付仲毅、李志鹏、张晓帆、郑好、周涵君、宗胜杰、凌天孝、管赛赛等。

本词汇可供广大从事烟草科学研究和教学的相关人员，以及烟草专业的本科生、研究生等参考使用，同时也可供对烟草农业感兴趣的广大读者阅读。为便于读者查询,书后附有汉文索引。

烟草农业涉及的学科领域众多,且本词汇涉及面较广,限于编者的学识和水平,书中有失偏颇乃至错误之处在所难免,恳请广大读者批评指正。

编　者

2016 年 8 月

使用说明

1. 本词汇英汉对照部分按英文字母次序编排。汉文索引部分按汉文拼音字母次序编排。词目中的数字、符号均不参加排序。

2. 缩略语后用(=)的形式表示出缩略语的全称。

3. 用(=)的形式表示出该词的同义词。

4. 释义中不同词义中间用";"分开。

5. 圆括号()表示词义中可省略的词。

6. <商>表示为商品名。

目 录

英汉对照

A	……… 1	J	……… 111	S	……… 192
B	……… 20	K	……… 111	T	……… 225
C	……… 28	L	……… 113	U	……… 236
D	……… 50	M	……… 126	V	……… 238
E	……… 64	N	……… 142	W	……… 241
F	……… 75	O	……… 151	X	……… 247
G	……… 87	P	……… 159	Y	……… 248
H	……… 94	Q	……… 180	Z	……… 249
I	……… 102	R	……… 181		

汉文索引

A	……… 250	J	……… 292	R	……… 323
B	……… 251	K	……… 300	S	……… 325
C	……… 256	L	……… 303	T	……… 337
D	……… 263	M	……… 308	W	……… 349
E	……… 271	N	……… 311	X	……… 353
F	……… 272	O	……… 315	Y	……… 359
G	……… 278	P	……… 315	Z	……… 374
H	……… 286	Q	……… 318		

英汉对照

A

A horizon 腐殖质层(淋溶层;A层)
a method for the determination of pyrethroids residues in tobacco leaves 烟叶中拟除虫菊酯杀虫剂残留量的测定方法
A. I. (= artificial insemination) 人工授精
AA (= allyl alcohol) 烯丙醇
AAFS (= atomic absorption flame spectrometer) 原子吸收火焰分光计
AAOCF(= Association of American Organic and Compound Fertilizers) 美国有机及合成肥料协会
AAPB (= American Association of Pathologists and Bacteridogists) 美国病理学家及细菌学家协会
AAS(= atomic absorption spectrometry) 原子吸收光谱测定法
AAT(= acetylanatabine) 乙酰基新烟碱
ABA(= abscisic acid) 脱落酸
abieninic acid 冷杉酸
abienol 冷杉醇
ability of germination 发芽能力
ability to retain moisture 保湿力
abiotic components 非生物组分
abiotic degradation 非生物降解
abiotic diseases 非生物性病害
abiotic environment 非生物环境
abiotic factor 非生物因素
abnormal development of tissues 植物组织发育不正常
abnormal early ripening 异常早熟
abnormal metabolism 异常代谢
abnormal seedling 非正常苗
abnormal sprout 非正常芽
ABOCF (= Association of British Organic and Compound Fertilizers) 英国有机及合成肥料协会
abolition of production 取消种植

· 1 ·

aborted 发育不全的
abortive 败育的;发育不全的
abortive ovule 败育胚珠;不完全胚珠
abortive pollen 败育花粉
abortive seed 败育种子;瘪籽
above sea level 海拔
above-ground pest 地上害虫
abridged 节略的
abridged drawing 略图
abridged general view 示意图
abridged spectrophotometer 滤光光度计
abscisic acid 脱落酸
abscissa 横坐标
abscission of leaves 叶子脱落
absolute age of soil 土壤绝对年龄
absolute alcohol 无水酒精
absolute difference 绝对差
absolute drought 绝对干旱
absolute dry weight 绝对干重
absolute duty of water 低限水量(指植物生长所必需的)
absolute ether 无水(乙)醚
absolute ethyl alcohol 无水乙醇
absolute frequency 绝对频数
absolute humidity 绝对湿度
absolute maximum fatal temperature 最高绝对致死温度
absolute maximum temperature 绝对最高温度
absolute minimum fatal temperature 最低绝对致死温度
absolute minimum temperature 绝对最低温度
absolute moisture content 绝对含水量
absolute moisture of soil 土壤绝对湿度
absolute vacuum 绝对真空
absolute volume 绝对容积
absolute water capacity 绝对持水量
absolute water content 绝对含水量
absolute yield 绝对产量
absorbance index 吸光系数;吸收系数
absorbance unit 吸光度单位;吸收度单位
absorbancy (= absorbance) 吸光度;吸收度
absorbed amount 吸收量
absorption ability of nutrient 吸肥力
absorption amount of nutrient 养分吸收量
absorption coefficient 吸收系数
absorption of nutrient 养分吸收
absorption of ultra-violet 紫外(线)吸收
abstain from smoking 戒烟
abstraction of moisture 水分排除
abstraction reaction 萃取反应
abundance of water 丰水
abundant and full body (香气)丰满
accelerated germination test 加速发

芽试验
accelerated weathering 加速风化（作用）
accelerating effect 促进效应
accelerating germination 催芽
accept 承兑;接受;认付;验收
acceptability 接受性;合格率;可接受性;可承兑性
acceptability of leaf grade （烟叶）等级合格率
acceptable quality level 合格质量水平
acceptable variety 合格品种
acceptance check 验收
acceptance region 接受域
accessory seed 混杂种子
accommodation of pollutants 污染物的容纳
accompanying cationic 相伴阳离子
accompanying weeds 伴生杂草
Accotab <商>除芽通
accumulated temperature 积温
accumulation 堆积（作用）
accumulation of nutrient 养分积累
accumulation of organic substance 有机质积累
accumulator plant 积累植物
accuracy of operation 操作准确性;操作精确性
accuracy of seeding 播种精度
acephate <商>高灭磷;杀虫灵

2-acetofuran 2-乙酰呋喃
2-acetopyrrole 2-乙酰吡咯
acescency 微酸味
acescent 微酸的
acetacetic acid 乙酰乙酸
acetacetic ester 乙酰乙酸乙酯;乙酰乙酸酯
acetal 乙缩醛;缩醛
acetaldehyde 乙醛
acetamide 乙酰胺
acetate 醋酸盐;乙酸盐
acetate fiber 醋酸纤维
acetate fiber filter 醋纤滤嘴
acetate filter rod 醋纤滤棒
acetate filter rod maker 醋纤滤棒成型机
acetate rayon process 醋酸纤维丝束工艺
acetate staple fiber 醋酸短纤维
acetate tow 醋纤丝束
acetate tow filter 醋酸纤维丝束型滤嘴
acetic acid 乙酸;醋酸
acetic aldehyde 乙醛
acetic fiber 醋酸纤维
acetidin (= ethyl acetate) 乙酸乙酯
acetone 丙酮
acetyl furan 乙酰呋喃
acetyl nornicotine 乙酰基去甲基烟碱;乙酰降烟碱

acid 酸
acid attack 酸侵蚀
acid brown soil 酸性棕壤
acid clay 酸性黏土
acid deposition 酸性沉积
acid humification 酸性腐殖化(作用)
acid humus 酸性腐殖质
acid insoluble nutrient 酸不溶性养分
acid ion 酸性粒子
acid oxide 酸性氧化物(成酸氧化物)
acid potassium sulfate 硫酸氢钾;酸式硫酸钾
acid purplish soil 酸性紫色土
acid rain 酸雨
acid resistant 抗酸性;耐酸的
acid sodium sulfate 硫酸氢钠
acid soil 酸性土壤
acid soluble nutrient 酸溶性养分
acid solution 酸性溶液
acid sulphate soil 酸性硫酸盐土;酸性土
acid tolerant 耐酸性
acid tolerant microorganisms 耐酸微生物
acid toxicity 酸害
acid-base balance 酸碱平衡
acid-base equilibrium 酸碱平衡
acid-base indicator 酸碱指示剂
acid-base reaction 酸碱反应
acid-base system 酸碱体系
acid-base titration 酸碱滴定法
acid-dichromate method 酸性重铬酸钾法
acid-fast 抗酸的
acid-forming bacteria 产酸细菌
acidic oxide 酸性氧化物(成酸氧化物)
acidification 酸化,酸化作用
acidification of land 土地酸化
acidification process 酸化过程
acidity 酸度(酸性)
acidity effect 酸度效应
acidophil(= acidophile) 嗜酸生物
acidophile bacteria 嗜酸菌
acidophiles 嗜酸微生物
acidophilic microorganisms 嗜酸微生物
acidoresistant 抗酸性
acidulate 酸化
acidulation 酸化
aciduric 耐酸性
aciduric bacterium 耐酸细菌
acid-washed quartz sand 酸洗净的石英砂
acquired resistance 获得抗病性;后天抗病性
acridity 辛辣;辛辣味
acronecrosis 向顶性坏死;顶端坏死
acrophyll 顶生叶

acrospire 初生叶;初生茎
across-slope ploughs 横坡耕作
Acta Tabacaria Sinica 中国烟草学报
actinomyces 放线菌
actinomycete 放线菌
activated acetic acid pathway 活性乙酸途径
activated carbon 活性炭
activated carbon fiber 活性炭纤维
activated charcoal 活性炭
active absorption 主动吸收
active acidity 活性酸度
active carbon 活性炭
active component 活性组分
active constituent 有效成分
active fertilizer 速效肥料
active humus 活性腐殖质
active ingredient 有效成分;活性成分
active ion uptake 离子主动吸收
active microflora 活跃性微生物区系
active nutrient 活性养分
active organic matter 活性有机质
active pore 活性孔隙
active porosity 有效孔隙度
active smoking 主动吸烟
active substance 活性物质
active transmitting porosity 有效透气孔隙度
active transport 主动运输
active unit 活力单位
active uptake 主动吸收

active uptake area （根系)活跃吸收面积
active volume 有效容积
activity of soil organism 土壤生物活性
actual capacity 有效容量
actual effect of control 实际防治效果
actual energy consumption 实际能耗
actual evapotranspiration 实际蒸散
actual fertility 实际肥力
actual significance level 实际显著性
aculeus 针刺;刺状产卵器;螯刺
acute hazard 急性危害
acute-toxic pesticide 剧毒农药
adaptability 适应性
adaptation 适应
adapted breed 适应品种
adapted soil 适应土壤
adaptive capacity 适应能力
added quantity 添加量
addicting drug 致瘾药物;成瘾剂
addition 添加;添加物
additional fertilizer 追肥
additional manuring 追肥
additive action of genes 基因加性作用
additive effect 加性效应;累加效应
additive gene 加性基因
additive genetic variance 加性遗传变量;加性遗传方差
adenosine triphosphatase 三磷酸腺苷酶

adenosine triphosphate(ATP)　三磷酸腺苷
adequate　充足
adequate amount　适量
adequate application　适量施肥
adequate fertilization　合理施肥
adequate irrigation　适宜灌溉;合理灌溉
adequate management　适当管理
adequate nutrient　足够养分
adhesion　黏附(现象)
adhesive water　黏附水
adjustment of soil acidity　土壤酸度调整
administration of tobacco monopoly　烟草专卖管理
administrative map　行政区划图
adnascent plant　寄生植物
adsorb　吸附(作用)
adsorbability　吸附性;吸附能力
adsorbed boron　吸附态硼
adsorbed copper　吸附态铜
adsorbed zinc　吸附态锌
adsorbent　吸附剂
adsorbing capacity　吸附容量
adsorption　吸附;吸附(作用)
adsorption ability　吸附力;吸附能力
adsorption attraction　吸附引力
adsorption dynamics　吸附动力学
adsorption water　吸附水
adult period　成虫期;成株期

adult plant resistance　成株抗性
adventitious root　不定根
adventive bud　不定芽
adverse factor　有害因素;不利因素
adverse growing condition　不利的生长条件
adverse soil conditions　不利的土壤条件
aeolian accumulation　风力堆积
aeolian deposit　风积物
aeolian erosion　风蚀(作用)
aeolian process　风积过程
aeolian soil　风积土
aerating cultivation　通气培养
aeration　换气;通风
aeration and agitation　通气搅拌
aeration cooling　通气降温;通风降温
aeration porosity　通气孔隙度
aeration status　通气状况
aeration-drying　通风干燥
aerial contamination　大气污染
aerial growth　地上部分的生长
aerial mycelium　气生菌丝
aerial part　地上部分
aerial remote sensing　航空遥感
aerial root　气生根
aerial spraying (= aerial application)　空中喷洒
aerobes　好氧微生物(好氧菌)
aerobic bacteria　需氧性细菌;好氧

性细菌
aerobic composting 需氧堆肥
aerobic condition 好氧条件,好气条件
aerobic decomposition 好气性分解
aerobic degradation 好氧降解
aerobic denitrification 好气反硝化作用(好气脱硝作用)
aerobic fermentation 好气性发酵
aerobic incubation 好气培养
aerobic nitrogen-fixing bacterium 好气固氮细菌
aerobic respiration 有氧呼吸;需氧呼吸
aerobic seed 好气性种子
aerobic treatment 好气性处理
aerobicorganism 好氧微生物
aero-bioreactor of organic garbage 有机垃圾好氧生物反应器
aeromicrobe 需氧微生物
aerophotogrammetric survey 航空摄影调查
aerosol 气溶胶
aerosol pesticide 气雾剂农药
aerotolerant 耐氧的
aerotolerant anaerobe 兼性厌氧菌
aerotolerant bacteria 耐氧细菌
aerotropism 向气性;向氧性
aetiolation 黄化现象;徒长
affected 受感染的;受侵害的;受影响的

afforestation 造林
after taste 余味
aftercrop 后作
after-effect 后效
after-fermentation 二次发酵;再次发酵;后发酵
after-mature 后熟
after-ripening seed 后熟种子
after-taste 余味;后味
after-treatment 补充处理;后处理
agar 琼脂
agar block method 琼脂块法
agar colony 琼脂菌落
agar culture media 琼脂培养基
agar diffusion test 琼脂扩散试验
agar layer method 琼脂层法
agar plate 琼脂平板
age 老化;陈化;年龄;醇化
age of leaf 叶龄
age of seed 种子年龄
aged leaf 醇化烟叶;陈化烟叶
aged seed 老化的种子;陈种子
aged tobacco 醇化烟草;陈化烟草
ageing(= aging) 老化;陈化;醇化
ageing time 醇化时间;老化时间
agglutinate 凝集;烧结
aggregate 团聚体
aggregate complex 复合团聚体
aggregate formation 团聚体形成(作用)
aggregate of soil 土壤团聚体

aggregate stability 团聚体稳定性

aggregate structure 团粒结构

aggregation 聚集(作用);群聚的;族聚;团聚作用

aggregation state 团聚状态

aging condition 醇化条件

aging of plant 植物老化

aging of seed 种子老化

aging process 醇化过程

aging time 醇化时间;老化时间

agitator 搅拌器

agrarian area 耕地面积

agric horizon 耕作淀积层

agriclimatic map 农业气候图

agricultural amelioration 农业土壤改良

agricultural area-source pollution 农业面源污染

agricultural chemistry 农业化学

agricultural drain 农田排水沟

agricultural ecology 农业生态学

agricultural ecosystem 农业生态系统

agricultural environment pollution evaluation information system 农业环境污染评价信息系统

agricultural experiment 农业试验

agricultural film 农用薄膜

agricultural greenhouse gas 农业温室气体

agricultural information science 农业信息科学

agricultural information system 农业信息系统

agricultural investment 农业投资

agricultural land evaluation 农田评价

agricultural land gradation 农用地分等

agricultural land resource information system 农耕地资源信息系统

agricultural land resources 农地资源

agricultural land suitability evaluation 农业用地适宜性评价

agricultural lime 农用石灰

agricultural map 农业地图

agricultural microbiology 农业微生物学

agricultural microclimate 农田小气候

agricultural output 农产品产量;农业产出

agricultural pesticide 农药

agricultural pollutant 农业污染物

agricultural pollution 农业污染

agricultural processing waste 农业加工废物

agricultural remote sensing 农业遥感

agricultural residue 农业残渣

agricultural resources 农业资源

agricultural soil moisture characteristics 农田土壤水分特性

agricultural test 农业试验

agricultural waste 农业废弃物

agricultural waste treatment 农业废

弃物处理
agricultural wastewater 农业废水
agricultural water 农业用水
agricultural water resources 农用水资源
agricultural water-saving 农业节水
agricultural yield forecast system 农业产量预报系统
agriculture and pasture interlaced region 农牧交错地
agriculture chemicals 农药
agriculture ecological zone 农业生态区
agriculture in greenhouse 温室农业
agriculture resource exploitation 农业资源开发
agriculture science and technology garden 农业科技园
agriculture soil 农业土壤
agriculture structure adjustment 农业结构调整
agriculture-animal husbandry ecotone 农牧交错地
agriplast 农用塑料
agritol (= bacillus thuringiensis) <商>苏云金杆菌
agro- 田野;农业;作物
agrobacterium 土壤杆菌;农杆菌
agrobiology 农业生物学
agrochemical map 农业化学图
agrochemical research methods 农业化学研究方法
agrochemicals 农用化学品(包括化肥、农药等)
agrochemistry 农业化学
agroclimatic map 农业气候图
agroecological pattern 农业生态模式
agroecology 农业生态学
agroecosystem 农业生态系统
agroecosystem ecology 农业生态系统生态学
agrogeological map 农业地质图
agromelioration 农业土壤改良
agro-meteorological 农业气象灾害
agronomic efficiency 农学效率
agronomic remediation 农艺修复
agronomic trait 农艺性状
agronomy 农艺;农学
agropedic degradation 农业土壤退化
agropedogenesis 农业土壤发生
agro-plastic film pollution 农业塑料薄膜污染
Agrotis ypsilon Rottemberg 小地老虎
agrotopoclimatology 农业地形气候学
agrotype 农艺类型
air blast high speed sprayer 高速喷雾器
air circulating bulk curing barn 空气循环密集烤房
air cleaning 空气净化
air contamination 大气污染

air current 气流
air dry weight 风干重
air drying 风干
air flow 气流
air flow redrying 气流复烤
air high speed sprayer 高速喷雾器
air humidity 空气湿度
air infection 空气侵染
air inlet 进风口
air leakage of barn 烤房漏气
air oven 烘箱;烤箱
air permeability （卷烟纸等的）透气度;透气性
air phase 气相
air phase volume 气相容积
air pollution 大气污染
air pollution control 空气污染控制
air porosity 通气孔隙度
air purification 空气净化
air quality 空气质量
air quality control standard 大气质量控制标准
air separation 风选,空气分离
air settlement 大气沉降
air stream 气流
air survey 航空测量(航空调查)
air-borne infection 空气传染
airborne pollution 空气污染
air-cured 晾制的
air-cured tobacco 晾烟
air-cured types 晾烟型

air-cured variety 晾烟品种
air-curing 晾制
air-curing barn 晾房
air-dried condition 风干状态
air-dry sample 风干样品
air-dry soil 风干土
air-filled porosity 通气孔隙度
airlift fermentor 气升式发酵罐
air-plant interface 大气－植物界面
air-seasoning 风干
air-tight pile fermentation 密封堆积发酵
Ala 丙氨酸;氨基丙酸
alanine 丙氨酸;氨基丙酸
Albic bleached soils 白浆土
albic bleaching 白浆化作用
albinism 白化(现象);白化病
albino chlorophyll deficiency 白化叶绿素缺乏症
albino tobacco tissue 白化烟草组织
alcali 碱;强碱
alcohol acid 醇酸;羟基酸
alcohol aldehyde 醇醛;羟基醛
alcohol fermentation 乙醇发酵;酒精发酵
alembic 蒸馏罐
alfalfa 苜蓿,紫花苜蓿
alfisol 淋溶土
alidade 照准仪
alkalescence 微碱性;弱碱性
alkalescency 微碱性

alkali fusion 碱熔融
alkali metal 碱金属
alkali tolerance 耐碱性
alkali-affected soil 碱化土壤
alkali-earth metal 碱土金属
alkali-hydrolyzable nitrogen 碱解氮
alkaline 碱性的
alkaline land 碱地
alkaline phosphatase 碱性磷酸(脂)酶
alkaline soil 碱性土
alkaline-calcareous soil 碱性石灰质土
alkalinity 碱性(碱度)
alkalinization 碱化作用
alkaliphiles 嗜碱微生物
alkaliphilic microorganisms 嗜碱微生物
alkali-resistance 抗碱性
alkalitolerant microorganisms 耐碱微生物
alkalization 碱化
alkalized soil 碱化土壤
alkaloid 植物碱;生物碱
alkaloid accumulations 生物碱积累
alkaloid biosynthesis 生物碱的生物合成
alkaloid content analyzer 生物碱测定仪;生物碱分析仪
alkaloid degradation 生物碱降解
alkaloidal 生物碱的
allele linkage analysis 等位基因连锁反应(分析)
allele specific restriction analysis 等位基因特异性限制内切酶分析
allelic complement 等位(基因)互补
allelic diversity 等位(基因)多样化
allelopathic substance 他感作用物质
allelopathy 化感作用
allelosomal 等位染色体(的)
alley cropping system 农林复合系统
allitic enrichment 富铝化(作用)
allitization 富铝化(作用)
allogamous 异花受精的;异体受精的
alluvial 冲积的
alluvial deposit 冲积物
alluvial soil 冲积土
alluvium 冲积物;冲积层;冲积土
allyl alcohol 烯丙醇
along-slope ploughs 顺坡耕作
alpine climate 高山气候
alternaria alternata 链格孢菌(烟草赤星病菌)
alternate freezing and thawing 冻融交替
alternate phyllotaxis 互生叶序
alternate yellow and green strips 黄绿相间条纹
alternating system of farming 农业轮作制
alternating temperature germination ch-

·11·

amber 变温发芽箱
alternation of drying and wetting 干湿交替(作用)
alternation of freezing and thawing 冻融交替(作用)
alternation of generations 世代交替
alternative hypothesis 对立假设
altimeter 高度计(高程计)
aluminium 铝
aluminium oxide 氧化铝
aluminosilicate 铝硅酸盐
aluminum toxicity 铝毒害
ambient air quality 环境空气质量
ambient air quality standard 环境空气质量标准
ambient conditions 环境条件
ameliorative effect 改良效应
ameliorative measure 改良措施
American blended-type cigarette 美式混合型卷烟
amidase 酰胺酶
amide 酰胺
amide nitrogen 酰胺态氮
amide nitrogen fertilizer 酰胺态氮肥
amination 胺化作用
amine 胺
amino acid 氨基酸
amino acid analyzer 氨基酸分析仪
amino acid composition 氨基酸组成;氨基酸成分

amino acid nitrogen 氨基酸态氮
amino nitrogen 氨基氮
amino sugar 氨基糖
amino sugar nitrogen 氨基糖氮
amino-N 氨基氮;胺氮
aminotransferase 转氨酶
ammonia 氨
ammonia emission 氨挥发
ammonia nitrogen 氨态氮
ammonia nitrogen fertilizer 氨态氮肥
ammonia oxidizing bacteria 氨氧化细菌
ammonia toxicity 氨毒害
ammonia volatilization 氨挥发
ammonia volatilization loss 氨挥发损失
ammonia water 氨水
ammoniacal nitrogen 氨态氮
ammoniacal nitrogen fertilizer 氨态氮肥
ammoniation 氨化(作用)
ammonification 氨化(作用)
ammonifiers 氨化细菌
ammonifying bacteria 氨化细菌
ammonifying capacity 氨化能力(氨化量)
ammonifying intensity 氨化强度
ammonium 铵
ammonium bicarbonate 碳酸氢铵
ammonium carbonate 碳酸铵

ammonium chloride 氯化铵
ammonium citrate soluble phosphorus 柠檬酸铵溶性磷
ammonium dihydrogen phosphate 磷酸二氢铵
ammonium fertilizer 铵态氮肥
ammonium fixation 铵固定
ammonium fixation by clay mineral 铵的黏土矿物固定作用
ammonium humate 腐殖酸铵
ammonium hydrogen carbonate 碳酸氢铵
ammonium in soil solution 土壤溶液中铵
ammonium molybdate 钼酸铵
ammonium nitrate 硝酸铵
ammonium nitrogen 铵态氮
ammonium nitrogen fertilizer 铵态氮肥
ammonium phosphate 磷酸铵
ammonium sulfate 硫酸铵
ammonium-containing fertilizer 含铵肥料
ammonolysis 氨解(作用)
amount of absorbed nutrient 养分吸收量
amount of effective temperature 有效积温
amount of nitrogen mineralization 矿化氮量
amount of nutrient supply 养分供应量
amount of precipitation 降雨量
amount of soil erosion 土壤侵蚀量
amount of supply water 供水量
amount of water required 需水量
amphoteric oxide 两性氧化物
ample rainfall 降水充沛
amplitude of annual variation 年较差
amylase 淀粉酶
amylolysis 淀粉分解
amylolytic enzyme 淀粉水解酶
amylopectin 支链淀粉
amylose 直链淀粉
amylum 淀粉
anabolism 合成代谢;同化作用
anaerobes 厌氧微生物(厌氧菌)
anaerobic bacteria 厌氧微生物;厌氧细菌
anaerobic biological treatment 厌氧生物处理
anaerobic composting 厌氧堆肥
anaerobic condition 厌氧条件
anaerobic culture 无氧培养
anaerobic culture dish 嫌氧培养皿
anaerobic decomposition 嫌气分解
anaerobic degradation 厌氧降解
anaerobic fermentation 厌氧发酵
anaerobic nitrogen-fixing bacterium 嫌气固氮细菌
anaerobion 厌氧菌;嫌气微生物
anaerooxidase 过氧化氢酶

analysis for soil soluble salts　土壤可溶盐分析
analysis of correlation　相关分析
analysis of covariance　协方差分析
analysis of regression　回归分析
analysis of variance　方差分析
analysis quality control　分析质量控制
analysis report　分析报告
analyte　待测成分
analytic hierarchy process　层次分析法
analytical apparatus　分析仪器
analytical balance　分析天平
analytical data　分析数据
analytical reagent　分析纯试剂
anaphase　后期
anatabine　新烟草碱
ando soil　暗色土(火山灰土)
andosol　火山灰土(暗色土)
angle of venation　叶脉角度
angular leaf spot　角斑病;(叶)角斑
angularity　倾斜度
angustifoliate　窄叶烟;细叶烟
animal dropping　动物粪便
animal ecology　动物生态学
animal excrement　动物粪便
animal excreta　动物粪便
animal feces　动物粪便
animal manure　动物性杂肥;畜肥
animal waste　动物粪便

anion　阴离子(负离子)
anion exchange　阴离子交换
anion exchange capacity　阴离子交换容量
annelid　环节动物
annual crop　一年生作物
annual mean　年平均
annual mean precipitation　年平均降雨量
annual mean temperature　年平均气温
annual plant　一年生植物
annual precipitation　年降水量
annual range　年较差
annual turnover　年总生物量;年营业额;年周转额
annual yield　年产量
annual yield of unit area　单位面积年产量
anorganotrophic　非有机营养的
ant　蚂蚁
antagonism　相克现象;拮抗作用
antagonism of nutrient absorption　养分吸收拮抗现象
antagonistic effect　拮抗效应
antagonistic soil　拮抗性土壤;抑制性土壤
anthesis　开花期(指单花)
anthracenone　蒽酮
anthracnose　炭疽病
anthracometer　二氧化碳计

anthranone 蒽酮
anthropic epipedon 人为表土层
anthropic erosion 人为侵蚀
anthropic soil 人为土壤
anthropogenic emission 人为排放
anthropogenic factors 人为因子
anthropogenic fertility 人为肥力
anthropogenic mellowing of soil 土壤的人为熟化作用
anthropogenic process 人为发生过程
anthropogenic soil 人为土壤
antibacterial 杀菌的;抗细菌的
antibiont 相克生物;对抗生物;拮抗生物
antibiotics 抗生素
antimony 锑
antitranspirant 抗蒸腾剂
apatite 磷灰石
apex 顶点;顶生组织;翅尖;顶端;叶尖
aphid 蚜虫
Aphidiidae 蚜茧蜂科
Aphidius gifuensis Ashmead 烟蚜茧蜂
aphosphorosis 缺磷症
apical dieback 顶梢枯死
apical dominance 顶端优势
apical growth 顶端生长
apical meristem 顶端分生组织
apical parts 烟株顶部
apical point 生长点

apoplast transport 质外体运输
apparent assimilation rate 表观同化率
apparent budget of soil nitrogen 土壤氮素表观盈亏
apparent cation exchange capacity 表观阳离子交换量
appearance quality 外观质量
application apparatus of fumigation 熏蒸应用装置
application dosage 施药量;使用剂量
application of base fertilizer 施基肥
application of base manure to the subsoil 施底肥
application of fertilizer by prescription 配方施肥
application rate 施用量
application time 施用时间
application together with watering 随水浇施
application width 施用幅宽
applications in concentrated bands 集中条施
applied ecology 应用生态学
apply manure in furrows 开沟条施
appraisal of quality 品质鉴定
approved variety 经过鉴定的品种
approximate calculation 近似计算
aquatic pesticide contamination 水中农药污染
aquic cinnamon soils 潮褐土
aquic moisture regime 潮湿水分状况

aquiculture 水培
arable land 可耕地
arable land area 耕地面积
arable land consolidation 土地整理
arable land improvement 土地改良
arable land readjustment 土地调整
arable land resources 土地资源
arable layer 耕作层
arable soil 可耕土壤
arable stratum 可耕层
arable system 耕作制
arachidic acid 花生酸
arbuscular mycorrhiza 丛枝菌根
arbuscule 丛枝
architecture of individual plant 个体株型
area charts 面积图
area of arable land 可耕地面积
area source of water pollution 水污染面源
area source pollution 面源污染
arene 粗沙
arenose 粗沙质
arescent 干燥
argentometric method 银量法
argillaceous bottom 黏质底土
arginine 精氨酸
arid 干燥的
arid land ecosystem 干旱地区生态系统
aridic moisture regime 干旱水分状况

aridity 干旱;干燥性;干旱性
arithmetic mean 算数平均
aroma 香气;芳香
aroma constituent 香气成分;香气组分
aromatic property 香味特性
aromatic tobacco(= oriental tobacco; Turkish tobacco) 香料烟(东方型烟;土耳其烟)
arrest of development 发育停滞
arsenic 砷;砒霜;三氧化二砷
arsenic contamination 砷污染
arsenium 砷(As)
artificial aggregate 人工团聚体
artificial atmosphere 人工气候
artificial binder 人造(雪茄烟)内包皮;人造黏合剂
artificial classification 人工分类
artificial climate 人造气候
artificial compost 人造堆肥
artificial culture 人工培养
artificial germination 人工催芽
artificial humic substance 人工腐殖物质
artificial intelligence 人工智能
artificial manure 堆肥;人造粪肥
artificial neural network 人工神经网络
artificial precipitation 人工降雨
artificial rain 人工降雨
artificial rainfall 人工降雨

artificial ripening　人工催熟
artificial shift erosion　人为侵蚀
artificial soil　人工土壤
artificial tobacco sheet　人造烟草薄片
ascending water　上升水
Ascochyta nicotianae Pass　烟草破烂叶斑病菌
ascorbate oxidase　抗坏血酸氧化酶
aseptic culture　无菌培养
aseptic operation　无菌操作
aseptic technic　无菌技术
aseptic technique　无菌操作
ash　灰
ash analysis　灰分分析
ash content　灰分(含)量
ash determination　灰分测定
ash element　灰分元素
ash rate　灰分比;含灰率
ashing　灰化
ashless filter paper　无灰滤纸
asiderosis　铁缺乏(症)
asparaginase　天冬酰胺酶
asparagine　天冬酰胺
asparagine synthase　天冬酰胺合酶
aspartic acid　天冬氨酸
aspect　坡向
aspergillus　曲霉属真菌
assimilation　同化;同化作用
assimilation efficiency　同化效率
assimilation of ammonium　铵盐同化作用

assimilation product　同化产物
assimilation substance　同化物
assimilatory coefficient　同化系数
assimilatory nitrate reduction　同化性硝酸盐还原作用
assimilatory quotient　同化系数
association analysis　关联分析
association coefficient　关联系数
associative nitrogen fixation　联合固氮作用
associative nitrogen fixing bacteria　联合固氮菌
astragalus sinicus　紫云英
astringency　涩味
asymbiotic azotification　非共生固氮作用
asymbiotic microorganism　非共生微生物
asymbiotic nitrogen fixation　非共生固氮作用
asymbiotic nitrogen fixer　非共生固氮生物
atalase　过氧化氢酶
atmosphere　大气圈
atmosphere pollution　大气污染
atmospheric composition　大气组成
atmospheric drought　大气干旱
atmospheric layer　大气层
atmospheric nitrogen deposition　大气氮沉降
atmospheric pollutant　大气污染物

atmospheric pollution 大气污染
atmospheric pollution source 大气污染源
atmospheric pressure 气压
atmospheric quality 大气质量
atmospheric temperature 气温;常温
atmospheric water 大气水
atomic absorption analysis 原子吸收分析
atomic absorption spectrometry 原子吸收光谱仪(AAS)
atomic absorption spectrophotometer 原子吸收分光光度计
atomic absorption spectrophotometry 原子吸收分光光度法
atomic absorption spectroscopy 原子吸收光谱分析
atomic emission spectrometry 原子发射光谱法
atomic fluorescence spectrometry 原子荧光光谱仪(AFS)
ATP synthase ATP 合酶
attemperation 温度调节
attractant 引诱剂;诱虫剂
attractant agent 引诱剂
attracting action 引诱作用
attracting method 诱捕法
attribute statistics 属性统计;品质统计
attrition mill 研磨机
auger method （根）钻法

auricle 叶耳
autecology 个体生态学;环境生态学
autoclave 高压灭菌锅;高压灭菌器
autoclaving 高压灭菌法
autocorrelation 自相关
autocorrelation analysis 自相关分析
automatic amino acid analyzer 氨基酸自动分析仪
automatic irrigation system 自动灌溉系统
automatic pipetting device 自动移液器
automatic priming harvester 自动烟叶采收机;自动采叶装置
automatic smoke sampler 自动烟气采样器
automatic smoking machine 自动吸烟机
automatic sprinkler 自动喷灌机
automatic titrator 自动滴定仪
automatic transplanter 自动移栽机
automatic transplanting 自动化移栽
automatic water feeding control 自动供水控制
autopurification 自净作用
autoregressive model 自回归模型
autotroph 自养生物
autotrophic nitrification 自养硝化（作用）
auxiliary root 侧根

auximone 植物生长素;发育激素
auxin 植物生长激素;植物生长素
auxin-type growth regulator 生长素型植物生长调节剂
auxoautotrophs 自养型微生物
auxoheterotrophs 异养型微生物
auxotroph 营养缺陷型
availability 有效性
availability index 有效性指标
availability of nutrient 养分的有效性
available capacity 有效容量
available carbon 有效碳
available coefficient 有效系数
available component 有效成分
available depth of soil 有效土层
available fertilizer 有效肥料
available field capacity 有效田间持水量
available moisture 有效湿度;有效水分
available nitrogen 有效氮
available nutrient 有效养分
available phosphorus 有效磷
available plant nutrient 有效植物养分
available potassium 有效钾
available potassium pool 有效钾库
available precipitation 有效降水
available soil moisture 土壤有效水分
available sulfur 有效硫
available water 有效水
available water capacity 有效水容量
available water holding capacity 有效持水量
average 平均值
average charts 均值图
average deviation 平均偏差
average moisture content 平均含水量
average output 平均产量
average production 平均产量
average run length 平均链长
average sample number 平均抽样个数
average total inspected 平均验查总数
average yield 平均产量
axenic culture 无菌培养
axial root 主根
axil 腋;叶腋
axillary bud 腋芽
axillary root 侧生根
axis 轴;轴线;坐标轴;坐标系;中心线;晶轴;主根;茎轴
azophoska 氮磷钾肥
azospirillum lipoferum 生脂固氮螺菌
azotate 硝酸盐
azotic acid 硝酸
azotification 固氮作用
azotobacter fertilizer 固氮菌肥料
azotobacteria 固氮菌
azotobacterin 固氮菌剂
azotobacterium 固氮细菌
azotometer 定氮仪

B

B (leaf) 上部叶组;上二棚
Bacillus 芽孢杆菌属
bacillus medium 芽孢杆菌培养基
bacillus thuringiensis <商>微生物杀虫剂;苏云金杆菌
background concentration 本底浓度;背景浓度
background content 背景含量
background level 本底水平
background value 背景值
background value of soil elements 土壤元素背景值
bacteria 细菌
bacteria density 细菌密度
bacteriafree culture 无菌培养
bacterial 细菌(性)的
bacterial alcoholic fermentation 细菌酒精发酵
bacterial colony counter 细菌菌落计数器
bacterial culture 细菌培养
bacterial diseases of tobacco 烟草细菌病害
bacterial fertilizer 细菌肥料
bacterial flora 细菌区系
bacterial lysis 细菌的溶解作用
bacterial manure 细菌肥料(菌剂)
bacterial necrosis 细菌性坏死
bacterial ooze 细菌分泌物;细菌渗出物
bactericide 杀菌剂
bacteriocidation 杀菌
bacteriolysis 溶菌作用
bacteriorhiza 菌根
bacteriostasis 抑菌作用
bacteriostat 抑菌剂
bacteriotrophy 细菌营养
bacterium 细菌
badland 劣地
bagging 套袋;装袋
Baijiang soil 白浆土
baking 烘烤;焙烧;干燥
balance of stock 资源平衡
balanced fertilization 平衡施肥
balanced nutrients fertilization 养分平衡施肥
balanced plant nutrition 植物营养平衡
balanced state 平衡状态
bale (烟)包;打包
bale press 打包机
band application 条施
band fertilization 条施肥
band placement 条施
band spreading 条状施肥
banded beside the row 行边条施

bar chart 柱状图表
bare soil 裸露土壤
barium 钡
barn 烤房;晾房;堆房;谷仓
barn loading method 装炕方法
barn maintenance 烤房维修
barn scald 挂灰;糊片;蒸片
barn sewage 畜舍污水
barn spot 烤房斑点病
barnyard manure 堆肥;厩肥
barophiles 嗜压菌
barophilic microorganisms 嗜压微生物
barotolerant microorganisms 耐压微生物
barren 贫瘠的
barren land 不毛之地
basal application 基施
basal dressing 施基肥
basal dressing for all 只施基肥
basal fertilization 施底肥
basal fertilizer 基肥;底肥
basal manure 基肥
basal root 主根
base 碱
base desaturation 盐基脱饱和(作用)
base eluviation 盐基淋溶
base exchange 盐基交换
base exchange capacity 盐基交换量
base saturation percentage 盐基饱和度

basic capacity 碱性(碱度)
basic farmland 基本农田
basic fertilizer 碱性肥料;基肥
basic ion 阳离子;碱离子
basic oxide 碱性氧化物(成碱氧化物)
basic soil 碱性土
basic solution 碱性溶液
basicity 碱度;碱性
basin 盆地
bast 韧皮部
batch fermentation 分批发酵
batching 分组
beaker 烧杯
bean aroma 豆香
bean cake 豆饼
bean cake fertilizer 饼肥
bed covers 苗床覆盖
bed irrigation 畦灌
bed rot 烂苗病
bedding 作垄
beef extract 牛肉浸膏
befouled water 污水
beginning stage of seedling 出苗始期
behavior signals 行为信号
bench irrigation 梯田灌溉;等高灌溉
bench terrace 水平梯田
benchtop fermentor 台式发酵罐

beneficial cycle 良性循环
beneficial effect 有益作用
beneficial element 有益元素
beneficial symbiont 有益共生物
benefit analysis 效益分析
benign circulation 良性循环
benzalcohol 苯甲醇;苄醇
benzaldehyde 苯甲醛
benzazole 吲哚
benzene 苯
benzene hexachloride 六六六;六氯化苯
benzidine 联苯胺
benzoanthracene 苯丙蒽
benzoic acid 苯甲酸
benzopyrene 苯并芘
3,4-benzopyrene (= benzo[α] pyrene) 3,4-苯并芘;苯并[α]芘
benzopyrrole 吲哚
benzyl alcohol 苯甲醇
best leaf 最好的上二棚(烟叶)
best management practice 最佳管理
best utilization of resources 资源最优化利用
beta distribution β分布
betaine 甜菜碱
better strains of seed 良种
BHC(= benzene hexachloride) 六六六;六氯化苯
bias 偏移;偏差

bias of estimator 估计量的偏差
bicarbonate 酸式碳酸盐;碳酸氢盐;重碳酸盐
big vein 巨脉病;巨脉
binary fertilizer 二元肥料
binary variable 二元变量
binder 雪茄内包皮;胶粘剂;黏合剂;捆扎机
binder group 内包皮叶组
binder leaf 雪茄内包皮烟叶
binomial distribution 二项分布
binomial test 二项式检验
bioaccumulation 生物富集
bio-agriculture 生物农业
bioassay 生物试验法
bioaugmentaion 生物增添作用
bioavailability 生物有效性
bioavailability concentration 生物有效性浓度
bioavailability of nutrient 养分生物有效性
bioavailability of soil nutrients 土壤养分生物有效性
bioavailable matter 生物有效性物质
bioavailable nutrient 生物有效养分
biochar 生物(质)炭
biochemical diagnosis 生物化学诊断
biochemical inhibition 生化抑制
biochemistry 生物化学(生化)

bioclimate 生物气候
biocoenosis 生物群落
biocommunity 生物群落
biocompatibility 生物适应性
bioconcentration 生物富集
biocontrol 生物防治
bioconversion 生物转化
biocycle 生物循环
biodegradability 生物降解能力
biodegradable 可生物降解的
biodegradable film 生物降解膜
biodegradation 生物降解
biodegradation of pollutant 污染物的生物降解(作用)
biodemethylation 生物脱甲基作用
biodiversity 生物多样性
biodiversity conservation 生物多样性保护
bioecology 生物生态学
bioelement 生命必要元素;生物元素
bioenergy 生物质能
bioenrichment 生物富集
bioerosion 生物侵蚀
bio-fertilizer 生物肥料
biofuel 生物燃料
biogas 沼气
biogas manure 沼气肥
biogas waste fertilizer 沼气肥
biogenic pesticides 生物源农药
biogeochemical cycle 生物地球化学循环
bioinformatics 生物信息学
bioinsecticide 生物杀虫剂
biologic products 生物制品
biological immobilization of nitrogen 氮的生物固持作用
biological accumulation 生物积累作用
biological activated carbon 生物活性碳
biological activity 生物活性
biological agent 生物因素;生物制剂
biological agriculture 生物农业
biological amplification 生物放大效应
biological balance 生物平衡
biological characteristics 生物学特征
biological chemistry 生物化学(生化)
biological cleaning 生物净化
biological concentration 生物富集;生物浓缩
biological control 生物防治
biological control measures 生物防治措施
biological cycle 生物循环
biological decomposition 生物分解(作用)
biological degradation 生物降解
biological enrichment 生物富集

biological equilibrium 生物平衡
biological erosion 生物侵蚀
biological factor 生物因素
biological fertility 生物肥力
biological fixation 生物固定(作用)
biological form 生物学类型
biological immobilization 生物固持（作用）
biological improvement 生物改良
biological inert carbon 生物惰性碳
biological insecticide 生物杀虫剂
biological interactions 生物相互作用
biological isolation 生物隔离
biological measures for soil and water conservation 水土保持生物措施
biological mineralization 生物矿化
biological monitoring 生物监测
biological nitrogen fixation 生物固氮
biological oxidation 生物氧化
biological pesticide 生物农药；生物杀虫剂
biological property 生物学性质
biological repair 生物修复
biological return ratio 生物归还率
biological selectivity 生物选择性
biological tie-up 生物固结
biological tolerance 生物学耐受性
biological transformation 生物转化作用
biological treatment 生物处理
biological weathering 生物风化

biological yield 生物产量
biology 生物学
biomarker 生物标志物
biomass 生物量
biomass of root system 根系生物量
biome 生物群落
biometrics 生物统计学
biomineralization 生物矿化
biomonitoring 生物监测
bionomics 生物学特性；生态学
bionomy 生态学
bio-organic fertilizer 生物有机肥
biopesticide 生物杀虫剂
bio-photodegradable film 生－光降解膜
biopiling 生物堆制
biopurification 生物净化
biorational insecticides 无公害杀虫剂
bioreactor 生物反应器
bioremediation 生物修复
biosequence 生物序列
biosphere 生物圈
biostimulation 生物刺激
biosynthesis 生物合成
biosystem 生物系统
biotest 生物试验
biotic activities of soil 土壤生物活性
biotic balance 生物平衡
biotic community 生物群落
biotic condition 生物条件
biotic factors 生物因子

biotic pesticide 生物农药
biotite 黑云母
biotransformation 生物转化作用
biotransport 生物迁移;生物转运
biotron 生物人工气候室
bioturbation 生物扰动作用
biting taste 辛辣味;刺激性;苦涩味
bitter 苦(味)的
bitterness 苦味
bivariate distribution 二维分布
bivariate Gauss distribution 二维高斯分布
bivariate normal distribution 二维正态分布
black carbon 黑炭
black earth 黑土
black fat 经过特殊处理的深色烟叶
black root rot 根黑腐病
black shade leaf 遮阴烟叶
black shank 黑胫病
black soil 黑土
blackleg 黑胫病;黑脚病
blank 空白
blank assay 空白试验
blank experiment 匀地试验
blank test 空白试验
bleaching 褪色
bleeding irrigation 渗灌
blended cigarette (= blended type cigarette) 混合型卷烟
blended fertilizer 掺合肥料
blended type cigarette 混合型卷烟
blending 调香;合香;掺和;配叶;配方
blending formula 配方
blending leaf groups 配方叶组
blending ratio 配方比例
blonde tobacco 黄色烟草;浅色烟草;淡色烟草
bloom 花,开花
blossoming 开花
blow 吹;吹风;送风;鼓风
blowing-downward type curing barn 气流下降式调制室;气流下降式烤房
blowing-upward type curing barn 气流上升式调制室;气流上升式烤房
blue algae 蓝藻
bluish grey 青灰色的
bodied leaf 有身份的烟叶
bodied taste 香味浓馥
body 物体;身体;身份;(香气)丰满度
bog soils 沼泽土
bogus soil 人为土壤
bonded nutrient 结合态养分
bone meal 骨粉
borax 硼砂
Bordeaux mixture 波尔多液
boric acid 硼酸

boric fertilizer 硼肥
borism 硼中毒
boron 硼(B)
boron deficiency 缺硼症
boron disturbance 硼素营养失调
boron fertilizer 硼肥
botanical bioremediation 植物生物修复
botanical insecticide 植物性杀虫剂
bottom application 施底肥
bottom horizon 底层
bottom layer 底层
bottom leaf 底脚叶
bottom soil 底土
bound water 束缚水
boundary row 保护行
bounded （烟气）抱团
bouquet 酒香
Bouyoucos hydrometer 鲍尧科斯比重计(甲种比重计)
box plots 箱线图
bract 苞;托叶
brand name 商标名称
breakage resistance 抗碎性;耐破度
break-even price 保本价格
breaks 叶位
breed 品种;培育
breeding 育种;繁育
breeding by crossing 杂交育种
breeding for pest resistance 抗虫育种
breeding for stress tolerance 抗逆育种

breeding material 育种材料
breeding process 育种过程;育种程序
breeding target 育种目标
bright leaf 亮色烟叶(指白肋烟)
bright leaf (= flue-cured, virginia) 淡色烟叶(烤烟)
bright tobacco 浅色烟;淡色烟;烤烟
brightness 亮度;光泽;鲜明度
brilliance 光泽
brittle 脆(性)的;易碎的
brittle fracture test 脆性破坏试验
broad cast 撒播;撒施;传播;广播
broad spectrum fungicide 广谱性杀菌剂
broad spectrum herbicide 广谱性除草剂
broad spectrum insecticide 广谱杀虫剂
broad spectrum pesticide 广谱杀虫剂
broadcast application 撒施
broadcast before plowing 耕翻前撒施
broadcast between rows 行间撒施
broadcast distribution 撒施
broadcast seeder 撒播(种)机
broadcast sowing 撒播
broadleaf tobacco 阔叶型烟草
broken 间断
broken bond water 断键水
broken ground 新开垦的地
bromocresol green 溴甲酚绿
bromomethane 溴代甲烷;溴甲烷

bromothymol blue 溴百里酚蓝
brown 褐色;棕色;褐色的;棕色的
brown calcic soil 棕钙土
brown earth 棕壤
brown forest soil 棕色森林土
brown leaf spot (= Ascohyta nicotianac Pass.) 烟草褐斑病
brown necrotic area 褐色坏死部分
brown necrotic spots 棕色坏死斑点
brown soil 棕壤
brown speck 褐色斑点
Brownian motion 布朗运动
Brownian movement 布朗运动
brownification 棕化(作用)
browning 棕(色)化;棕化;挂灰
browning reaction 棕化反应;棕色化反应
brown-red soil 棕红壤
Buchner funnel 布氏漏斗
bud inhibition 抑芽
bud stage 蕾期
budding period 现蕾期
budworms 蚜虫
buff color 浅黄色(晾烟颜色)
buffer 缓冲(缓冲剂)
buffer action 缓冲(作用)
buffer capacity 缓冲容量(缓冲力)
buffer system 缓冲体系
buffering 缓冲(作用)
buffering capacity 缓冲能力
buffering power of soil 土壤缓冲能力
building up fertility 培养肥力;提高肥力
bulk barns 堆积烤房;密集烤房
bulk blended fertilizer 散装掺混肥料(BB肥)
bulk curer 堆积烤房;密集烤房
bulk curing 密集烘烤
bulk curing barn 密集烤房
bulk density 堆积密度;体积密度,单位体积重量
bulk fermentation 堆积发酵
bulk fermenting 堆积发酵
bulk turning 翻堆
bulk yellowing 堆积变黄
bulking 膨胀
bunch planting 穴播;点播
bundle loosening 解把;松把
bundling 打捆;扎把
buret 滴定管
buried fertilization 深施
burlap wrapper for bale 打包麻袋布
burley 白肋烟
burley tobacco 白肋烟
burley type 白肋烟型
burned lime 生石灰;氧化钙
burning capacity 燃烧能力;持火力
burning characteristics 燃烧特性
burning holding property 持火力;阴燃持火力
burning of leaf margins 叶缘烧伤

burning of leaf tips 叶尖烧伤
burning quality 燃烧性;燃烧质量
burning rate 燃烧速率
burnt aroma 焦香
burnt symptom 焦灼症(缺钾)
burnt-sweetness aroma 焦甜香
burnt-sweetness type 浓香型
burozem 棕壤
burying of manure 耕埋厩肥
bushy habit of plant 植株丛生

C

C (= Cutters) 中部叶组,中部叶,腰叶
C horizon 母质层(C层)
c.v (= coefficient of variation) 变异系数
C/N ratio 碳氮比
cadmium 镉(Cd)
cadmium pollution 镉污染
cadmium-polluted fertilizer 镉污染肥料
cake fertilizer 饼肥
cake manure 饼粕堆肥
caked mass 硬块
caking 结块(结块性)
calcareous 石灰的;含钙的
calcareous paddy soil 石灰性水稻土
calcareous soil 石灰土
calcic cinnamon soils 石灰性褐土
Calcic lime concretion black soils 石灰性砂姜黑土
Calcic purplish soils 石灰性紫色土
Calcisol 钙积土
calcium 钙
calcium bicarbonate 碳酸氢钙;酸式碳酸钙;重碳酸钙
calcium carbonate 碳酸钙
calcium chloride 氯化钙
calcium deficiency 缺钙
calcium dihydrogenphosphate 磷酸二氢钙
calcium disturbance 钙素失调症
calcium fertilizer 钙肥
calcium hydrogencarbonate 碳酸氢钙;重碳酸钙
calcium magnesium phosphate 钙镁磷肥
calcium nitrate 硝酸钙
calcium oxide 氧化钙;生石灰
calcium soil 钙质土
calcium sulfate 硫酸钙
calcium superphosphate 过磷酸钙
calcium-magnesium ratio 钙镁比
calibration curve 标准曲线
calmodulin 钙调素

caloristat 恒温器;恒温箱
Calvin cycle 卡尔文循环
calyptra 根冠
calyx 花萼
cambium 形成层
canonical distribution 典型分布
canonical analysis 典范分析
canonical correlation 典型相关
canonical correlation analysis 典范相关分析
canopy 株冠;冠层
capacity of biological productivity 生物物质生产的能力
capillarity 毛细(管)作用(毛细现象)
capillary 毛细管的
capillary adsorbed water 毛管吸附水
capillary electrophoresis 毛细管电泳
capillary gas chromatography 毛细管气相色谱法
capillary interstice 毛管孔隙
capillary phenomenon 毛管现象
capillary pore 毛管孔隙
capillary tube 毛细管
capillary water 毛管水
capillary water holding capacity 毛管持水量
carbamide 尿素(脲)
carbendazim 多菌灵
carbendazol 多菌灵
carbohydrate 碳水化合物;糖类

carbohydrate metabolic pathways 碳水化合物代谢途径
carbon 碳
carbon and nitrogen partitioning 碳氮分配
carbon balance 碳平衡
carbon capture and storage 碳捕获和封存
carbon content 碳含量
carbon cycle 碳循环
carbon density 碳密度
carbon dioxide 二氧化碳
carbon dioxide application 二氧化碳施肥
carbon dioxide assimilation 二氧化碳同化作用
carbon dioxide fertilization 二氧化碳施肥
carbon dioxide fixation 二氧化碳固定(作用)
carbon dynamics 碳动态
carbon emission reduction 碳减排
carbon emissions trading 碳排放交易
carbon metabolism 碳代谢
carbon monoxide 一氧化碳
carbon nitrogen ratio 碳氮比
carbon nutrition 碳营养
carbon pool 碳库
carbon pool management index 碳库管理指数
carbon sequestration 固碳

carbon sink 碳汇
carbon source 碳源
carbon-sulphur ratio 碳硫比
carbon/nitrogen 碳氮比
carbonaceous matter 含碳物质
carbonaceous oxidation 碳质氧化
carbonate bounded form 碳酸盐结合态
carbonate horizon 碳酸盐层
carbonhydrate 碳水化合物;糖类
carbonic acid 碳酸
carbonic acid system 碳酸体系
carbonification 碳化作用
carbonifying 碳化的
carbonization 碳化(作用)
carbon-nitrogen ratio 碳氮比
carbon-rich waste 富碳有机废物
cardboard holder 纸质烟嘴
cardinal temperatures （植物生长的三种）基本温度（即最低、最高、最适）
carotene(= carotin) 胡萝卜素;叶红素
β-carotene β-胡萝卜素
carotenoid 类胡萝卜素;类叶红素
carrier gas 载气
carry-over effect (= residual effect) 剩余效应;残余效应
carryover fertilizer 残留肥料
cartographic data bank 地图数据库
cartographic digital data bank 地图数字化数据库
carton （卷烟）条
case 箱;盒;套;
cased tobacco 加过料的烟叶
casting plowing 外翻耕
catabolism 分解代谢
catabolite 分解(代谢)产物
catalase 过氧化氢酶(触酶)
catalysis 催化作用
catalytic action 催化作用
catalytic degradation 催化降解
catch crop cultivation 填闲作物栽培
catechol 儿茶酚;邻苯二酚
category 分类
category approach 分类法
category axis 分类轴;尺度轴
catharometer 气体分析仪
cathetometer 测高计(高差表)
cation 阳离子;正离子
cation exchange 阳离子交换
cation exchange capacity 阳离子交换量
cation exchange capacity of root 根系阳离子交换量
cation exchange constant 阳离子交换常数
cation exchange site 阳离子交换(位)点
cation fixation 阳离子固定
cation interchange 阳离子交换

cation-exchange chromatography 阳离子交换色谱(法)
cattle droppings 畜粪
cattle dung 牛粪
cattle feces 牛粪
cattle manure 牛粪
cattle waste 牛粪
cause and effect diagram 因果分析图
caustic potash 苛性钾;氢氧化钾
causticity potash 碱性钾;氢氧化钾
cell wall 细胞壁
cellar irrigated agriculture 窖灌农业
cellar type 水窖类型
cellular regulation 细胞调节作用
cellulolytic microorganism 纤维素降解菌
cellulose 纤维素
cellulose decomposer 纤维素降解菌
cellulose decomposition 纤维素分解(作用)
cellulose-decomposing bacterium 纤维素分解细菌
cellulosic biomass 纤维素类生物质
cembratriendiol 西柏三烯二醇
centigrade degree 百分温度;摄氏温度
central distance 中心距
central tendency 中心趋势
centrifugal filter 离心过滤器
centrifuge 离心机

centrifuge tube 离心管
cercobin M 甲基托布津
cereal 禾本科作物;谷类作物;谷香
certified seed 合格种子;认可种子
change of land quality 土地质量变化
channeling 作穴
Chao soil 潮土
char 炭
characteristic appearance 外部特征
characteristic odor 特征香气
characteristic symptom 特征性症状
characteristics 生物学特征
characteristics of moisture content 含水特征
characteristics of soil erosion 土壤侵蚀特征
characteristics of soil fertility 土壤肥力特征
characteristics of stalk position 部位特征
characteristics of textural classes 各组质地特征
characters of maturity 成熟特征
charcoal 木炭;炭;活性炭
charcoal filter 加活性炭复合滤嘴
charge 电荷
charged surface 带电表面(荷电表面)
charing (=charring) 炭化

·31·

charming odor 美好香气;优雅香气
charts 统计图形
characteristics of mineralization 矿化特征
check analysis 对照分析
check experiment 对照试验
check irrigation 畦灌
check plant 对照植株
check plot 对照区
check sample 对照样本
check test 对照试验
check variety 对照品种
chelate compound 螯合物
chelate fertilizer 螯合肥料
chelate ring 螯合环
chelated copper 螯合铜
chelated iron 络合态铁
chelated nutrient 螯合态养分
chelate-induced phytoremediation 螯合剂诱导的植物修复
chelating agent 螯合剂
chelation 螯合(作用)
chemical adsorption 化学吸附
chemical attack 化学侵蚀
chemical contamination 化学制品污染
chemical decomposition 化学分解
chemical defence 化学防御
chemical degradation 化学降解
chemical denitrification 化学反硝化作用;化学脱氮

chemical element 化学元素
chemical erosion 化学侵蚀
chemical extraction 化学提取
chemical extraction technology 化学浸提技术
chemical fertilizer 化肥
chemical fertilizer contamination 化肥污染
chemical fertilizer pollution 化肥污染
chemical fixation 化学固定(作用)
chemical fixation of phosphorus in soil 土壤磷素化学固定
chemical growth control agents 化学生长调节剂
chemical herbicide 化学除草剂
chemical injuries 农药伤害
chemical insecticide 化学杀虫剂
chemical pesticide 化学农药
chemical pollution 化学制品污染
chemical precipitate 化学沉淀
chemical pure 化学纯
chemical remediation 化学修复
chemical remediation of contaminated soil 污染土壤的化学修复
chemical resistance 抗药性
chemical signals 化学信号
chemical soil test 土壤化学测定
chemical sterilization 化学灭菌
chemical sucker control 化学抑芽
chemical tank 化学试剂罐;药剂桶

chemical topping 化学剂打顶
chemical toxicant 化学毒物
chemical weathering 化学风化
chemical weed control 化学防除杂草
chemically acidic fertilizer 化学酸性肥料
chemically alkaline fertilizer 化学碱性肥料
chemically basic fertilizer 化学碱性肥料
chemically neutral fertilizer 化学中性肥料
chemicals dissolving tank 药液罐
chemicking 漂白
chemisorption 化学吸附
chemoattractant 化学引诱物;化学吸引物;趋化物
chemotaxis 趋化性;趋药性
chemotroph 化能自养生物
chernozem 黑钙土
chewing plug 嚼烟饼
chewing tobacco 嚼烟
chicken droppings 鸡粪
chicken manure 鸡粪
chief constituent 主要成分
chief of graders 主分级员
chilli 辣椒
chilling injury 低温冷害
chilling point 冷冻温度
chilling resistance 抗寒性

chilling tolerance 耐寒性
chilling treatment 冷冻处理
China Cigarette Sales and Marketing Corporation 中国卷烟销售公司
China Ecological Research Net 中国生态研究网络
China Leaf Tobacco Corporation 中国烟叶生产购销公司
China National Import & Export Group Corporation 中国烟草进出口（集团）公司
China National Tobacco Corporation 中国烟草总公司
China Tobacco 《中国烟草》（杂志）
China Tobacco Machinery Group Corporation 中国烟草机械（集团）公司
China Tobacco Materials Corporation 中国烟草物资公司
China Tobacco Museum 中国烟草博物馆
China Tobacco Quality Supervision and Test Center 国家烟草质量监督检验中心
China Tobacco Science 《中国烟草科学》（杂志）
China Tobacco Science and Technology information center 中国烟草科技信息中心
China Tobacco Society 中国烟草学会

China Tobacco Standardization Research Center 中国烟草标准化研究中心
Chinese milk vetch 紫云英
chisel planting 松土播种法
chiseling 深松耕
chi-square chiasma 卡方交叉
chi-square distribution 卡方分布
chitinase 几丁质酶
chloraniformethan 双胺灵
chlorate 氯酸盐;氯化
chlorethylene 氯乙烯
chloride 氯化物
chloride of lime 漂白粉
chloride toxicity 氯害
chloride-containing fertilizer 含氯化肥
chlorine 氯
chlorine deficiency 缺氯
chlorine disturbance 氯素失调症
chlorine free fertilizer 无氯肥料
chlorine hydride 氯化氢
chlorine-containing fertilizers 含氯化肥
2-chloroallyl-N-diethyl-dithiocarbamate 草克死(除草剂)
2-chloroethylphosphonic acid 乙烯利;2-氯乙基磷酸(CEPA)
chloroform 三氯甲烷;氯仿
chloroform fumigation 氯仿熏蒸法
chlorogenic acid 绿原酸

chloromethane 氯(代)甲烷
chlorophyll 叶绿素
chlorophyll defect 叶绿素缺乏
chlorophyll deficiency 缺叶绿素症
chlorophyll deficient type 叶绿素缺乏型
chlorophyll degradation 叶绿素降解
chlorophyll fluorescence 叶绿素荧光
chlorophyll levels 叶绿素含量
chlorophyllin 叶绿酸
chloropicrin 氯化苦
chloroplast 叶绿体
chlorosis 白化;褪绿(病);缺绿;萎黄病
chlorosis effect 失绿效应
chlorotic disease 失绿病
chlorotic spot 褪绿斑点
chlorotic tip 褪绿叶尖
chocolate spot 褐色斑点
cholesterin 胆甾醇;胆固醇
cholesterol 胆甾醇;胆固醇
chroma 色泽;色度
chromatogram map 色谱图
chromatogram scanning 色谱图扫描
chromatograph 色谱仪;用色谱分析
chromatographic column 色谱柱
chromatography 色谱分析法
chromium 铬(Cr)
chromium pollution 铬污染
chromometer 比色计

CID（=cigarette inspection device）卷烟检测器；卷烟检测装置
cig（=cigarette） 卷烟；香烟
cigar 雪茄；雪茄烟
cigar aroma type 雪茄香型
cigar binder 雪茄内包皮
cigar binder tobacco 雪茄内包叶
cigar filler 雪茄芯烟
cigar filler tobacco 雪茄芯烟烟叶
cigar tobacco 雪茄烟草
cigar type cigarette 雪茄型卷烟
cigar wrapper tobacco 雪茄外包烟叶
cigar wrapper type 雪茄外包皮烟叶类型
cigarette 卷烟
cigarette acetate tow 烟用醋酸纤维丝束
cigarette burning rate meter 卷烟静燃仪
cigarette maker 卷烟机
cigarette making machine 卷烟机
cigarette manufacturing technology 卷烟工艺
cigarette paper 卷烟纸；盘纸
cigarette physical characteristics 卷烟物理性能
cigarette size 卷烟规格
cigarette smoke 卷烟烟气
cigarette tip 烟嘴
cigar-filler type 雪茄芯烟类型

cigarillo 小雪茄
cinnamal 肉桂醛
cinnamaldehyde（=cinnamic aldehyde）肉桂醛
cinnamic acid 肉桂酸
cinnamon soil 褐土
circling fertilization 环形施肥
circuit 循环
circular economy 循环经济
circular trench fertilization 环沟施肥
circular use 循环利用
circulating application 环施
circulation 循环
circulation economy 循环经济
cis-abienol 顺式冷杉醇
citrate acid soluble phosphatic fertilizer 枸溶性磷肥
citrate or citric acid soluble phosphorus 枸溶性磷
citrate-insoluble phosphorus 枸不溶性磷
citrate-soluble phosphorus 枸溶性磷
citric acid 柠檬酸
citric acid cycle 柠檬酸循环
citric acid soluble phosphorus 枸溶性磷
citrulline 瓜氨酸
CK（=check） 对照；核对
class 组
class limit 组限

class of soil particles　土壤颗粒分级
class width　组距
classes of pesticide toxicity　农药毒性等级
classification of soil texture　土壤质地分类
classify　分类；分级
classing　分类；分级
clay　黏粒；黏土
clay content　黏土含量
clay group　黏土组
clay loam　黏壤土
clay mineral　黏土矿物
clay particle　黏粒
clay sand　黏质沙土
clay silty loam　黏质粉沙壤土
clay soil　黏土
clayed bottom　黏质底土
clayey　黏质的(含黏土的)
clayey fine sand　黏质细沙土
claying horizon　黏化层
claying of soil　土壤黏化
clayish　黏土质
clay-sized mineral　黏粒矿物
clean bench　超净工作台
clean graded seed　清选出的分级种子
clean stubble　灭茬
clean tillage　除草
cleaness　（烟气）干净
cleaning efficiency　净化效率

cleanup effect　净化作用
cleanup of pollutants　污染物的净化
clean-up of seed　种子清选
clear　清晰
clear liquid fertilizer　清液肥料
click beetle　叩头虫
climagram　气候图
climagraph　气候图
climate change　气候变化
climate resources information system　气候资源信息系统
climate sensitivity　气候敏感性
climatic anomaly　气候异常
climatic chamber　气候室；人工气候室
climatic element　气候因素
climatic factor　气候因子；气候因素
climatic fertility　气候肥力
climatic potential productivity　气候生产潜力
climatic resources　气候资源
climatological atlas　气候图集
climatophytic soil　气候性土壤
climax leaves　完熟叶
clinograph　坡度计
clinometer　坡度计
clinosol　坡积土
cloddy-pulverescent structure　碎块状结构
close grown　密植
close planting　密栽；密植

close spacing 密植株行距
close texture 致密质地
cloud-burst 暴雨
cloudy weather 阴天;多云天气
clump 土块
clump statistics 聚类统计
cluster 团聚体
cluster analysis 聚类分析法
cluster sampling 分组取样;整群抽样
clustering 聚类
CMV（=cucumber mosaic） 黄瓜花叶病
CNTC（=China National Tobacco Corporation） 中国烟草总公司
CNTIEGC（=China National Tobacco Import & Export Group Corporation） 中国烟草进出口(集团)公司
CO_2 assimilation CO_2同化作用
CO_2 cycle 二氧化碳循环
CO_2 discharge 二氧化碳排放
CO_2 fertilization 二氧化碳施肥
CO_2 fertilizer 二氧化碳肥料
CO_2-equivalent 二氧化碳当量
CO_2-equivalent concentration 二氧化碳当量浓度
CO_2-equivalent emission 二氧化碳当量排放
coacervation 团聚(作用)
coarse 粗的;粗糙的
coarse fragment 粗碎片(粗碎块)

coarse grained soil 粗粒质土壤
coarse gravel 粗沙砾
coarse humus 粗腐殖质(地面枯叶层)
coarse sand 粗沙粒
coarse sand soil 粗沙土
coarse sandy loam 粗沙质壤土
coarse sandy soil 粗沙质土
coarse screen 粗孔筛
coarse sieve 粗筛
coarse soil 粗沙土
coat thickness 种皮厚度
coated controlled release 包膜控释
coated fertilizer 包膜肥料;包衣肥料
coated seed 包衣种子
coating 包膜
coccus 球菌
co-composting 混合堆肥
coefficient of correlation 相关系数
coefficient of dispersion 分散系数
coefficient of regression 回归系数
coefficient of respiration 呼吸系数;呼吸商
coefficient of variation 变异系数;偏差系数
coefficients 系数
coenzyme 辅酶
coevolution 协同进化
coexistence 共存
co-immobilization 共固定化作用

cold air advection 寒流
cold damage 低温冷害
cold digestion 冷浸提;冷消解
cold endurance 耐寒性
cold hardiness 耐寒性
cold in the late spring 倒春寒
cold injury 寒害;冷害;冻害
cold manure 冷性肥料
cold mortality 冻死率
cold point 冷点;冻点
cold resistance 耐寒性;抗寒性
cold resistant plant 抗寒植物
cold tolerant 耐寒性
cold wave 寒潮
collapse of chloroplast 叶绿体分解
collection and preparation of fertilizer sample 肥料样品的采集与制备
colloform 胶体
colloid 胶体
colloid solution 胶体溶胶
colloidal 胶体的
colloidal form 胶体形态
colloidal material 胶体物质
colloidal matter 胶体物质
colloidal property 胶体性质
colloidal sol 溶胶
collosol 溶胶
colony 菌落
colony forming unit 菌落形成单位
colony-counting method 菌落计数器
color fixing 定色

color fixing period (= color stabilizating stage) 定色期
color fixing stage (= color stabilizating stage) 定色期
color intensity 颜色强度;(叶片)色度
color setting 定色
color stabilizating stage 定色期
color symbols 颜色代号
colorimetry 比色法
colority 颜色;色度
colour 颜色
colour of ash 灰色
colour shade 色度;色泽
colour symbols 颜色代号
column diagram 直方图;方形图
columnar structure 柱状结构
colza cake 菜籽饼
combinated fertilizer 配合肥料;复合肥料
combine ripeness 联合收割机
combined form humic acid 结合态胡敏酸
combined nutrient 结合态养分
combined pollution 复合污染
combined pollution of metals and organic compounds 金属、有机复合污染
combined remediation 联合修复
combined water 束缚水
combustibility 燃烧性

combustibility of tobacco 烟叶燃烧性
combustible 可燃的;易燃的
combustion method 燃烧法
commanded land 自流灌溉农田
commensalism 互惠共生(现象);共生(现象)
commercial fertilizer 商品肥料
commercial humate 工业用腐殖酸盐
commercial organic fertilizer 商品有机肥
commercial value 经济价值;交换价值
commercialized agriculture 商业化农业
comminution 粉碎;磨细
comminutor 粉碎机
commodity pesticide regulation for acceptance 商品农药验收规则
common factor 公共因子
common fertilizer application technique 通用施肥技术
common mosaic 普通花叶病
common vetch 箭舌豌豆
community ecology 群落生态学
compact crop 密植
compact curing (= bulk curing) 仓式调制;堆积调制
compact layer 紧实层;致密层
compact tractor 小型拖拉机

compactibility 压实性
compacting 镇压
compaction 压实
compactness (卷烟)松紧度;紧密度;坚实度;紧实度
companion cropping 间作
company standard 企业标准
comparative advantage 比较优势
comparative test 比较试验
compass 罗盘
compensation ion 互补离子(补偿离子)
compensation of ecological factor 生态因子补偿作用
compensational exploitation 有偿使用
competing ion 竞争离子
competition equilibrium 竞争平衡
competitive advantage 竞争优势
competitive inhibition 竞争性抑制
complementary action 互补作用
complementary ion 互补离子
complementary ion principle 陪补离子原则
complementary technique 配套技术
complementation 互补作用
complete analysis 全量分析
complete association 绝对关联
complete block design 完全区组设计
complete cultivation 全面翻耕;全

面中耕
complete culture solution 完全营养液
complete fertilizer 完全肥料
complete flower 完全花
complete leaf 完全叶
complete treatment 完全处理
completely biodegradable 完全生物降解的
completely degradable 可彻底降解的
completely randomized design 完全随机设计
complex compound 络合物
complex experiment 复合试验;复因子试验
complex fertilizer 复合化肥;复合肥料
complex pollution 复合污染
complex reaction 络合反应
complex stress 复合胁迫
complexation 络合作用
complexation reaction 配位反应
complexed iron 络合态铁
complexed organic material 复合的有机物质
complexing agent 络合剂
complexity 复杂性
complexity of ecosystem 生态系统复杂性
complexometric titration 络合滴定法
component analysis 组分分析

component of variance 方差分量
composite aggregate 复合团聚体
composite benefit 综合效益
composite hypothesis 复合假设
composite sampling 混合取样
composite soil sample 混合土样
compositional analysis 组成分析
compost 堆肥
compostable refuse 可制堆肥垃圾
composted straw 秸秆堆肥
composting 堆制;堆沤;堆肥法
composting promoter 腐熟促进剂
compound determiner 复因子
compound fertilizer 复合肥料
compound manure 混合类肥;复合类肥
compound ped 复合土壤结构体
compound pollution 复合污染
comprehensive amelioration 综合改良
comprehensive benefit 综合效益
comprehensive evaluation of land productivity 土地综合生产力评价
comprehensive experiment 综合试验
comprehensive index of soil pollution 土壤污染综合指数
comprehensive nutrient management planning 综合养分管理规划
comprehensive test 综合试验
comprehensive utilization 综合利用
computational science 计算科学
computer fertilization system 计算机

施肥系统
concentrated 浓集的；浓缩的；浓的
concentrated fertilization 集中施肥
concentrated fertilizer 浓肥；浓缩肥料
concentrated nitric acid 浓硝酸
concentrated sulfuric acid 浓硫酸
concentrated superphosphate 重过磷酸钙
concentration 浓度；浓缩；蒸浓；提浓
concentration gradient 浓度梯度
concentration gradient measurement 浓度梯度测定
concentration of nutrient 养分浓度
concentration-mortality curve 死亡浓度曲线
concurrent parasitism 兼性寄生
concurrent saprophytism 兼性腐生
condensate pipe 冷凝管
conditional distribution 条件分布
conductance cell 电导池
conductibility 导电性（传导性）
conductivity 导电性（传导性）
conductivity apparatus 电导仪
conductivity meter 电导仪
conductivity method 电导分析法
conductivity of soil solution 土壤溶液电导率
conductivity-cell constant 电导池常数
conductometry 电导分析法
confidence band 置信区域

confidence interval 可靠区间；置信区间
confidence interval for regression line 回归线的置信区间
confidence level 置信水平（置信概率）
confined water 束缚水
conical flask 三角烧瓶；锥形瓶
conical type plant 圆锥形株型
conidium 分生孢子
connectivity 连通性
Conoderus vespertinus 烟草金针虫
conservation （菌种）保藏
conservation agriculture 保护性农业
conservation farming 保土耕作（保护性耕作）
conservation of soil and water 水土保持
conservation tillage 保土耕作（保护性耕作）
conservatory 温室
conserving cultivation 保土耕作（保护性耕作）
conserving seedling 保苗
conserving use of resources 节约地使用资源
consistence 结持性（稠性）
constant fertilization 经常施肥
constant temperature and humidity container 恒温恒湿箱
constant temperature bath 恒温浴

constant temperature oven 恒温烘箱
constant temperature oven method 恒温烘箱法
constants 常数
constituent 组分；成分
construction of seed bed 苗床结构
construction of technology 技术结构
consumption and compensation of soil water 土壤水分消耗与补偿
contact absorption 接触吸收
contact action （接）触(使)杀作用
contact area 接触面积
contact herbicide 接触性除草剂
contact insecticide 触杀剂
contaminated soil 污染土壤
contaminated soil remediation technology 污染土壤修复技术
contamination by pesticide 农药污染
contamination hazard 污染危害
contaminative source 污染源
content of detrimental substances 有害物质数量
content of nutrients 养分含量
continual improvement 持续改进
continuity analysis 连续性分析
continuous 连续的
continuous absorption 连续吸收
continuous analysis 连续分析
continuous broth out system 培养液连续排出设备

continuous crop rotation system 连续轮作制
continuous cropping 连作
continuous cropping obstacle 连作障碍
continuous cultivation 连作
continuous culture method 连续培养法
continuous distribution 连续分布
continuous double cropping 两熟连作
continuous fermentation 连续发酵
continuous flooding 连续漫灌(持续淹没)
continuous flow analysis 连续流动分析
continuous flow analyzer 连续流动分析仪
continuous flow culture 连续流动培养
continuous flow method 连续流动法
continuous horizon 连续(土)层
continuous irrigation 连续灌溉
continuous moderate rain 连续性中雨
continuous monitoring 连续监测
continuous phytoextraction 连续的植物提取
continuous rain 连续雨；绵雨
continuous random variable 连续随机变量
contour cultivation 等高耕作
contour farming 等高耕作；梯田农业
contour irrigation 等高灌溉

contour line 等高线
contour map 等高线图
contour planting 等高种植
contour ploughing 等高耕作
contour plowing 等高耕作
contour strip intercropping 等高带状间作
contour terrace 等高梯田
contour tillage 等高耕作
contoured land 坡地
contouring field 等高农地
contract farming 合同制农业
contrasting 异色(与叶脉颜色有差别)
contribution margin 边际贡献;边际利润;边际所得
control 对照(物);控制;调节;操纵;管理
control experiment 对照试验
control of insects 虫害防治
control of plant disease 植物病害防治
control of pollution sources 污染源控制
control of sucker 抑制腋芽
control plot 对照区
control sample 对照样品
control test 对照试验
control variety 对照品种
controllable cost 可控制费用
controlled environment chamber 人工气候室
controlled moist condition irrigation 湿润灌溉
controlled release 控释
controlled release fertilizer 控释肥料(长效肥)
controlled release mechanism 控释机理
controlled release pesticide 控释农药
controlled release fertilizer 长效肥料
conventional agrotechnique 常用农业技术(传统栽培技术)
conventional application 常规施药(肥)
conventional assay 常规检测
conventional barn 老式烤房;传统烤房
conventional chemical fertilizer 常规化学肥料
conventional curing 传统调制;常规烘烤;一般烘烤;习惯烘烤
conventional test 普通试验;标准试验
conversion from farmland to forest or grasslands 退耕还林还草
conversion of cropland to forest 退耕还林
convert the land for forest 退耕还林
convertible husbandry 轮作
converting farmland to forestland 退耕还林
conveyance technologies of highyield cultivation 高产栽培配套技术

Convolvulaceae　旋花科
Convolvulus　旋花属
cool damage　冷害
cool house　低温室;冷藏室
co-operative seedling　联合育苗
coordinate　坐标
coordinated development　协调发展
copper　铜
copper deficiency　缺铜症
copper disturbance　铜素营养失调
copper fertilizer　铜肥
copper fungicide　铜素杀菌剂
copper pollution　铜污染
copper shortage　缺铜
copper sulfate　硫酸铜
copper sulphate　硫酸铜
coprogenous aggregate　粪粒团聚体
coprophage members of the soil fauna　食粪性土壤动物
copy sample　仿制样品
corcle　胚;胚芽;胚根
CORESTA (= Cooperation Centre for Scientific Research Relative to Tobacco)　烟草科学研究合作中心
corolla　花冠
corrected retention time　校正保留时间
correction of soil acidity　土壤酸度中和(酸性土壤改良)
correlated character　相关性状
correlation　相关

correlation analysis　相关分析
correlation and regression　相关与回归
correlation coefficient　相关系数
correlation function　相关函数
correlogram　相关图
corrugation infiltration　垄沟渗漏
corrugation irrigation　垄沟灌溉
cosmid　黏粒
cost-benefit analysis　成本效益分析
costs of agricultural products　农产品成本
cotransport　协同转运
cotyledon　子叶;绒毛叶
cotyledon petiole　子叶柄
coumaraldehyde　香豆醛
coumaric acid　香豆酸
coumarin　香豆素
coumarinic lactone　香豆素内酯
covariance　协方差
covariance design　协方差设计
covariance function　协方差函数
cover crop　覆盖作物;肥田的农作物
cover glass　盖玻片
cover plant　覆盖植物
covered by rice straw　覆盖稻草
covered culture　覆盖栽培
covering area per curing barn　每个烤房适用种烟面积
covering effect　覆盖效应
covering with plastic film　地膜覆盖
cow droppings　牛粪

cow dung 牛粪
cow manure 牛粪
cowshed manure 牛粪
crack of stem 茎破裂
crasher 粉碎机
crest 峰值;峰顶;顶点;波峰;螺纹牙顶;牙顶;齿顶
crest of bands 谱带顶
crest of peak 峰顶
crinkle 皱褶;皱叶病
crinkle mosaic 皱缩花叶
crinkling 起皱
criss-cross method 方格计数法
criteria of soil fertility grade 土壤肥力等级标准
critical concentration 临界浓度
critical concentration of element 元素的临界值浓度
critical humidity 临界湿度(临界含水量)
critical level 临界水平
critical moisture content 临界含水率
critical moisture point 水分临界点
critical nutrient range 养分的临界值范围
critical period of nutrition 营养临界期
critical point 临界点
critical potassium concentration 钾的临界浓度
critical region 临界域
critical shearing stress 临界剪应力

critical soil moisture 土壤临界含水量(指植物最低需水量)
critical stage and maximum efficiency stage of nutrition 营养临界期和最大效应期
critical stage of nutrition 营养临界期
critical stage of water requirement 需水临界期
critical temperature 临界温度
critical value 临界值
critical value of nutrient 养分临界值
critical value of soil pollution 土壤污染临界值
critical zone 临界区间
crop condition 作物长势
crop alternation 轮作
crop climate adaptation 作物气候适应型
crop climate ecotype 作物气候生态型
crop composition and distribution 作物布局
crop duty 作物灌水率
crop ecophysiology 作物生态生理学
crop growing 作物栽培
crop growth rate 作物生长率
crop index 农作物指数;作物指标
crop monitoring 作物监测
crop nutrient requirement 作物营养需求
crop of depletion soil fertility 耗地作物

crop of improving soil fertility　养地作物
crop of rotation　轮作作物
crop price/fertilizer cost ratio　作物/肥料价格比
crop protection　作物保护
crop recovery efficiency　作物回收率
crop removal efficiency　作物移走率
crop residue(s)　作物残体;作物残茬
crop response curve to fertilizer　肥料反应曲线
crop response function to fertilizer　肥料效应函数
crop response to fertilizer　肥料效应
crop rotation　轮作
crop rotation succession　作物轮作顺序(茬口顺序)
crop rotation system　作物轮作制
crop season　种植季节
crop space application　株间施药
crop succession　轮作顺序
crop uniformity　品种单一性;作物单一性
crop vulnerability　作物脆弱性
crop water requirement　作物需水量
crop water stress index　作物水分胁迫指数
crop yield　农作物产量
cropland　耕地;农田
cropland retirement　退耕

cropper　收割机(收获机)
cropping　种植
cropping arrangement　茬口安排
cropping index　种植指数;复种指数
cropping pattern　种植模式
cropping season　栽培季节
cropping sequence　种植顺序
cropping structure　种植结构
cropping system　种植制度;耕作制度
cropping-pastoral ecotone　农牧交错地
cross　横向的;杂交;杂种
cross breed　杂种
cross breeding　杂交繁育;杂交育种
cross combination　杂交组合
cross compatibility　杂交亲和性
cross sterile　杂交不育
cross-pollinated cultivated plants　异花传粉栽培作物
cross-pollinated plant　异花传粉植物
crown　冠;根茎;副(花)冠
crown canopy　根冠层
crown rot　根冠腐烂
crowning　顶生的
crucible　坩埚
crucible tongs　坩埚钳
cruciferous type　十字花科型
crude　天然的;粗的;生烟;生叶
crude ash　粗灰分
crude cellulose　粗纤维素
crude density　相对密度;粗密度
crude extract　粗提出物

crude fruit 未熟果
crude leaf bundle 自然烟叶扎把
crudeness 生硬
Crueiferae 十字花科
crumb structure 团粒状结构
crumpled leaf bundle 自然烟叶把［非平坦状烟叶把］
crushing strength 抗碎强度
crust and mulch structure 上结下松结构
cryic layer 永冻层
cryostat 低温恒温器
cryoturbation 融冻扰动(作用)
crystalline water 结晶水
crystallization water 结晶水
CTQSTC (= China Tobacco Quality Supervision and Test Center) 国家烟草质量监督检验中心
CTS (= China Tobacco Society) 中国烟草学会
CTSRC (= China Tobacco Standardization Research Center) 中国烟草标准化研究中心
cube 立方体,(育苗的)营养钵
cubic content 容量;体积
cucumber mosaic 黄瓜花叶病
cucumber mosaic virus 黄瓜花叶病毒
Cu-deficient 缺铜症
cultigen 栽培种
cultimulcher 耙地松土镇压器
cultispecies 栽培种

cultivability 耕性(可耕度、熟化度)
cultivable area 可耕地区
cultivar 栽培品种;栽培品系
cultivar purity field plot test 品种纯度田间小区试验
cultivar specificity 品系特异性
cultivar 品种
cultivate 中耕;耕作
cultivated acreage 耕地面积
cultivated character 栽培性状
cultivated form 栽培类型
cultivated horizon 耕作层
cultivated land 耕地;农田
cultivated land loss 耕地流失
cultivated land protection 耕地保护
cultivated land quantity 耕地数量
cultivated land resources 耕地资源
cultivated lands productivity 耕地生产力
cultivated layer 耕作层
cultivated soil 耕作土壤;熟土
cultivated species 栽培种
cultivated variety 栽培品种
cultivated vegetation type 栽培植被型
cultivating period 培土期;栽培期
cultivating practice 栽培措施
cultivation 中耕;栽培;培土
cultivation and banking 中耕培土
cultivation and weeding 中耕除草
cultivation by direct seeding 直播栽培
cultivation depth 耕深

cultivation fallow 中耕休闲
cultivation in protecting condition 保护地栽培
cultivation layer 耕作层
cultivation management 栽培管理
cultivation model 栽培模式
cultivation system 栽培体系;栽培制度
cultivation technique 栽培技术
cultivation test 耕作试验
cultivation time 培养时间
cultivator 中耕机
cultural condition 培养条件
cultural control 耕作防治
cultural eutrophication 人为富营养化
cultural fallow 栽培休闲
cultural practices 栽培措施
cultural system 培养系统
culture 培养;培养物;菌种;栽培
culture bottle 培养瓶
culture dish 培养皿
culture flask 培养瓶
culture fluid 培养液
culture in glass 离体培养
culture in the field 露地栽培
culture in vitro 离体培养
culture in vivo 活体培养
culture medium 培养基;培养液
culture method 培养法
culture on the hill 垄上栽培;垄作栽培
culture packet 培养袋
culture pan 营养钵
culture preservation 菌种保藏
culture season 栽培季节
culture solution 培养液;营养液
culture solution of element deficient 缺素培养液
culture technique 培养技术
culture tube 培养管
culture vessel 培养容器
culture without rotation 连作
cultured soil 耕作土
cumulative distribution function 累积分布函数
cumulative frequency 累积频数
cumulative sum chart 累积和图
cumulative temperature 积温
cumuli 堆垫
cumulic 堆垫
cupping of leaves 叶片翘曲
cupric sulfate monohydrate 一水硫酸铜
curability of leaf (烟叶)调制性能;烘烤性能
cured 调制后的(烟叶);调制
cured leaf 初烤烟叶;原烟
cured leaf weight 干叶重
cured tobacco shred 烟丝
curing 调制
curing barn 调制室;烤房;晾房
curing condition 调制条件

curing control　调制管理
curing facility　调制设施
curing frame　调制架
curing hours　调制时间
curing matter loss　调制损失
curing method of tobacco　烟叶调制方法
curing module　烘烤组件;调制组件
curing of compost　堆肥熟化
curing operation　调制操作
curing period　调制期
curing rate　调制速度;烘烤速度
curing system　调制系统
curing technique　调制技术
curing temperatures　调制温度
curl downwards　向下卷曲
curl of leaf margins　叶子边缘卷曲
curl upwards　向上卷曲
curled leaves　卷叶
curling of leaves　叶片卷缩
curly leaf　卷叶病毒病
current land utilization map　土地利用现状图
curve fitting　曲线拟合
customary　惯例
customary fertilizer application　习惯施肥法
cut rag　烟丝
cut rolled stem　梗丝
cut tobacco　烟丝
cut tobacco drier　烘丝机

cut tobacco feeding　喂丝
cuticle　表皮;护膜;角质层
cuticular wax　表皮蜡质
cutis　表皮(复)
cutters　中部叶组;腰叶
cutters of lugs　中下部(叶):代号CX
cutters　中部(叶):代号C
cutworm　地老虎
cyanide　氰化物
cyanide pollutant　氰化物污染物
cyanide pollution　氰化物污染
cyanobacteria　蓝细菌
cycle　循环
cycle coefficient　循环系数
cycle of material　物质循环
cycle of substances in agriculture　农业的物质循环
cycling of soil nutrient in agroecosystem　农田生态系统土壤养分循环
cyme　聚伞花序
cymose inforescence　头状聚伞花序
cysteine　半胱氨酸
cystine　胱氨酸
cyto-anatomy　细胞解剖学
cytoarchitecture　细胞构造
cytochrome　细胞色素
cytochrome oxidase　细胞(呼吸)色素氧化酶
cytolysosome　细胞溶酶体

cytomembrane 细胞膜
cytomixis 细胞融合

cytoplasm 细胞质

D

daily maximum temperature 日最高温度
daily mean temperature 日平均温度
daily minimum temperature 日最低温度
daily range 日较差
daily variation 日变化
damage 伤害;病害;病伤;损坏;受伤
damage by coldness 冻害
damage by disease and pest 病虫害
damage by drought 枯死
damage by excess fertilization 肥料过量伤害
damage by insect 虫害
damage by low temperature 低温灾害
damage by ozone 臭氧害
damage by storm 暴风雨灾害
damage by unusual weather 异常气象灾害
damage by wind 风害
damage caused by low temperature 冷害
damaged 受害的
damaged degree 受害程度

damaged leaf 损伤叶;伤残叶
damascone 二氢大马酮
damp 潮湿
damper of curing barn 调制室风门;烤房
damping off 猝倒病
dampness 水分(含水量)
dappled 有斑点的
dark 暗色
dark air-cured 深色晾制
dark air-cured tobacco 深色晾烟
dark brown 深褐色
dark brown leaf 褐片(烟叶)
dark brown soil 暗棕壤
dark cigarettes 深色卷烟
dark fire cured 深色烟熏烟
dark germination 遮光发芽
dark germinator 黑暗发芽器
dark green 暗绿色
dark lusterless brown leaf 褐糟叶
dark lusterless leaf 黑糟叶
dark mahogany （烟叶）深棕色
dark metabolism 暗代谢
dark mildew 烟霉;煤烟病
dark red 深红色;暗红色
dark respiration 暗呼吸作用

dark sun-cured tobacco 深色晒烟
dark tobacco 深色烟草
Dark-brown soils 暗棕壤
dark-fire tobacco 明火烤烟
data base management system 数据库管理系统
data base system 数据库系统
data mining 数据挖掘
data preparation 数据预处理
date of maturity 成熟期
date test 播种期试验
day breeze 昼风;日风
day length 昼长;日长
day light 白天光照
day without frost 无霜日
days after transplanting 移栽后天数
days for germination 发芽天数
days for yellowing 变黄天数
days in field 大田生育期
days of flowering 花期天数
DDT (= dichloro-diphenyl trichloro-ethane) 滴滴涕;二氯二苯三氯乙烷
2-DE (= two-dimensional electrophoresis) 双向电泳
dead heart 枯心
dead leaf 枯叶
dead soil covering 枯枝落叶层
deadly temperature 致死温度
dealkalization 脱碱(作用)
deamination 脱氨作用

deaquation 脱水(作用)
decapitation 打尖(去顶)
decay 腐烂;衰变;衰减
decay process 腐解过程
decayed stable manure 腐熟厩肥
dechlorination degradation 脱氯降解
deciduous materials 脱落物
decision function 判定函数;决策函数
decision rules 判决规则
decision support system 决策支持系统
decision tree 决策树
declarable content 标明量
decline of soil fertility 地力衰退
decoated seed 去皮种子
decomposed dung 腐熟厩肥
decomposer 分解者
decomposition 分解
decomposition course 分解过程
decomposition degree 分解程度
decomposition product 分解产物
decomposition rate of soil organic matter 土壤有机质分解率
decontamination factor 净化因素;去污系数
deep 浓(色度)
deep brown (= dark brown) 深褐色
deep color 深色
deep cultivating 深耕

deep digging 深松耕
deep lemon (=dark lemon) 正黄
deep orange (=dark orange) 深橙(橘)黄
deep placement 深施
deep ploughing 深耕
deep ploughing and careful cultivation 深耕细作
deep plowing 深耕;深松;深翻
deep scarification 深松耕
deep seeding 深播
deep tillage 深耕
deep tillage with fertilizer application 深耕施肥法
deep tillage work 深耕作业
deeper transplanting 深栽
defect cigarette 残烟;次烟;疵烟
deficiency 缺乏
deficiency disease 营养不足症
deficiency symptom (养分)缺乏症状
deficient 缺乏的
deficient element 亏缺元素
definite inflorescence 有限花序
deflation 风蚀(作用)
deforestation 毁林
degenerate (品种)退化
degenerated form 退化类型
degenerated soil 退化土壤
degenerating stage 衰退期
degeneration 退化(作用)

degenerative process 衰退过程
degradable 可降解的
degradable carbon 降解碳
degradable plastic film 可降解地膜
degradable plastic mulching film 可降解地膜覆盖
degradable pollutant 可降解污染物
degradation 退化
degradation mechanism 降解机制
degradation of herbicide 除草剂降解
degradation of pollutants 污染物的降解
degradation product 降解产物
degradation succession 退化演替
degree 度;程度;等级
degree of acidity 酸度
degree of aggregation 团聚度
degree of base saturation 盐基饱和度
degree of breakage 烟叶破损度
degree of browning 变褐率
degree of compaction 压实度
degree of dispersion 分散度
degree of horizon differentiation 层次分化(程)度
degree of humification 腐殖化程度
degree of salinization 盐渍化程度
degree of saturation 饱和度
degree of spoilage (烟叶)损伤程度
degrees of freedom 自由度

dehydrating of excess water （烟叶调制）初期凋萎
dehydration 脱水(作用)
dehydrogenases 脱氢酶
dehydrohumic acid 脱水胡敏酸(脱氢胡敏酸)
dehydrolysis 脱水(作用)
deionized water 去离子水
dejection 排泄物
delayed germination 延期萌发;发芽迟缓
delayed pollination 延迟授粉
deleterious element 有害元素
delicate 灵敏的;准确的;优美的;细致的;细腻的;(烟叶)组织细致
delicious 芬芳;芳香的;美味的
demineralized water 纯水;去矿质水;软化水
demonstration area 示范区;试验区
demonstration effect 示范效应;示范作用
demonstration experiment 示范试验
demonstration farm 示范农场
demonstration field 示范田
denitration 脱硝(作用)
denitrification 反硝化作用(脱氮作用)
denitrification inhibitor 反硝化抑制剂
denitrification loss 反硝化损失
denitrification potential 反硝化势
denitrifying bacteria 脱氮菌;反硝化菌
denitrobacteria 反硝化细菌(脱氮细菌)
denitrogenation 脱氮
dense （稠)密的;浓厚的
density 密度
density function 密度函数
density of crop 种植密度;播种密度
density of smoke 烟气浓度
denudation 剥蚀
denutrition 营养不良(营养不足)
depelleted seed 去除包衣的种子
depletion technique 耗竭法
depollution 去污染;除污染
deposit 沉淀
deposition 堆积(作用)
depotassication 脱钾作用
depressant 抑制剂
depression 洼地
depth of leaf layer 叶层厚度
depth of plowing 耕深
depth of root 根深
depth of runoff 径流深度
depth of sampling 取样深度
depth of sowing 播种深度
depth of tillage 耕深
depth regulating device 耕深调节装置
desalinization 脱盐(作用)

desaturation 去饱和(作用)
descending order 顺序下降
descending water 渗透水;下降水
description of soil 土壤描述
descriptive statistic(s) 描述性统计
desertification 荒漠化
desertification of land 土地荒漠化
desertification process 荒漠化过程
desertization 荒漠化
desiccation 干燥
desiccator 干燥器
desiccator method 干燥法(测水分)
desorption 解吸;脱附;解吸作用
desorption process 解吸过程
destroy of ecosystem 生态破坏
destroying stubble 灭茬
destructing damage 毁灭性灾害
destruction of biodiversity 生物多样性的破坏
destruction of vegetation 植被破坏
destructive disease 毁灭性病害
destructive distillation 分解蒸馏
destructive exploitation 掠夺式开采
desuckering agent 抑芽剂
desulfuration 脱硫作用
detached leaf 离体叶片
detailed reconnaissance survey 详细的勘测
detailed soil survey 土壤详查
detection limit 检测限
deterioration 退化

determinand 被测定物
determinant 决定因素
determinant factor 决定因素
determinate inflorescence 有限花序
determination coefficient 决定系数
determination method of pH value for pesticides 农药 pH 值的测定方法
determination of alkaloids 总植物碱的测定
determination of fertilizer moisture content 肥料水分测定
determination of nitrogen content in chemical nitrogenous fertilizer 化学氮肥含氮量测定
determination of phosphorous content in chemical phosphatic fertilizer 化学磷肥含磷量测定
determination of plant coarse ash 植物粗灰分测定
determination of plant moisture and dry matter 植物水分和干物质测定
determination of plant total nitrogen 植物全氮测定
determination of plant total phosphorus 植物全磷测定
determination of plant total potassium 植物全钾测定
determination of potassium content in chemical potassic fertilizer 化学

钾肥含钾量测定

determination of residual stem content 含梗率的测定

determination of soil ammonium nitrogen 土壤铵态氮测定

determination of soil arsenic 土壤砷测定

determination of soil available phosphorus 土壤有效磷测定

determination of soil available potassium 土壤速效钾测定

determination of soil cadmium 土壤镉测定

determination of soil cation exchange capacity 土壤阳离子交换量测定

determination of soil copper 土壤铜测定

determination of soil density 土壤密度测定

determination of soil exchange acidity 土壤交换性酸测定

determination of soil hardness 土壤坚实度测定

determination of soil humus composition 土壤腐殖质组成测定

determination of soil hydrolysable nitrogen 土壤水解性氮测定

determination of soil lead 土壤铅测定

determination of soil lime requirement 土壤石灰需要量测定

determination of soil mercury 土壤汞测定

determination of soil moisture content 土壤含水量测定

determination of soil nickel 土壤镍测定

determination of soil nitrate nitrogen 土壤硝态氮测定

determination of soil organic matter 土壤有机质测定

determination of soil organic pollutants 土壤有机污染物测定

determination of soil particle density 土粒密度测定

determination of soil pH 土壤pH测定

determination of soil porosity 土壤孔隙度测定

determination of soil redox potential 土壤氧化还原电位测定

determination of soil reducing substances 土壤还原性物质测定

determination of soil selenium 土壤硒测定

determination of soil slowly available potassium 土壤缓效钾测定

determination of soil temperature 土壤温度测定

determination of soil total exchangeable bases 土壤交换性盐基总量

测定

determination of soil total exchangeable calcium 土壤交换性钙测定

determination of soil total exchangeable magnesium 土壤交换性镁测定

determination of soil total exchangeable potassium 土壤交换性钾测定

determination of soil total nitrogen 土壤全氮测定

determination of soil total phosphorus 土壤全磷测定

determination of soil total potassium 土壤全钾测定

determination of soil zinc 土壤锌测定

determination of strip particle size 叶片大小的测定

determination of total nitrogen 总氮的测定

determination of total volatile bases 总挥发碱的测定

determination of water nitrate 水样硝酸根测定

determination of water pH 水样 pH 测定

determination of water phosphate 水样磷酸根测定

determination of water potassium 水样钾离子测定

determination of water soluble sugars 水溶性糖的测定

determination of water sulphate 水样硫酸根测定

determination of water total ammonium 水样铵离子测定

determination of water total nitrogen 水样全氮测定

determination of water total phosphorus 水样全磷测定

determination range 测定范围

detrimental effect 有害作用

detritus 枯枝落叶层

devegetation 植被破坏

development of reproductive organ 生殖器官的发育

development of seed 种子发育

development stage 发育期

developmental condition 发育条件

developmental phase 发育期

deviation 离差

devolution 退化

dew point 露点

dewing in barn 烤房水珠凝结

dew-point hygrometer 露点湿度计

dew-point recorder 露点记录仪

dew-point transducer 露点传感器

dextrose 葡萄糖;右旋糖

diagnosis 诊断

diagnosis and recommendation integrated system 诊断推荐(施肥)综

合法
diagnosis approach to nutrient deficiency 养分缺乏的诊断措施
diagnosis index 诊断指标
diagnosis method of crop root vigour 作物根系活力诊断法
diagnosis method of fertilization 施肥诊断法
diagnosis method of foliar color 叶色诊断法
diagnosis method of leaf analysis 叶片分析诊断法
diagnosis method of plant chemistry 植物化学诊断法
diagnosis method of rapid tissue measurement 组织速测诊断法
diagnosis of biological incubation 生物培养诊断
diagnosis of method by critical value 临界值诊断法
diagnosis of microstructure 显微结构诊断
diagnosis of plant nutrition 植物营养诊断
diagnosis of remote sensing technique 遥感诊断
diagnostic characteristics 诊断特性
diagnostic technique 诊断技术
dialyzing paper 渗析纸
diametrical ratio 长宽比
diammonium phosphate 磷酸氢二铵（DAP）
diastase 淀粉酶;淀粉酶制剂
diazinon 二嗪农;氯化苦
diazotroph 固氮生物
diazotrophic bacterium 固氮细菌
diazotrophic biocoenosis 联合共生固氮
diazotrophic system 固氮体系
diazotrophy 氮素营养(吸氮作用)
dibble seeding 点播
dibbling 点播
dibbling implement 开穴机
dibbling of fertilizer 穴施
dibutyl phthalate 邻苯二甲酸二丁酯
dicalcium phosphate 磷酸二钙
dicalcium phosphate dehydrate 二水磷酸二钙
dichloroethane 二氯乙烷
dichloromethane 二氯甲烷
dichromate 重铬酸盐
dicot 双子叶植物
dicotyledon 双子叶植物
dieback of plant 植株枯萎
dietary disease 营养性疾病
difference between dry and wet bulb 干湿球温度差
differential precipitation 差别沉淀法
difficultly soluble nutrient 难溶性养分

difficultly soluble phosphatic fertilizer 难溶性磷肥
diffuse 扩散
diffusion 扩散(作用)
diffusion coefficient 扩散系数
diffusion condition 扩散条件
diffusion path 扩散路径
diffusion process 扩散过程
diffusion transport 扩散运输
diffusion-controlled process 扩散控制过程
diffusiveness 透发性
digeneration 退化
digenesis 世代交替
digestion 消化;蒸煮;浸提
digging plant hole 开穴
digging stubble 刨茬
digital agriculture 数字农业
digital earth 数字地球
digital elevation model(DEM) 数字高程模型
digital geomorphic model 数字地形模型
digital terrain model 数字地形模型
dihydroactinidiolide 二氢猕猴桃内酯
dihydrocoumarin 二氢香豆素
dihydroionone 二氢紫罗兰酮
dilatate 膨胀
dilatation 膨胀
dilution effect 稀释效应

dilution plate count 稀释平板计数
dilution rate 稀释率
dilution-plate method 稀释平板分离法
diluvial deposit 洪积物
diluvial soil 洪积土壤
dimethoate 乐果
dimethyl furan 二甲基呋喃
2,5-dimethylfuran 2,5-二甲基呋喃
dimethyl phthalate 邻苯二甲酸二甲酯
dimethyl pyrrole 二甲基吡咯
diminishing marginal return 边际报酬递减
diminishing marginal utility 边际效用递减
diminishing returns 报酬递减
diminution of fertility 肥力减退
dingy 光泽暗;(叶色)很暗
2,4-dinitrophenol 2,4-二硝基苯酚
dipotassium hydrogen phosphate 磷酸氢二钾(磷酸二钾)
dipped leaf 水浸叶
dipping seeds 浸种
dipterex (= trichlorfon) 敌百虫
direct counting method 直接计数法
direct drilling 直接播种
direct seeding 直播
direct solar radiation 直射阳光;太

阳直接辐射
direction for safe use of agricultural chemical 农药安全使用说明书
directional selection 定向选择;定向育种
directive breeding 定向培育
directive variation 定向变异
dirty （烟叶）污脏;不洁净
disaccharide 二糖;双糖
disagreeable odor 不愉快气息
disaster caused by windstorm 风灾
discarding plant 间苗
discharge furrow drain 排水沟
discharging barn （烟叶）出炕
discing 圆盘耙耕作
discoloration 褪色;脱色;变色
discoloring 褪色的;褪色（作用）;脱色
discounted variety 淘汰品种
discrete 离散
discrete altitude matrix 离散高程矩阵
discrete distribution 离散分布
discrete random 离散随机变量
discriminant analysis 判别分析
discriminant function 判别函数
discriminant test 判别试验
disease caused by weather extreme （烟草）气候性病害
disease control 病害防治
disease cycle 病害循环

disease diagnosis 病害诊断
disease endurance 耐病性
disease escaping 避病性
disease forecasting 病害预测
disease index 病情指数
disease intensity 病害强度
disease management 病害治理
disease monitoring 病害监测
disease outbreaking field 发病烟田
disease potential 病势
disease rating 病害分级
disease resistance 抗病性;抗病力
disease resistance breeding 抗病育种
disease resistant cultivar 抗病品种
disease spot 病斑
disease steady variety 抗病品种
disease susceptibility 感病性
disease symbiosis complex 病害共生复合体
diseased plant 病株
diseased seed 带病种子
disease-escape mechanism 避病机制
disease-producing germ 病原菌
diseases control of seed bed 苗床病害防治
diseases of seedling stage 苗期病害
diseases of tobacco field 烟草大田病害
disinfect 消毒

disinfecting action 消毒作用
disinfection 消毒
disinfestation 杀虫
disintegration of soil 土壤结构破坏
disintegrator 粉碎机
disk (disc) plow 圆盘犁
disk harrow 圆盘耙
disk type transplanter 圆盘式移栽机
disorder 失调
dispersal 传播
dispersant 分散剂
dispersion 离差
dispersion variance 离散方差
dissimilarity 不相似
dissimilatory nitrate reduction 异化性硝酸盐还原作用
dissipation of pesticide 农药分解代谢
dissolubility 溶解性
dissolution method 溶解法
dissolved inorganic phosphorus 溶解态无机磷
dissolved organic carbon 溶解有机碳
dissolved organic matter 可溶性有机物;可溶性有机质(DOM)
dissolved organic phosphorus 溶解态有机磷
dissolved reactive phosphorus 溶解态活性磷

dissolving tank 溶解罐
distance between plants 株距
distance between rows 行距
distance interval of hanging garlands (sticks) 挂烟间隔
distance isolation 间距分离
distance of planting 行株距
distant crossing 远缘(有性)杂交
distant hybrid 远缘杂种
distant hybridization 远缘杂交
distil 蒸馏
distillable 可蒸馏的
distillate 馏出液
distillation 蒸馏(作用)
distillation cut 馏分
distilled water 蒸馏水
distribution function 分布函数
distribution of crops 作物分布
distribution of resource 资源分布
disturbance horizon 扰动层
disturbance in transpiration 蒸腾作用失调
disturbance of enzyme activity 酶活性失调
disturbed soil 扰动土壤
ditch irrigation 沟灌
ditching 开沟
dithane manganese 代森锰
diurnal range 日较差
diurnal temperature and humidity change 温湿度昼夜变化

diurnal variation 日变化
divergent survey method 分散调查法
diversity 多样性
diversity index 多样性指数
dividends on land 土地报酬
division of agricultural region 农业区划
DNA amplification in vitro DNA体外扩增
DNA bending DNA转折;DNA弯曲
DNA blotting DNA印迹(法)
DNA body 脱氧核糖核酸体
DNA catenation DNA 连环
DNA circle DNA 环(环状DNA)
DNA cleavage DNA 裂解;DNA切割
DNA cloning DNA克隆(化)
DNA connected device DNA连接仪
DNA cross-linking 脱氧核糖核酸交联
DNA ligase 脱氧核糖核酸连接酶
DNA methylation 脱氧核糖核酸甲基化
DNA nicking DNA切口形成
DNA pitch DNA螺距
DNA polymerase 脱氧核糖核酸聚合酶
DNA recombination 脱氧核糖核酸重组
DNA reiteration 脱氧核糖核酸重复;脱氧核糖核酸复制
DNA restriction 脱氧核糖核酸限制
DNA sealase 脱氧核糖核酸连接酶
DNA sequence analysis DNA 序列分析
DNA sequencer DNA 序列测定仪;DNA 测序仪
DNA sequencing DNA 序列测定;DNA 测序
DNA sizing DNA 大小筛分
DNA sizing gene DNA 大小决定基因
DNA synthesizer DNA 合成仪
DNA typing DNA 分型
DNAase DNA 酶
DNAase I footprinting DNA 酶足迹法
DNA-binding protein 脱氧核糖核酸结合蛋白质
DNA-DNA Hybridization 脱氧核糖核酸-脱氧核糖核酸杂交
DNA-like RNA 类似脱氧核糖核酸的核糖核酸
DNP derivative 2,4-二硝基苯衍生物
dodecyl aldehyde 十二醛;月桂醛
dolomite 白云石
dolomite limestone 白云石质;石灰石
Dolycoris baccarm Linnaeus (蝽类的)斑须蝽象
domestic animal 家畜

domestic animals manure 畜粪尿
domestic tobacco leaf 国产烟叶
dominance index 优势度指数
dormancy of seed 种子休眠
dormancy stage 休眠期
dormancy-breaking 打破休眠；破除休眠
dormancy-inducing factor 诱导休眠因素
dormant stage 休眠阶段
dosage 用量
dosage effect 剂量效应
dosage for application 施药量；使用剂量
dosage form 剂型
dose 剂量；(用)药量
dose rate 施肥量
dot distribution map 散点图
dot mapping 散点制图
double auxotroph 双重营养缺陷型
double calcium superphosphate 重过磷酸钙
double cropping 一年两熟(两茬复种)
double phase erosion 两相侵蚀
double row sowing 双行播种
double sampling inspection 二次抽样
double superphosphate of lime 重过磷酸钙
double tube method 双管法(测土壤导水率)
double-degradable plastic film 双降解膜
down-draft (curing) barn 气流下降式烤房
downland 丘陵地
downslope cultivation 顺坡耕作
downward curling of leaf tip 叶尖下卷
downy mildew 霜霉；霜霉病
downy mildew blue mould 霜霉病
drab soil 褐土
draft equipment 通风设备
draft oven 鼓风干燥箱
drain ditch 排水沟
drainage 排水
drainage ditch 排水沟
drainage technique 排水技术
draw resistance 吸阻
draw resistance meter (tester) 吸阻测定仪
drench 浸润；浸湿
dress 整理；种子拌药(消毒)
dressed seed 拌药的种子
dressing 种子拌药(消毒)；拌种；施肥；追肥
dried cow manure 干牛粪
dried fruit type 干果香型
dried matter 干物质
dried poultry manure 干禽粪
dried sheep manure 干羊粪
drift 偏差

drill 条播；条播机；钻头
drill box 播种机种子箱
drill fertilization 条施(肥)
drill seeder 条播种机
drill sowing 条播
drilling 条施；条播
drilling seed 条播种子
drilling width 播幅
drip irrigation 滴灌
drip irrigation tape 滴灌带
drip irrigation under plastic mulch 膜下滴灌
drip system 滴灌系统
droop 低垂
drop off 脱落
drop-by-drop irrigation 滴灌
drought 干旱
drought calamity 干旱灾害
drought damaged tobacco 受旱烟草
drought disaster 干旱灾害
drought hardiness 耐旱性
drought monitoring 旱情监测
drought period 干旱周期
drought resistance 抗旱性；耐旱性
drought resistance line 抗旱品系
drought resistant and water saving 抗旱节水
drought season 枯水期；旱季
drought spot 旱斑；干斑
drought tolerance 耐旱性
drought-enduring 耐旱性

drought-resistant and water-absorbent polymer 抗旱保水剂
drowning 水害，淹涝，水淹，涝害
drug resistance 抗药性
drug-tolerance 耐药性
dry and wet bulb thermometer 干湿球温度计
dry bulb temperature 干球温度
dry combustion method 干烧法
dry cropland 旱耕地，旱作农田
dry damage 旱灾
dry deposition 干沉降
dry farming agriculture 旱作农业
dry hardiness 耐旱性
dry land 旱地
dry loam sand 干旱壤质沙土
dry matter 干物质
dry matter accumulation 干物质积累
dry matter partition ratio 干物质分配率
dry matter production 干物质产量
dry matter yield 干物质产量
dry method 干法
dry season 干季
dry soil 干旱土壤
dry through 烟气刺喉
dry weight 干重
dry-farming land 旱作农田
drying 干燥
drying apparatus 干燥装置；干燥设备
drying cabinet 干燥箱

63

drying degree 干燥程度
drying oven 烘箱;干燥炉;干燥箱
drying sample 样品干燥
drying-down 干燥
dryland farming 旱地农业
dryland wireworm 旱地金针虫
dryness 干燥感
duff 粗腐殖质
duff horizon 粗腐殖质层
duff mull 半腐解腐殖质
dull 暗(光泽);香气沉闷;无光的;单调的
dull color (烟叶)暗色
dull greenish 死青(烟叶);青暗色
dullness 沉闷
Dumate 代森锰锌
dung 粪;粪肥
dunghill 堆肥
duo-trio test 三角试验评吸法
durability of insecticide 药效持续法
durability of the aggregate 团聚体稳定性
durable aroma 持久香气
duration of day 日照长度;日长
duration of sunshine 日照时间;日照长度
dusky (叶色光泽)稍暗
duty of water 灌溉需水总量;作物需水量
dwarfing 矮化的
dwarfing agent 矮化病
dwarfism 矮态;矮生性;矮化现象
dwelling-curing 农家调制;农家烘烤
dynamic balance 动态平衡
dynamic characteristic 动态特征
dynamic equilibrium 动态平衡
dynamic model 动态模型
dynamic property of soil 土壤动力学性质
dysbiosis 生态失调
dystrophy 营养不良(营养不足)

E

earliness of germination 提早发芽
early bearing 早果
early budding 早花
early crop tobacco 早烟
early cropping 早熟
early flower stage 开花早期
early frost 初霜
early harvesting 早期采收
early high topping 早期高打顶
early leaf senescence 叶片早衰
early low topping 早期低打顶
early maturing variety 早熟品种

early period resistance　早期抗病性
early planted tobacco　早栽烟
early senescence　早衰
early sowing　早播
earth humus　土壤腐殖质
earth odor　壤香;泥土气味
earth temperature　地温
earth thermometer　地温计
earthworm　蚯蚓
earthworm cast　蚯蚓粪
earthworm compost　蚯蚓堆肥
earthy odour　土腥气
easy curing property　易烘烤性
eco-agriculture　生态农业
ecobiotic adaptation　生态环境区适应
ecoclimatic adaptation　生态气候适应
ecocycle　生态循环
ecocycling　生态循环
eco-demonstration area　生态示范区
eco-dumping　生态倾销
eco-engineering　生态工程
eco-environment recovery　生态环境恢复
eco-environment system　生态环境系统
eco-environmental degradation　生态环境退化
eco-environmental issues　生态环境问题

ecofactor　生态因子
ecofarming　生态农业
eco-interception　生态拦截
ecologic fitness　生态适应性
ecological adaptability　生态适应性
ecological adaptation　生态适应性
ecological agriculture　生态农业
ecological amplitude　生态幅
ecological and environment water requirements　生态环境需水量
ecological and physiological water requirement　生理生态需水
ecological appraisal　生态评价
ecological background　生态背景
ecological balance　生态平衡
ecological benefit　生态效益
ecological benefit of soil and water conservation　水土保持生态效益
ecological biodiversity　生态生物多样性
ecological breakdown　生态破坏
ecological capacity　生态承载力;生态容量
ecological carrying capacity　生态承载力
ecological collapse　生态破坏
ecological community　生态社区;生态群落
ecological compensation　生态补偿
ecological conservation　生态保护
ecological control　生态控制;生态

防治

ecological crisis 生态危机
ecological cycle 生态循环
ecological damage 生态破坏
ecological degeneration 生态退化
ecological deterioration 生态恶化
ecological disturbance 生态失调
ecological dominant 生态优势种
ecological economic equilibrium 生态经济平衡
ecological effect 生态效应
ecological engineering 生态工程
ecological environment 生态环境
ecological equilibrium 生态平衡
ecological evaluation 生态评价
ecological factor 生态因子
ecological field 生态场
ecological form 生态型
ecological fragility 生态脆弱性
ecological function 生态功能
ecological function division 生态功能区划
ecological function region 生态功能区
ecological harmfulness 生态危害
ecological hazard 生态危害
ecological heterogeneity 生态异质性
ecological impact 生态影响
ecological impact assessment 生态影响评价
ecological indicator 生态指示种;生态指示植物;生态表征

ecological integrity 生态完整性
ecological invasion 生态入侵
ecological menace 生态威胁
ecological model 生态模型
ecological monitoring 生态监测
ecological niche 生态小生境
ecological optimum 生态最适度
ecological plasticity 生态可塑性
ecological preservation 生态保护
ecological pressure 生态压力
ecological resources 生态资源
ecological restoration 生态恢复
ecological risk assessment 生态风险评价
ecological safety 生态安全
ecological security 生态安全
ecological setting 生态背景
ecological similarity 生态相似性
ecological simulation 生态模拟
ecological strategy 生态对策
ecological stress 生态压力
ecological suitability 生态适宜性
ecological suitability assessment 生态适宜性评价
ecological system 生态系统
ecological system of paddy field 稻田生态系统
ecological type 生态型
ecological vulnerability 生态脆弱性
ecological water quantity 生态用水

量
ecological water requirement 生态需水
ecological zonation of soil water 土壤水分生态分区
ecologically damaged 生态破坏
ecologically sustainable development 生态可持续发展
ecology 生态学
ecology harnessing 生态治理
ecology of biosphere 生物圈生态学
eco-mass 生态质量
economic benefit 经济效益
economic costs 经济成本;节约的成本
economic crop 经济作物
economic effect 经济效益
economic fertility 经济肥力
economic fertilization 经济施肥
economic optimum application rate 经济最佳施肥量
economic plant 经济作物
economic yield 经济产量
economical benefit 经济效益
economical character 农艺性状
economical crop 经济作物
economical use of resources 节约地使用资源
economics of fertilization 施肥的经济学
economics of fertilizer 肥料经济学

ecopedology 生态土壤学
ecoregulation 生态调节
ecosphere 生物圈;生态层
ecosystem 生态系统
ecosystem carbon balance 生态系统碳平衡
ecosystem component 生态系统要素
ecosystem degradation 生态退化
ecosystem destruction 生态破坏
ecosystem diversity 生态系统多样性
ecosystem function 生态功能区
ecosystem health 生态系统健康
ecosystem integrity 生态系统完整性
ecosystem modeling 生态建模
ecosystem sensitivity 生态系统敏感性
ecosystem-type 生态系统类型
ecotone 生态过渡带;生态交错带
ecotope 生态环境
ecotype 生态型
ectendotrophic mycorrhiza 内外生菌根
ectocommensal relationship 共生关系
ectomycorrhiza 外生菌根
ectoparasitism 外寄生(现象);外寄生性
ectosymbiosis 外共生(现象)
ectotrophic 体外营养的,外生的
ectotrophic mycorrhiza 外生菌根
edaphic community 土壤群落
edaphic drought 土壤干旱

edaphic ecotype 土壤生态型
edaphic factors 土壤因子
edaphic niche 土壤生态位
edaphology 土壤生态学
edaphon 土壤微生物(群)
edatope 土壤环境
eelworm 线虫;蠕虫
effect of control 防治效果
effect of fertilizer 肥效
effect of fertilizer on crop yield 肥料的产量效应
effect of interception 截留效应
effect of soil drying 干土效应
effect of soil freezing 冻土效应
effect on animal health (土壤)对动物健康的影响
effect on human beings health (土壤)对人类健康的影响
effect on plant health (土壤)对植物健康的影响
effective accumulated temperature 有效积温
effective aperture 有效孔径
effective area 有效面积
effective cation exchange capacity 有效阳离子交换量
effective component 有效成分
effective concentration 有效浓度
effective diameter of soil particle 土粒有效直径
effective fertility 有效肥力

effective fertility of soil 土壤有效肥力
effective pore space 有效孔隙
effective porosity 有效孔隙度
effective precipitable water 有效降水
effective precipitation 有效降水量;有效降雨量
effective radiation 有效辐射
effective rainfall 有效降雨量
effective root length 有效根长
effective rooting depth 有效扎根深度
effective soil fertility 土壤有效肥力
effective temperature 有效温度
effective utilization evaluation of resources 资源高效利用评价
effective volume 有效容积
efficiency for light utilization 光利用效率
efficiency for solar energy 太阳能利用率
efficiency of water application 水分利用率
efficient rotation 合理轮作
efficient use of fertilizer 肥料的有效利用
efflorescence 开花
effluent from farmland 农田排水
egesta 粪
eggplant 茄子

Eh value　Eh 值
eigenvalue　特征值
eigenvector　特征向量
eigenwert　特征值
elasticity　弹性;伸缩性
electric charge　电荷
electric conductance　电导
electric conductivity　导电性;电导率
electric conductivity method　电导法
electric furnace　电炉
electric heat seed bed　电热苗床
electric oven　电烘箱
electric plate　电热板
electric potential　电位(电势)
electrical conductivity　电导率;导电性
electrochemical analysis　电化学分析
electrochemical potential　电化学位
electrochemical property of soil　土壤电化学性质
electrode　电极
electron microscope　电子显微镜
electron microscopy　电子显微技术
electron scanning microscope　电子扫描显微镜
electron theory of acid and base　酸碱电子理论
electron transfer chain　电子传递链
electronic probe　电子探针

electronic scanning microscope　电子扫描显微镜;扫描电镜
electrophoresis　电泳
electroprobe microanalyser diagnosis method　电子探针诊断法
element analyzer　元素分析仪
element geochemistry　元素地球化学
element migration　元素迁移
element of cost　成本要素
element synergism　元素协同作用
element toxicity　元素毒害
elemental composition　元素组成
elemental nitrogen　元素氮
elementary charge　基本电荷
elementary constituent　基本组分
elementary microstructure　基本微结构
element-deficient culture solution　缺素营养液
elevation　海拔
ELISA(=enzyme-linked immunosorbent assay)　酶联免疫吸附实验;酶联免疫吸收分析,酶结合抗体法
elite　原种;良种
elite plot　原种圃
elite seeds　精选种子;优良种子
elongation zone(region)　伸长区
eluvial horizon　淋溶层
eluvial layer　淋溶层

eluvial soil 淋溶土
eluviation 淋溶作用;残积作用
eluvium 淋溶层
Embden-Meyerhof-Parnas pathway EMP途径
embedding 埋生的;包埋的
embryo 胚;胚胎
embryo axis 胚轴
embryo buds 胚芽
embryo stem 胚茎
embryonal axis 胚轴
embryonal structure 原始结构
embryonic bud 胚芽
embryophyta 有胚植物
embryotrophy 胚体营养
emergence 萌发;突出体
emergence of seedling 出苗
emigration 迁移
emission spectrometer 发射光谱仪
empirical constants 经验常数
empirical distribution 经验分布
empirical model 经验模型
empirical regression coefficient 经验回归系数
empirical regression equation 经验回归方程
emptier （烟味）平淡
empty 空的;(烟味)平淡
empyreumatique 焦香
end point 终点
end point of titration 滴定终点

endomycorrhiza 内生菌根
endoparasite 内寄生;内寄生物;内寄生虫
endosperm 胚乳
endotrophic mycorrhiza 内生菌根
endurance 耐性
enemy crop 天敌作物
energy crops 能源作物
energy element 能量元素
energy flow 能流
energy resource consumption 能源消耗
energy utilization 能源利用
energy-saving 节约能源的
energy-saving technology 节能技术
engineering erosion 工程侵蚀
English cigarette 英式卷烟;烤烟型卷烟
English-type straight virginia 英式纯烤烟
enid 草乃敌(除莠剂)
enol 烯醇
enovit 甲基托布津
enriched flavour （香气）浓郁
enrichment 富集
enrichment coefficient 富集系数
enrichment horizon 富积层
ensilage 青贮饲料
entisol 新成土
entomochorus 虫传的
entomochory 昆虫传播

entomogenous nematodes 昆虫病原线虫体
entomophily 虫媒
environment 环境
environmental assessment 环境评价
environmental background value 环境背景值
environmental bearing capacity 环境承受力
environmental benefit 环境效益
environmental biotechnology 环境生物技术
environmental capacity 环境容量
environmental capacity of soil 土壤环境容量
environmental challenge 环境挑战
environmental contamination 环境污染
environmental decay 环境衰退;环境破坏
environmental degeneration 环境退化
environmental degradation 环境退化
environmental deterioration 环境恶化
environmental effectiveness 环境有效性
environmental factor 环境因子
environmental friendly sustainable technology 环境友好型可持续发展技术
environmental humidity 环境湿度
environmental indicator 环境指示物
environmental issues 环境问题
environmental pollutant 环境污染物
environmental pollution 环境污染
environmental pollution assessment 环境污染评价
environmental pollution loss 环境污染损失
environmental problem 环境问题
environmental purification mechanism 环境净化机制
environmental quality 环境质量
environmental quality standard for soils 土壤环境质量标准
environmental renovation 环境修复
environmental standard of pollutant 污染物环境标准
environmental stress 逆境
environmental sustainability 环境可持续性
environmental temperature 环境温度
environmental toxic chemicals 环境有毒化学物质
environmental worsening 环境恶化
environmentally friendly 环境友好
environmentally friendly soil quality 友好的土壤环境质量
enzymatic degradation 酶促降解
enzymatic reaction 酶促反应

enzyme 酶
enzyme activity 酶活性
enzyme reaction 酶反应
enzyme selectivity 酶选择性
enzyme specificity 酶专一性
enzymology 酶学
enzymology diagnosis 酶学诊断
eolian 风积的(风成的)
eolian erosion 风蚀(作用)
eolian soil 风成土
ephebic 有翅成虫
epidemic 流行病;病害流行
epigeal cotyledon 出土子叶
epiphytic microorganism 附生微生物
episperm 种皮
epoxy ethane 环氧乙烷
equilibrium moisture content 平衡水分;平衡含水率
equilibrium water content 平衡水分;平衡含水率
ergamine 组胺
Erlenmeyer flask 锥形烧瓶
erode 腐蚀;侵蚀
eroded land 被侵蚀土地
eroded soil 被侵蚀土壤
erosion 侵蚀
erosion action 侵蚀作用
erosion by human activities 人为侵蚀
erosion by water 水蚀

erosion factor 侵蚀因素
erosive action 侵蚀作用
error 误差;错误
error analysis 误差分析
error bar charts 误差条形图
error control 误差控制
eruca 幼虫
ESM (= eletronic scanning microscope) 电子扫描显微镜;扫描电镜
essence 香精
essential amino acid 必需氨基酸
essential component 主要成分
essential element (生命)必需元素
essential elements of plant 植物必需元素
essential fatty acid 必需脂肪酸
essential mineral element 必需矿物质元素
essential nutrient 主要养分
essential plant element 植物必需营养元素
essential plant nutrient 植物必需养分
establishing (= seedlings) 定苗
establishment agriculture 设施农业
estimation 估计
ethanal 乙醛
ethanol 乙醇
ethanolamine 乙醇胺
ethene 乙烯

ether extract 乙醚萃取物
ethereal oil 芳香油类;精油
ethrel 乙烯利
ethyl acetate 醋酸乙酯;乙酸乙酯
ethyl alcohol 乙醇
ethyl aldehyde 乙醛
ethylenediamine tetraacetic acid 乙二胺四乙酸(EDTA)
ETS (= environment tobacco smoke) 环境烟气
eubiosis 生态平衡
eutric 肥沃的
eutrophication 富营养化
Euxoa intracta Walker 暗黑地老虎
Euxoa oberthuri Leech 白边地老虎;白边切夜蛾
Euxoa segetum schiffer Uller 黄地老虎
Euxoa trifarca Eversmana 三叉地老虎
evaluation index system 评价指标体系
evaluation method 评价方法
evaluation of site quality 立地质量评价
evaluation principle 评价原则
evaluation system 评价系统
evaporation 蒸发(作用)
evaporation capacity 蒸发量
evaporation curve 蒸发曲线
evaporation from land surface 地面蒸发
evaporation from vegetation 植物蒸腾(植物蒸发)
evaporation from water surface 水面蒸发
evaporation index 蒸发指数
evaporation loss 蒸发耗损
evaporation pan 蒸发皿
evaporation rate 蒸发速率
evapotranspiration 蒸发(作用)
even 平滑的;平展的;(烟叶)平展
evenness 均匀度
ever frozen layer 永冻层
excavation 挖掘
exceeding deficiency 极缺
excellent variety 优良品种
excess 过量
excess and deficiency 过剩与不足
excess fertilization 施肥过量
excess moisture 水分过多
excess nitrogen 过量施氮
excess of nitrogen 氮过量
excessive application of fertilization 施肥过量
excessive grazing 过度放牧
excessive moistening 过度湿润(过湿)
excessive vegetative growth 疯长
exchange adsorption 交换吸附
exchange capacity 交换容量
exchange diffusion 交换扩散
exchange phenomenon 交换现象
exchangeable acidity 交换性酸度

exchangeable adsorption 交换吸附
exchangeable ammonium 交换性铵
exchangeable anion 交换性阴离子
exchangeable base 交换性盐基
exchangeable calcium 交换性钙
exchangeable cation 交换性阳离子
exchangeable cation percentage 交换性阳离子百分率
exchangeable copper 交换态铜
exchangeable form 可交换态
exchangeable hydrogen 交换性氢
exchangeable ion 交换性离子
exchangeable iron 交换态铁
exchangeable magnesium 交换性镁
exchangeable manganese 交换性锰
exchangeable metal cation 交换性金属阳离子
exchangeable potassium 交换性钾
exchangeable sodium 交换性钠
exhausted soil 耗竭土壤
existing available moisture 实际有效水量
expanded cut rolled stem 膨胀梗丝
expanded cut tobacco 膨胀烟丝
experimental project 试验项目
experimental result 实验结果
experimental station 试验站
experimental value 试验值
expert knowledge 专家经验
expert system 专家系统
exploitability 可开发性;可利用性
exploitative agriculture 掠夺性农业
exploration 勘测(查勘;考察)
exploratory soil survey 土壤考察
exponential curve 指数曲线
exponential distribution 指数分布
exponential equation 指数方程
exponential function 指数函数
exponential growth 指数增长
exponential model 指数模型
exposure 坡向
ex-root fertilization 根外施肥
ex-root nutrition 根外施肥;根外营养
extension of new technology 新技术推广
extensive agriculture 粗放农业
extensive farming 粗放农业
extensive pattern 粗放型
external flavour type cigarette 外香型卷烟
external property 外部特征;外观特性
external property on appearance 外观性状
external standard method 外标法
external surface 外表面
external surface area 外表面积
extinguish 熄火;(烟支)熄火
extra pure 高纯试剂
extractable humus 可浸提腐殖质
extraction 萃取;浸提;提取

·74·

extractor 萃取器;浸提器
extra-root feeding 根外施肥
extreme arid 极端干旱
extreme halophiles 极端嗜盐菌
extreme thermophilic bacterium 极端嗜热菌
extreme value 极值
extremely hard 超硬实
extremum 极值

F

facilitated transport 协同运移
facility agriculture 设施农业
factor 因素;因子
factor analysis 条件分析;因素分析;因子分析
factor of soil fertility 土壤肥力因子
factor variance 因子方差
factorial fertilizer trial 肥料因子试验
factorial Kriging analysis 因子克立格分析法
factory nursery 工厂化育苗
facultative aerobe 兼性需氧菌
facultative anaerobe 兼性厌氧菌
facultative denitrification 兼性反硝化(作用)
facultative thermophile 兼性嗜热菌
fading out 褪色(分级)
faint odor 微弱香气
faintly colored 暗淡色
fall sowing 秋播
fallout of pesticide 农药的沉降
fallow 休(闲)耕
fallow cultivation 休闲地耕作
fallow culture 休闲栽培
fallow land 休闲地
fall-sown type 秋播型
family farming 小农经济;家庭式农业
fancy name 品名
far infrared 远红外
far ultra violet 远紫外
farm compost 农家肥
farm drainage 农田排水
farm economy 农业经济
farm efficiency 农业效率
farm implements 农具
farm manure 厩肥;农家肥
farm operation system 农事操作体系
farm pollution 农田污染
farm soil conservation 农地土壤保持
farm tools and implements 农具

farm water requirement 农田需水量
farm water supply engineering 农田给水工程
farm work 农活
farmer's leisure season 农闲期
farmers' grades and manufacturers' classification 农场等级和卷烟加工分级
farming 耕作
farming plastics membrane 农用塑料薄膜
farming system 耕作制;农作制度(农耕制度)
farming system of paddy field 水田耕作制
farming system of rainfed area 旱地耕作制
farming technique 耕作技术
farming with rainwater harvesting technology 集水农业
farming-grazing ecotone 农牧交错地
farmland 耕地;农田
farmland classification 农用地分等
farmland ecosystem 农田生态系统
farmland grading 农用地分等
farmland irrigation 农田灌溉
farmland mulch 农田覆盖
farmland protection 耕地保护
farmland quality assessment 农田质量评价
farmland resources 耕地资源

farmyard manure 农家肥
farmyard manure layer 厩肥层
farmyard manure spreader 撒堆肥机
fast growing period 旺长期
fat acid 脂肪酸
fat soil 肥土(沃土)
fat soil layer 肥土层
fat synthesis 脂肪合成
fatal dryness 致死干旱度
fatal high temperature 致死高温
fatal humidity 致死湿度
fatal low temperature 致死低温
fatal temperature 致死温度
fate of fertilizer nitrogen in agroecosystem 农田生态系统中肥料氮的去向
fatigue of soil 土壤疲乏(土壤耗竭)
fats 脂肪
fat-soluble pesticide 脂溶性农药
fatty acid 脂肪酸
fatty acid composition 脂肪酸成分
fatty stems 调制中烟梗水分过大的烟叶
favorable days for field work 田间作业适宜天数
favorable season for field work 田间适宜作业季节
favorable time for harvest 采收适宜期
F-distribution F-分布

Fe protein 铁蛋白
fecal manure 粪肥
fecal mass 粪
fecal material 粪
fecal pollution 粪便污染
feces 粪
fecund soil 肥土(沃土)
female parent 母本
femina 雌性
Fe-Mn oxide combined form 铁锰氧化物结合态
fenaminosulf ＜商＞敌克松
fermentation 发酵
fermentation chamber 发酵室
fermentation process 发酵过程
fermentation reaction 发酵反应
fermentation time 发酵时间
fermentative heat source 酿热物
fermenter 发酵罐
fermentor 发酵罐
ferrallitic soil 铁铝土
ferredoxin 铁氧还蛋白
ferric sulfate 硫酸铁
ferrous sulfate 硫酸亚铁
ferrous sulphate 硫酸亚铁
fertigation 灌溉施肥;肥水灌溉;加肥灌溉
fertile 肥沃的
fertile farmland 良田
fertile land 沃地;肥地
fertile land forming 熟地化

fertile soil 肥沃土壤
fertility 肥力
fertility capability classification (土壤)肥力分类
fertility characterization 肥力鉴定
fertility degree 肥力等级
fertility element 肥力要素
fertility evaluation 肥力评价
fertility grade 肥力等级
fertility index 肥力指标
fertility monitoring 肥力监测
fertilization 受精作用;施肥;传粉
fertilization based on formulation 配方施肥
fertilization diagnosis method 施肥诊断法
fertilization information system 施肥信息系统
fertilization model 施肥模式
fertilization pattern 施肥方式;施肥模式
fertilization planning 施肥计划
fertilization recommendation 推荐施肥
fertilization system 施肥制度
fertilizer 肥料
fertilizer analysis 肥料分析
fertilizer analytic formula 肥料分析式
fertilizer application 施肥
fertilizer application rate 施肥量

fertilizer application scheme 施肥方案
fertilizer application techniques for high yield 高产施肥技术
fertilizer applied before setting 栽前施肥
fertilizer at seeding stage 苗肥
fertilizer availability 肥料有效性
fertilizer burn 肥料烧伤
fertilizer component 肥料成分
fertilizer component and formula 肥料成分和配合式
fertilizer constituent 肥料成分
fertilizer deep applicator 深层施肥机
fertilizer deficiency 肥料缺乏
fertilizer demand 需肥量
fertilizer dressing 撒施(底)肥
fertilizer drilling 条施
fertilizer economics 肥料经济学
fertilizer effect 肥料效应
fertilizer efficiency 肥料效率
fertilizer element 肥料元素
fertilizer enhancers 肥料增效剂
fertilizer experiment 肥料试验
fertilizer experiment design 肥料试验设计
fertilizer experimental design 肥料试验设计
fertilizer formula 肥料配方
fertilizer grade 肥料品位

fertilizer in bands 带状施肥
fertilizer injury 肥料伤害
fertilizer input 肥料投入
fertilizer intolerant 不耐肥的
fertilizer levels 施肥水平
fertilizer loss 肥料流失
fertilizer management 肥料管理
fertilizer marking 肥料标识
fertilizer nutrient 肥料养分
fertilizer placement 施肥;施肥位置
fertilizer pollution 化肥污染;肥料污染
fertilizer pollution on environment 肥料对环境的污染
fertilizer practice 施肥技术
fertilizer rate 肥料用量;施肥量
fertilizer ratio 肥料配合比例
fertilizer recommendation 推荐施肥;肥料推荐
fertilizer recommendation based on soil testing 测土施肥;测土配方施肥
fertilizer regime 施肥制度
fertilizer requirement 需肥量
fertilizer response function 肥料效应函数
fertilizer response model 肥效模型
fertilizer salt injury 肥料盐害
fertilizer solitary applicating 施单一肥料
fertilizer source 肥源
fertilizer spreading 撒施肥料

fertilizer test 肥料试验
fertilizer tolerant variety 耐肥品种
fertilizer toxicity 肥害
fertilizer type 肥料种类
fertilizer unit 肥料单位
fertilizer with amino acid 含氨基酸类肥料
fertilizer-salt damage 肥料盐害
fertilizer-water irrigation 肥水灌溉
fertilizing structure 施肥结构
fertillizer pollution 肥料污染
fertiphos 磷肥
ferulic acid 阿魏酸
fiber(=fibre) 纤维;叶脉;筋脉
fiber color （叶脉）脉色
fibric soil material 纤维有机土壤物质
fibrous root 纤维根;须根
fibrous root system 须根系
field(moisture)capacity 田间持水能力
field capacity 田间持(水)量
field coefficient of permeability 田间渗透系数
field control 田间防治
field crop 农作物;大田作物
field cropping 大田栽培
field culture 露地栽培
field curing 田间调制
field development 大田生长发育
field diagnosis 田间诊断
field diseases 大田病害
field drainage system 田间排水系统
field ecosystem 农田生态系统
field experiment 田间试验
field experiment design 田间试验设计
field experiment interpretation 田间试验解释
field growth duration 大田生长期
field growth stage 大田生长期
field infection 田间侵染
field inspection 田间检验
field inspection of seedlings 查苗
field management 田间管理;现场管理
field management method 田间管理法
field maturing process 耕地熟化过程
field maximum moisture capacity 田间最大持水能力
field measurement 田间测定
field microclimate 农田小气候
field moisture deficiency 田间水分亏缺
field moisture equivalent 田间水分当量
field moisture regime 田间水分状况
field observation 田间观察
field of investigation 调查范围;研究领域

field plan 田间试验计划
field plot experiment 田间小区试验
field preparation 整地
field release 田间释放
field research 田间研究;实地研究;实地调查
field soil 田间土壤
field stratum 地面植被层
field strip cropping 带状耕种
field survey 田间调查;现场调查
field technique 田间技术
field test 田间试验;大田试验;肥料试验
field trial 田间试验
field water balance 农田水分平衡
field water capacity 田间持水量
field water consumption 农田耗水量
field water requirement 田间需水量
field water use efficiency 田间用水效率
field water-holding capacity 田间持水量
field wilting 田间萎蔫
fields for autumn sowing 秋地
fields for spring sowing 春地
field-specific nutrient management 定位养分管理
filamentous fungi 丝状真菌
filler ratio 填充料比例
filler tobacco 芯叶;填充料烟叶
filling capacity of cut tobacco 烟丝填充能力
filling the gaps with seedling 补苗
film-covering cultivation 地膜覆盖栽培
filter-tipped cigarette 滤嘴卷烟
filtration irrigation 渗灌
fine earth 细土
fine grain 细粒
fine granule 微粒
fine mature 适熟
fine odor 优美香气
fine powder 细粉
fine sand 细沙(细沙土)
fine sand soil 细沙土
fine screen 细筛
fine sieve 细筛
fine soil 细土
finely sandy 细沙的
fingerprint 指纹;指纹结构;指纹技术
fingerprint map 指纹图
fingerprinting 指纹法;指纹分析
finish furrow 开垄
finite resources 有限的资源
fire curing 明火调制
fire damp 沼气
fire damp fertilizer 沼气肥
fire-cured 明火烘烤的;明火烘烤
fire-cured tobacco 明火烤烟;熏制烟
fire-curing 明火烘烤;明火烤制

first flower 初花;第一朵花
first frost 初霜;早霜;霜降
fitness of environment 环境适应性
fixation 固定
fixation of ammonium by clay minerals 铵的黏土矿物固定作用
fixed ammonium 固定态铵
fixed atmospheric nitrogen 固定(的)大气氮素
fixed cost 固定成本
fixed potassium 固定态钾
fixing 固定;(烤烟过程)定色;卷烟纸、水松纸的上色处理
flame photometer 火焰光度计
flame photometric detector 火焰光度检测器
flame photometry 火焰光度法
flame spectrophotometer 火焰分光光度计
flame-emission spectrophotometer 火焰发射分光光度计
flask 烧瓶
flat land 平地;平原
flattened leaf bundle 平摊把
flavor component 香味组分
flavor composition 香味组分;(食用)香精
flavor notes 香味特征;香韵
flavor precursor 香气前体物质
flavor substance 香味物质;食用香料;调料

flavor type 香型
flavoring tobacco 香味型烟草;主料烟
flavouring ratio 加香率
flavouring tobacco 主料烟
flavourous 有香味的;有滋味的;味浓的
flavoursome 香气足
fleshy (烟叶)肥厚;稍厚
float bed 悬浮苗床;漂浮苗床
float system (育苗)漂浮系统
floating culture 漂浮栽培
flood 洪水
flood injury 涝害
flood irrigation 漫灌;淹灌
flood land 河滩地
flood-freezing injury 冻涝害
flooding 漫灌;洪水;溢流
flooding irrigation 漫灌;淹灌
flooding pipe 溢流管
flooding resistance 抗涝性
floor area of curing barn 调制室面积
floral aroma 花香
floral axis 花轴
floral bud 花(序)芽
floral differentiation 花芽分化
floral leaf 花叶
floral odour 花香
floral sterility 花而不实
florescence 开花期

· 81 ·

flow 流量
flow injection analysis 流动注射分析
flow injection analyzer(FIA) 流动注射分析仪
flow velocity 径流强度;流速
flower abscission 花蕾脱落
flower axis 花轴
flower axis length 花轴长度
flower branch 花枝
flower bud 花芽;花蕾
flower bud differentiation 花芽分化
flower colour 花色
flower drop 落花
flower induction 开花诱导
flower thinning 疏花
flower-bud appearing stage 现蕾期
flowering 开花
flowering date 开花日期
flowering inducing 诱导开花
flowering initiation 花蕾初期
flowering period 开花时期
flowering phase 开花阶段;花期
flowering stage 开花期
flowering without setting seed 花而不实
flow-off 径流
flue (烤房)火管;烟道
flue curing 火管烘烤;火管烤制;烘烤
flue curing barn 火管烤房;炕房;烤房

flue curing technique 烘烤技术
flue-cured 火管烘烤(的);烘烤(的)
flue-cured tobacco 烤烟
flue-cured type 烤烟型
flue-curing 烘烤
flue-curing barn 烤房
fluffy 指组织疏松、稍有身份、填充力好的烟叶
fluffy consistency 蓬松结持(度)
fluffy leaf 破碎叶
fluid fermentation 液态发酵
fluid fertilizer 液体肥料;流体肥料
fluid filter 液体过滤器;流体过滤器
fluid medium 液体培养基;流体介质
fluid suspension culture 液体悬浮培养
fluorescence analysis 荧光分析
fluorescence characteristic 荧光特性
fluorescence spectrophotometer 荧光分光光度仪
fluoride 氟化物
fluorimetric analysis 荧光分析
fluorine 氟
flush period 降雨期
fluvial erosion 河流侵蚀
fluviatile loam 冲积壤土
fluviogenic soil 冲积土壤

fluvisol 冲积土
fluvo-aquic soil 潮土
fly 脚叶
flying 白肋烟、烤烟最底部脚叶；脚叶；飘叶（分级，指淡晾烟）
foeniculin 茴香苷
foliage application 叶面施用（药）
foliage disease 叶面病害
foliage dressing 叶面喷洒
foliage spray 根外追肥；叶面施肥；叶面喷洒
foliage transpiration 叶面蒸腾作用
foliage treatment 茎叶处理
folial 小叶
foliar absorption 叶部吸收
foliar analysis 叶片分析
foliar application 叶面喷洒；茎叶喷洒；叶面施肥；叶面施药；根外追肥
foliar base 叶基
foliar diagnosis 叶片诊断
foliar feeding 根外追肥；叶面施肥
foliar fertilization 根外追肥；叶面施肥
foliar fertilizer 叶面肥料
foliar microelement fertilizer 微量元素叶面肥料
foliar N application 叶部施氮
foliar nutrition 叶片营养
foliar pathogen 叶部病原菌
foliar water absorption 叶部水分吸收
following crop 后作；后茬作物

Food and Agricultural Organization （FAO） 联合国粮农组织
food packet 营养袋
food security 食物安全
foot rot 根腐病；基腐病
forced aging 人工醇化
forced circulation 强制循环
forced sweating 人工发酵
forced ventilation 强制通风
forecast 预测；预报
foreign odor 不适的气味；异味
forest 森林
forest crop inter-cropping 林农间作
forest soil 森林土壤
form 叶态；叶状；（类）型；形态；形状；方式
form of nutrient 养分形态
form of soil consistency 土壤结持类型
form of soil nutrient 土壤养分形态
formaldehyde 甲醛
formalin ＜商＞福尔马林；甲醛水溶液
formation 形成（生成）
formative process 形成过程
formic acid 甲酸
formic ether 甲酸酯；甲酸甲酯
forms of soil boron 土壤硼形态
forms of soil calcium 土壤钙形态
forms of soil magnesium 土壤镁形态

forms of soil nitrogen 土壤氮素形态
forms of soil nutrient 土壤养分形态
forms of soil organic matter 土壤有机质的存在形态
forms of soil potassium 土壤钾素形态
formula fertilization 配方施肥
formulated fertilization 配方施肥
foster 培养
foul land 污染的土壤
foul water 污水
Fourier magnitude 傅里叶值
Fourier transform 傅里叶变换
Fourier transform infrared spectrometer 傅里叶变换红外光谱仪(计)
Fourier transform infrared spectroscopy 傅里叶变换红外光谱(法)
fowl manure 禽粪肥
fractional distillation 分馏;分馏作用
fractional precipitation 分级沉淀
fragility 脆性
fragrance 芳香;香味;香气;香料
fragrance material 香料
fragrant 香的;芳香的
frame culture 温床栽培
frame seeding 温床播种
Framework Convention on Climate Change 气候变化框架公约
frangibility 脆性
Frankliniella fusca Hinds 烟褐花蓟马;烟褐蓟马
free acid 游离酸
free alkali 游离碱
free amines 游离胺
free amino acid 游离氨基酸
free ammonia 游离氨
free burning rate 自由燃烧速度
free combustion (卷烟的)自由燃烧
free diffusion 自由扩散
free living nitrogen fixation 自生固氮作用
free oxide 游离氧化物
free particulate organic matter 游离态颗粒有机物
free sampling 自由采样
free survey 任意调查
free water 自由水
free-living nitrogen fixing bacteria 自生固氮细菌
free-living nitrogen-fixer 自生固氮菌
freeze injury 冻害
freeze point 冰点;凝固点
freeze tolerance 耐冻性;抗冻性
freeze-drying 冷冻干燥
freeze-thaw action 冻融作用
freeze-thaw erosion 冻融侵蚀
freeze-thaw process 冻融过程
freeze-thawing 冻融
freeze-thawing erosion 冻融侵蚀

freezing and thawing soil　冻融土壤
freezing damage　冻害
freezing injury　冻害
freezing tolerance　抗冻性;耐冻性
frenching　焦灼病(缺锰)
frequency　频率
frequency curve　频率曲线
frequency distribution　频率分布
frequency of application　施肥次数
frequency of sampling　采样频率
fresh and graceful aroma　香气清雅
fresh leaf　鲜烟叶
fresh leaf selection　鲜叶挑选
fresh leaf weight　鲜叶重
fresh manure　新鲜厩肥
fresh organic matter　新鲜有机物质
fresh seed　新鲜种子
fresh soil　荒地土壤(未垦地土壤)
fresh tobacco　鲜烟(叶)
fresh weight　鲜(叶)重
freshly　(香气)清新
freshly harvested　新采收的(烟叶)
fresh-sweetness aroma　清甜香
fresh-sweetness type　清香型
friability　松散性(酥性;脆性)
friability of soil　土壤酥度;土壤松散性
friable　松散的(酥性的;脆性的)
friable consistency　松散持(度)
friable soil　松散土壤
friction　摩擦(力)

frictional coefficient　摩擦系数
frictional resistance　摩擦阻力
frog eye（ = Cercospora nicotianae Ellis et Everhart）　烟草蛙眼病
frog-eye spot　蛙眼病斑
frond fracture　叶片破裂
frost　霜冻
frost bite　冻伤(霜害)
frost damage　冻害;霜害
frost hardiness　耐寒性
frost injury　冻害;霜害;霜冻害
frost period　有霜期
frost resistance　耐霜性;抗霜性
frost season　霜期
frost tolerance　耐寒性
frostbite　冻伤;霜害
frostbitten　霜害;霜打
frosted　霜冻(叶)
frosted leaf　霜冻烟叶
frost-free day　无霜日
frost-free growing season　无霜生长期
frost-free period　无霜期
frost-free season　无霜季;无霜期
frosting　降霜
frost-killing　霜冻致死
frostless season　无霜期
frozen　冻伤(烟叶)
frozen ground　冻土;冻地
frozen injury　冻害
frozen rain　冻雨

frozen soil 冻结土壤
frozen temperature 结冰温度
fructification 结实;结果;子实体
fructify 结果;结实
fructosan 果聚糖
fructose 果糖
fruit（ing）body 子实体
fruit acids （水）果酸
fruit bearing percentage 结果率
fruit set 结实;坐果;着果
fruit setting habit 坐果习性
fruit thinning 疏果
fruitfulness 丰收;多实;多产
fruity aroma 果香
full aroma 饱满香气
full bodied （叶态）丰满;身份充实
full bodied smoke 烟量充足;烟气饱满
full matured 完熟
full rich flavour 烟味丰富
full ripe stage 完熟期
full ripened seed 全熟种子
full seed 饱满种子
full seeding 撒播
full seedling stage 全苗期
full stand 全苗
full-flavored 浓郁
fullness （香气）充实
fullness and thickness （香气）浑厚
full-riped seed 完熟种子
full-time tobacco farmer 专业烟农

fully flower stage 盛花期
fully stocked 完全郁闭
fully-fermented compost 腐熟堆肥
fulvic acid 富啡酸(黄腐酸)
fumaric acid 富马酸;反式丁烯二酸;延胡索酸
fumigant 熏蒸剂
fumigating condition 熏蒸条件
fumigating technique 熏蒸技术
fumigation 熏蒸(消毒)法
fumigation injury 熏蒸药害
fumigation method 熏蒸法
fumigation operation 熏蒸操作
fumigation procedure 熏蒸步骤
function of fertilizer effect on crop 作物肥料效应函数
functional group 官能团
functional nutrient 功能养分
functional value 函数值
functions of alkaloids 生物碱功能
fungal disease 真菌病害
fungal remediation 真菌修复
fungicides effects 杀菌剂的效果
fungilytic action 溶菌作用
fungistatic action 抑真菌作用
fungitoxicity 真菌毒性
fungus 真菌
fungus disease 真菌病害
fungus disease of tobacco 烟草真菌病害
funky tobacco 霉味烟叶

funnel 漏斗
furaldehyde 糠醛
2-furaldehyde 2－呋喃甲醛;2－糠醛
furan(e) 呋喃
furancarbinol 呋喃甲醇;糠醇
furfural residue 糠醛渣
furrow 沟;畦;垄沟;槽;开沟
furrow and ridge tillage 沟垄耕作
furrow application 沟施;条施
furrow dressing 沟施
furrow drilling 沟播
furrow irrigation 沟灌
furrow planting 沟植
furrowing machine 开沟机
further outlook 短期天气展望
Fusarium 镰刀菌属;镰孢菌属
fused calcium-magnesium phosphate 钙镁磷肥
fused magnesium phosphate 镁磷混合肥
fusion method 熔融法
fuzzy mathematics 模糊数学方法
fuzzy theory 模糊理论

G

G (=green) 青色;绿色;青烟
gamma distribution γ分布
gang plow 多铧犁
gap of supply and demand 供需缺口
garbage compost 垃圾堆肥
gas analysis meter 烟气分析仪;气体分析仪
gas chromatogram 气相色谱图
gas chromatograph 气相色谱仪;气相色谱分析
gas chromatograph-mass spectrometer 气相色谱－质谱联用仪
gas chromatography 气相色谱;气相色谱法;气相色谱仪(GC)
gas chromatography-infrared technique 气相色谱－红外联用技术
gas chromatography-mass spectrometry 气相色谱－质谱分析法(气质联用)
gas exchange 气体交换
gas fertilizer 气体肥料
gas permeability 通气性;透气性
gas sampling 气体采集
gas toxicity 气体毒害
gas toxicity of protected land 保护地气体毒害
gaseous cycle 气体型循环
gaseous exchange 气体交换
gaseous fertilizer 气体肥料
gaseous fumigant 气体熏蒸剂

gaseous loss 气态损失
gaseous nitrogen loss 气态氮损失
gas-liquid chromatography 气-液色谱法
gas-solid chromatography 气-固色谱法
gathering plowing 内翻耕
gault 重黏土
gault clay 重黏土
Gauss distribution 高斯分布
Gaussian distribution 高斯分布
GC(＝gas chromatography) 气相色谱法
GC-IR(＝gas chromatography-infrared technique) 气相色谱-红外联用技术
GC-MS(＝gas chromatography-mass spectrometer) 气相色谱-质谱仪
geitonogamy(＝geitongamy) 同株异花受精
gel 凝胶
gel electrophoresis 凝胶电泳(法)
gelatin 凝胶
gem 叶芽
gene bank accession 基因库材料
gene chip hybridization system 基因芯片杂交系统
gene conversion 基因转换
gene diversity 基因多样性
gene duplication 基因复制;基因重复
gene engineering 基因工程学
gene expression 基因表达
gene location 基因定位
gene manipulation 基因操作
gene pool 基因库;基因源
gene recombination 基因重组
gene substitution 基因替代
genecology 遗传生态学;物种生态学
genera 属
general detailed soil survey 土壤普查
general fertility 总肥力
general fertilization model 通用施肥模型
general linear model 一般线性模型
general soil survey 土壤普查
generalized soil survey 土壤概查
generating season of pests 虫害发生季
generating tissue 分生组织
generation 一代;世代;时代
generous manure 优质肥料;优质厩肥
generous soil 肥土(沃土)
genesis and evolution of soil 土壤发生与演变
genesis of *Nicotiana* plants 烟草属植物起源
genetic diversity 遗传多样性

genetic drift 遗传漂变;遗传漂移
genetic factor 遗传因子
genetic immunization 基因免疫接种
genetic information 遗传信息
genetic marker 遗传标记
genetic pollution 基因污染
genetic potential 遗传潜力
genetic profile 基因档案
genetically modified agriculture 转基因农业
genetically modified cell 基因修饰细胞;基因改造细胞
genetically modified tobacco 基因修饰烟草;基因改造烟草
genotype 基因型
genotype-environment correlation 遗传型－环境相关
genotype-environment interaction 遗传型－环境互作
gentle slope 缓坡
genuineness of strain 品系纯度;纯种性
genus 属;类
genus hybrid 属间杂交
geochemical element 地球化学元素
geochemistry cycle 地球化学循环
geochory 土壤传播
geographic information 地理信息
geographic information science 地理信息科学
geographic information system（GIS） 地理信息系统
geoinformatics 地理信息科学
geometric growth 几何级数增长
geomorphic map 地貌图
geomorphological map 地貌图
geo-spatial data base 地理空间数据库
geosphere 陆圈
geostatistic mapping 地统计制图
geostatistics 地统计学
geothermal gradient 地温梯度
geothermometer 地温计
geotome 取土钻;取土器
geranyl acetone 香叶基丙酮
germ plasm 种质
germinating energy 发芽势
germinating power 发芽能力;萌发力
germination rate 发芽率
germination speed 发芽速度
germination temperature 发芽温度
germination test 发芽试验
germination test in blotter 吸水纸发芽试验法
germination test in soil 土壤床发芽试验
germplasm resource 种质资源
gibberellin 赤霉素
glandular hair 腺毛
glandular trichome 腺毛体;腺毛状体

glass electrode 玻璃电极
glass electrode potentiometry 玻璃电极电位计法
glass house culture 温室栽培
glass house effect 温室效应
glass rod 玻璃棒
glass slide 载玻片
glasshouse 温室
glasshouse cultivation 温室栽培
glasshouse culture 温室栽培
glasshouse effect 温室效应
glassware 玻璃器皿;玻璃仪器
gley 潜育土
gley horizon 潜育层
gley paddy soil 潜育水稻土
gleyed paddy soils 潜育水稻土
gleyed soil 潜育化土壤
gleyization 潜育作用
gleysol 潜育土
gliotoxin 木霉素;胶(霉)毒素
global assessment of soil degradation 全球土壤退化评价
global carbon cycle 全球碳循环
global sustained development 全球可持续发展
global warming 全球变暖
glow duration 发光持续时间
GLT(= green leaf threshing) 原烟打叶
glucose 葡萄糖
glucose oxidase 葡萄糖氧化酶

glutamate 谷氨酸盐
glutamate synthase 谷氨酸合酶
glutamic acid 谷氨酸
glutaminase 谷氨酰胺酶
glutamine 谷氨酰胺
glutamine synthase 谷氨酰胺合成酶
glutamine synthetase 谷氨酰胺合成酶
glutaraldehyde 戊二醛
glutathione 谷胱甘肽
glutathione peroxidase 谷胱甘肽过氧化物酶
Gly(= Glycine) 甘氨酸
glyceric acid 甘油酸
glycerol adsorption method 甘油吸附法(测比表面)
glycine 甘氨酸
glycolysis 糖酵解
glyoxal 乙二醛
glyoxylate cycle 乙醛酸循环
glyphosate 草甘膦
GMP(= genetically modified plant) 基因修饰植物;基因改造植物
Gnorimoschema operculella Zeller 马铃薯块茎蛾;烟潜夜蛾
gnotobiotically grown plant 无菌栽培植物
good cycle of agroecosystem 农业生态良性循环
goodness of fit 拟合优度;吻合度

goodness-of-fit test 拟合优度检验
gradational difference of fertility 肥力级差
grade and investigating method of tobacco disease 烟草病害分级及调查方法
grade of leaf quality elements 烟叶品质因素等级
graded bench 缓坡梯田
grades mixed 混级
grades of cultivated land 耕地等级
grades of tobacco leaf 烟叶等级
grading 分级
grading analysis 粒度分析
grading and handling 分级和整理
grading exercises 分级练习
grading factor 分级因素
grading leaf sample 分级烟叶样品
grading room 验级室
grading service 分级指导
gradual 渐变
graduated cylinder 量筒
graduated flask 量瓶
graft 嫁接
grain field 粮田
grain for green 退耕还林还草
grain safety 粮食安全
grain weight 粒重
grainy (= open grained) 组织疏松、吸湿性好的烟叶
gramineous green manure 禾本科绿肥

granite 花岗岩
granular fertilizer 颗粒肥料
granular pesticide 粒状农药
granular structure 团粒结构(粒状结构)
granville wilt (= Ralstonia solanacearum) 烟青枯病
grape sugar (= glucose) 葡萄糖
grass and cropping rotation 草田轮作
grass and wood ashes 草木灰
grassland 草地
grass-like odor 青草气
grassy odor 生青味
grassy smell 生青气
gravel 沙砾
gravel ground 砾质土
gravel soil 砾土(沙砾土)
gravelly 砾质
gravelly clay 砾质黏土
gravelly coarse sand soil 砾质粗沙土
gravelly fine sand soil 砾质细沙土
gravelly loam 砾质壤土
gravelly sand soil 砾质沙土
gravelly soil 砾质土
gravimeter 比重计;重力仪;重力计
gravimetric 重量的;重量分析的
gravitation erosion 重力侵蚀
gravitational erosion 重力侵蚀
gravitational potential 重力势

· 91 ·

gravitational sedimentation 重力沉降

gravitational water 重力水

gravity bottle 比重瓶

gravity irrigation 自流灌溉

gravity test 比重测定

gray correlation analysis 灰色关联分析

gray leaf 灰叶

gray spot 灰斑病

great circulation 大循环

great drought 大旱

great group of soil 土类

green 绿色的;新鲜的;未成熟的;青(叶);青色;青黄(分级术语)

green agriculture 绿色农业

green blight disease 青枯症(缺锌)

green economy 绿色经济

green fallow 绿肥休闲地

green farm produce 无公害农产品

green food 绿色食品

green gray 青灰色

green leaf 原叶;青烟;鲜(烟)叶

green leaf processing factory 原烟加工厂

green leaf threshing 原烟打叶

green leaf threshing and redrying 原叶打叶复烤

green manure 绿肥

green manure crop 绿肥作物

green manuring 压青;施绿肥

green note 青香韵

green nuance 青香

green odour 青杂气

green sickness 失绿症(萎黄症)

green spotty 青痕

green tips 顶叶

green tobacco 鲜烟叶;青烟叶;原烟(初烤烟叶)

green weight 鲜重;复烤前重量

green(A) 叶耳带绿

greenhouse 温室

greenhouse culture 温室栽培

greenhouse effect 温室效应

greenhouse gas 温室气体(GHG)

greenhouse management 温室管理

greenhouse nursery 温室育苗

greenhouse research 温室研究

greenhouse seedling 温室育苗

greenhouse test 温室试验

greenhouse type 温室类型

greenhouse ventilation 温室通风

greenish 黄带浮青;(烟叶)带青色;微带青

greenish leaf 青烟

greenish odour 青杂味

greenish white 青泛白;浮青

greenish-yellow 青黄;黄绿色

greenish-yellow leaf group 青黄叶组

greenish 微带青:代号V

green-leaf scent 青叶子香;鲜叶子香

greenness 绿色;新鲜
green-yellow (烟叶)青黄;青黄色(烟);青(烟);代号GY
grey desert soil 灰漠土
grey-brown desert soil 灰棕漠土
grid sampling 网格采样法
grid survey 网格调查
grind 粉碎;磨细
grinder 粉碎机;砂轮机;研磨机
grinding 研磨;研磨作用
gritty soil 沙砾质土
gross soil loss 土壤流失总量
grouan 粗沙
ground application 地面施用(药)
ground covered culture with plastic membrane 地膜覆盖栽培
ground fertilizer 基肥
ground information 地面信息
ground leaf 底(部)脚叶
ground litter 枯枝落叶层
ground making 整地
ground manure 基肥
ground phosphate rock 磷矿粉
ground resolution 地面分辨率
ground runoff 地表径流
ground seed bed 露地苗床
ground stratum 地面植被层
ground temperature 地面气温;地温
ground vegetation 地面植被
ground water 地下水
ground water erosion 地下水侵蚀

ground water resource 地下水资源
groundplasm 基质
groundwater 地下水
groundwater contamination 地下水污染
groundwater depth 地下水埋深
groundwater level 地下水位
groundwater table 地下水位
grouping 分组
grouping by colours of leaf 按烟叶颜色分组
grouping symbols 分组代号
groups 分组
growing area 种植面积
growing box 培养箱
growing character 栽培特性
growing medium 培养基
growing period 生长期
growing point 生长点
growing potentiality of seedling 苗期生长势
growing season 生长期;生长季节
growing wheat after rice 稻茬麦
growth analysis 生长分析
growth and development 生长发育
growth and management 种植管理
growth cabinet 生长室;人工气候箱
growth chamber 生长室;人工气候箱
growth curve 生长曲线
growth delay 生长发育迟缓

growth diagnosis 生长诊断
growth duration after transplant 大田生育期
growth environment 生长环境
growth factor 生长因素
growth hormone 生长激素
growth medium 生长介质
growth monitoring 长势监测
growth of early stage 早期生长发育
growth period 生长期
growth periodicity 生长周期性
growth phase 生长期
growth promoter 生长促进剂
growth rate 生长速率;增长率
growth regulating substances 生长调节物质
growth regulation factor 生长调节因子

growth regulator 生长调节剂
growth regulator and modifiers 生长调节剂与改良剂
growth rhythm 生长规律;生长节律
growth season 生长季节
growth substance 生长素;生长物质
Gryllotalpidae 蝼蛄科
guaranteed high-yielding field 高产稳产田
guaranteed reagent 优级纯试剂
guard row 保护行
guide to quality 品质导向
guideline for safety application of pesticides 农药合理使用准则
gum arabic 阿拉伯树胶
gutter-plough 开沟犁
gypsum 石膏

H

habitat 生境
habitat amelioration 生境改善
habitat classification 生境分类
habitat complex 生境综错作用;生境总体
habitat factor 生境因素
habitat stability 生境稳定性
hail 雹;冰雹
hail damage 雹害;雹灾

hail disaster 冰雹灾害
hail injury 雹害
hail-cut 雹伤(害)
hailed 雹伤的(烟叶)
hairs drop 茸毛脱落
hairy root 须根;毛壮根
hairy vetch 毛叶苕子
half leaf method 半叶法
half-leaf test 半叶测定

halophile 嗜盐微生物;喜盐性细菌
halophilic microorganisms 嗜盐微生物
halophytes 盐生植物
halotolerant 耐盐性
hand feeling 手感
hand loosening 解把
hand of tobacco 烟把
hand sorting 手选
hand tying 人工扎把
handed-leaf 把烟;扎把烟叶
hand-primed 手工采收;分批摘叶
hands loosening 解把
hand-suckered 人工抹杈
hand-tied 人工扎把
handy transplanter 手工移栽器
hanged retention water 悬着水
hanger 挂烟工
hanging density 挂(烟)密度
hanging leaf 挂叶
hanging sprayer 肩负型喷雾器
hanging tier number 挂烟层数
hanging water 悬着水
hanging weight of leaf 挂烟重量
haplic andosol 普通火山灰土(普通暗色土)
haplic gleysols 普通潜育土
haplic luvisol 普通淋溶土
haplic phaeozem 普通黑土
haplic solonetz 普通碱土
hard 硬实

hard tight leaf 坚实叶
hard to dissolve phosphatic fertilizer 难溶性磷肥
hardening 紧实
hardening of seedling 锻苗;蹲苗
hardening of the soil 土壤硬化
hardiness 抗性;耐性
hardiness physiology 抗性生理
hardly soluble nutrient 难溶性养分
hardness (烟气)粗糙,硬度
harmful element 有害元素
harmful substance 有害物质
harmless 无害的
harmonious development 协调发展
harmony (香气)谐调
harrow 耙
harrow plow 碎土犁
harrowing 耙地
harsh 粗糙(指烟叶、烟气)
harsh odor 粗糙气息
harsh taste 吃味粗糙;刺激味
harshness 粗糙性;刺激性
harvest index 收获指数
harvest maturity 采收成熟度
harvest method 收获方法
harvest period 收获期
harvest time 采收期
harvest yield 收获量
harvested area 收获面积
harvested leaf 采叶
harvester 采收机;收割机

harvesting 收割
harvesting aid 采收辅助机(设备)
harvesting on stalk(= stalk-cut harvesting) 砍株收割
harvesting operation 采收操作
hastening germination 促进萌发
hay incense 干草香
hazard of crop 农作物受害
hazardous substance 有害物质
heart rot 腐心症(缺硼)
heat conductivity 导热率
heat effect 热效应
heat efficiency 热效率;热能利用率
heat energy 热能
heat energy use efficiency 热能利用率
heat exchange 热交换
heat exchanger 热交换器;散热器;换热器
heat injury 热害
heat insulation 绝热;保温;热绝缘
heat insulator 隔热材料;保温材料;热绝缘体
heat interchange 热交换
heat radiation 热辐射
heat resource 热量资源
heat stress 热害;热逆境
heat tolerance 耐热性
heat wave 热浪
heated air circulating barn 热风循环式烤房

heated air curing 热风调制
heater cartridge 加热器套管
heater control 加热器控制
heater tube 加热管
heating equipment 加热设备
heating of manure 厩肥发酵
heating oven 烘箱;加热炉
heating power 燃烧热;热值
heating tube 加热管
heat-treated phosphate fertilizer 热法磷肥
heavier-textured soil 黏重土壤
heavy 浓的;重的;浓味;厚;大型的
heavy body 身份厚实;组织紧密;(烟叶分级)身份重
heavy clay 重黏土
heavy clay soil 重黏土
heavy crop 丰收;高产
heavy fertilization 重施肥法
heavy fraction of humic substance 重组腐殖质
heavy fraction of soil organic matter 重组土壤有机质
heavy fragrant aroma 香气浓馥
heavy green 深绿色
heavy leaf 厚叶(分级)
heavy loam 重壤土
heavy loam soil 重壤土
heavy metal 重金属
heavy metal accumulation 重金属富集

heavy metal bioavailability 重金属生物有效性
heavy metal combined contamination 重金属复合污染
heavy metal ion pollution of water body 水体重金属离子污染
heavy metal pollutant 重金属污染物
heavy metal pollution index 重金属污染指标
heavy metal stress 重金属胁迫
heavy odor 浓厚香气;沉厚香气
heavy rainstorm 暴风雨
heavy sandy clay 重沙质黏土
heavy sandy loam 重沙质壤土
heavy silt soil 重粉沙土
heavy soil 黏重土壤
heavy texture 黏重质地
heavy yield 丰收
height 高;海拔;顶点
height of ridge 垄高;畦高
height of seedling 苗高
height of topping 打顶高度
Heilu soil 黑垆土
heliogreenhouse 日光温室
heliophiles 喜阳植物;适阳植物
heliophilous seed 喜光性种子
Heliothis assulta Guenee 烟青虫(烟草夜蛾)
helpful element 有益元素
hemic soil material 半分解土壤有机物质

hemicellulose 半纤维素
hemicellulose decomposition 半纤维素降解
herbaceous 草本的;草质的;药草香的;干草气的
herbal aroma 药草香
herbal odour 药草气
herbicide 除草剂
herbicide contamination 除草剂污染
herbicide tolerance 耐除草剂
herby note 干草气息;药草香
heterochronogenous soil 次生土壤
heterogeneous combustion 不均匀燃烧
heterosis 杂种优势
heterotroph 异养生物
heterotrophic nitrification 异养硝化作用
heterotrophic nitrifier 异养硝化细菌
heterozygous plant 杂交后代植株
hexadecanoic acis 棕榈酸
hexoic acid 己酸
hexoic aldehyde 己醛
hexose 己糖
hexose diphosphate pathway 己糖二磷酸途径
hexyl caproate 己酸己酯
hibernacle 越冬场所;越冬巢
hibernal 冬季的

hibernate diapause 冬眠状态;冬眠滞育
hidden deficiency 潜在缺乏
hidden hunger 潜在饥饿(潜在缺素)
hierarchical cluster 聚类分析
hierarchy analytical method 层次分析法
hierarchy scheme 等级图
hierarchy theory 层级系统理论
high 高
high analysis fertilizer 高成分肥料
high and stable yield 高产稳产
high class leaf 上等烟
high concentration medium 高浓度培养基
high condensed complex fertilizer 高浓度复合化肥
high correlation 高度相关
high efficient and intensified use 高效集约化利用
high efficient utilization of resources 资源高效利用
high fatty acid 高级脂肪酸
high fertility 高肥力
high mountain region 高山地区
high mountain soil 高山土壤
high nutrient level 高营养水平
high performance liquid chromatogramphy masss pectrometer 高效液相色谱－质谱仪
high performance liquid chromatography(HPLC) 高效液相层析;高效液相色谱(法)
high pressure steam sterilizer 高压蒸汽灭菌器
high production 高产
high production land 高产田
high quality 优质
high resolution gel imaging system 高分辨率凝胶成像系统
high ridge 高垄;高畦
high risk rate 高发病率
high stress 重度胁迫
high temperature fermentation 高温发酵
high temperature injury 高温伤害
high temperature stress 高温胁迫
high yield 高产
high yield cultivating technique 高产栽培技术
high yield cultivation 高产栽培
high yield culture 高产栽培
high-analysis fertilizer 高浓度肥料
high-benefit eco-agriculture 高效生态农业
high-efficiency sustainable agriculture 高效持续农业
high-efficient eco-agriculture 高效生态农业
highland agriculture 山地农业
high-low charts 高低图

highly acidic 强酸性的
highly biodegradable 很容易生物降解的
highly decomposed organic horizon 高分解有机层
highly effective pesticide 高效农药
highly significant 高度显著性
high-speed centrifuge 高速离心机
high-temperature fermentation process 高温发酵法
high-yield and fine quality 高产优质
high-yield and high benefit 高产高效
high-yield field 高产田
high-yield grain area 高产粮区
high-yield plot 丰产田
hill 小山;丘陵
hill and mountain 丘陵山地
hill farm 丘陵农地
hill farming 山区农业
hill land 丘陵地
hill placement 穴施
hill seeder 穴播机
hill seeding 点播
hill sowing 穴播
hill spacing 穴距
hilling 培土
hillside farm 坡地农田
hillside land 坡地
hillslope runoff 坡面径流

hilly land 丘陵地
histidine 组氨酸
histogram 直方图
histogram breakpoint 直方图断点
histogram match 直方图匹配
histone 组蛋白
histosol 有机土
hoeing 锄地
hog manure 猪粪
holard 土壤总含水量
hole application 穴施
hole diameter 孔径
hole irrigation 穴灌(点浇)
hole seeding 穴播
hole treatment 穴施
hollow kernel 空籽;不饱满籽粒
hollow slide 凹玻片
hollow stalk 空茎病
hollow-ground slide 凹载玻片
holocoen 生态系统
holoside 多糖
homeostasis 动态平衡
homocysteine 高半胱氨酸
homogenized leaf curing 均质烟叶调制
homoserine 高丝氨酸
horizon C 母质层(土壤)
horizontal pathogenicity 水平致病性
horizontal resistance 水平抗性
horizontal transmission 水平传播
hormone 激素

hormone balance   激素平衡
host plant   寄主植物
host-specific   寄主专门化；寄主特异化
host-specificity   寄主专一性
hot damage   热害
hot fermentation   高温发酵
hot house   温室
hotbed farming   温床栽培
hot-house effect   温室效应
human and animal excreta   动物性废弃物有机肥；人畜粪
human disturbance   人为干扰
human excreta   人粪尿
human feces and urine   人粪尿
humate   腐殖酸盐
humi-   腐殖
humic   腐殖化
humic acid   腐殖酸；胡敏酸
humic acid compound fertilizer   腐殖酸复合肥
humic acid fractions   胡敏酸组分
humic acids / fulvic acids ratio   胡敏酸－富啡酸比（HA/FA）
humic fertilizer   含腐殖酸类肥料
humic materials   腐殖物质
humic substance   腐殖物质
humics   腐殖质
humid farming   灌溉农业
humidified air exhausting   排湿
humidifier   加湿器（机）；润湿器

humidity reducing by raising temperature   升温降湿
humidity reducing by ventilation   通风排湿
humification   腐殖化（作用）
humification coefficient   腐殖化系数
humification process   腐殖化过程
humified organic matter   腐殖化有机物质
humin   胡敏素（HM）
humo-fulvic acid   胡敏－富啡酸
humus   腐殖土；腐殖质
humus accumulation   腐殖质积累作用
humus analog   腐殖质类似物
humus coal   腐殖煤；褐煤；腐质煤
humus content   腐殖质含量
humus enriched horizon   腐殖质富集层
humus form   腐殖质组型
humus fractional composition   腐殖质组成
humus fractionation   腐殖质分级
humus horizon   腐殖质层
humus layer   腐殖质层
humus quality   腐殖质品质
humus reserve   腐殖质储量
humus soil   腐殖质土壤；腐殖土
humus-clay complex   腐殖质－黏粒复合体
humus-forming process   腐殖质形成

过程
hungry soil 瘠薄土壤
hurricane 飓风
hybrid variety 杂交品种
hybrid vigor 杂种优势
hydragric horizon 潴育层
hydragric paddy soil 潴育水稻土
hydrated lime 熟石灰;消石灰
hydrated oxide 氢氧化物
hydration compensation 水合补偿点
hydration water 结合水
hydraulic lift （根系）提水作用
hydraulic lift of root system 根系提水作用
hydride 氢化物
hydrocarbon 烃;碳氢化合物
hydrochloric acid 盐酸
hydrocinnamic acid 苯丙酸;氢化肉桂酸
hydro-culture 水培
hydrocyanic acid 氢氰酸
hydrogen 氢
hydrogen bromide 溴化氢
hydrogen cyanide 氰化氢
hydrogen ion concentration 氢离子浓度
hydrogen ion exponent 氢离子指数
hydrogen ion index 氢离子指数
hydrogen ion indicator 氢离子指示剂
hydrogen peroxide 过氧化氢

hydrogen sulfide 硫化氢
hydrogenic plant culture 水培养
hydrolases 水解酶
hydrologic cycle 水循环
hydrological balance 水分平衡
hydrolysable nitrogen 水解氮
hydrolysate 水解产物;水解液
hydrolysis 水解
hydrolysis constant 水解常数
hydrolytic reaction 水解反应
hydrolyzable nitrogen 水解氮
hydrometer （液体）比重计
hydrometer method 比重计法
hydromorphic paddy soil 潴育水稻土
hydromorphic soil 水成土
hydrophilic colloid 亲水胶体
hydrophilic polymer 亲水聚合物
hydrophobic colloid 疏水胶体
hydrophobic substances 厌水物质
hydroponic culture 水培
hydroponics 水培法
hydro-regime 水分状况
hydroscopic 吸湿的
hydroscopicity 吸湿性;吸湿度
hydrosol 水溶胶
hydrosphere 水圈
hydroxide ion 氢氧离子(羟基离子)
hydroxybutanedioic acid 羟基丁二酸;苹果酸

101

hydroxyl 羟基
hydroxyl group 羟基
hydroxyl ion 氢氧离子(羟基离子)
hygrometer 湿度计
hygroscopic 吸湿的
hygrothermograph 温湿度自动记录仪;温湿计
Hymenoptera 膜翅目
hyper spectral data 高光谱数据
hyper spectral remote sensing 高光谱遥感
hyper spectral sensor 高光谱传感器
hyperaccumulation 超富集
hyperaccumulator 超积累植物(超富集体)
hyperaccumulator plant 超积累植物
hyperfine structure 超细微结构;超精细结构
hyperspectral remote sensing 高光谱遥感
hyperthermophiles 超嗜热菌
hypertrophy 畸长;疯长
hypha 菌丝
hyphosphere 菌丝际
hypoplasia 细胞减生;抑生(现象)
hypoplastic 发育不良的;细胞减生的
hypoplasy 发育不全
hypothesis 假设
hypothesis testing 假设检验
hypotrophys 营养不良(营养不足)
hypsographic map 高程地图
hysteresis 滞后现象(平衡阻碍)
hythergraph 温度雨量图;温湿图

I

I. D. (= infective dose) 侵染剂量
IAA Oxidase 吲哚乙酸氧化酶
IAA(= indole acetic acid) 吲哚乙酸;异苗长素
ice box 冰箱
ice storm 冰爆
ichneumonid parasite of tobacco budworm 烟草夜蛾姬蜂
Ichneumonidae 姬蜂科;地老虎寄生姬蜂
IDA(= isotope dilution analysis) 同位素稀释分析(法)
ideal climate 理想气候
ideal cultural system 理想栽培系统
ideal system architecture 理想构型
ideal texture 理想质地
identification 鉴定
identification of disease resistance in tobacco cultivar 烟草品种抗病性鉴定

idioecology　个体生态学
idiotroph　专性营养
idle land　休闲地
IFR（＝infrared）　红外线（的）
igneous rock　火成岩
ill-aerated　通气性差
illite　伊利石
illumination culture　光照栽培
illumination degree　照度
illustrated chart　图示曲线；图表
illustration　插图；图解；例证说明；实例；解说
illuvial horizon　淀积层
illuvial soil　淀积土壤
illuviation　淀积作用
illuvium　淀积层
imagines　成虫
imago　成虫（复，imagines，imagos）
imbedding　覆盖；埋藏
imbibing seed　吸胀种子
imbibitional moisture　吸入水
imidole（＝pyrrole）　吡咯
immature　未熟；未成熟；成熟度不足
immature embryo　未成熟胚
immature seed　未成熟种子；未成熟籽
immature soil　生土；未成熟土
immature（＝incompletely riped）　成熟度不足；未成熟的（烟叶）
immature　未熟

immaturity　未成熟
immigration　移入
immobile nutrient　非活性养分
immobility　固定性
immobilization　固定
impact　劲头
impact of nitrogen to the environment　氮对环境的影响
impaired development　发育受阻
impaired drainage　排水不良
impart depth　增浓香气
impeded drainage　排水不良
imperfect grain　不完整种子；不饱满籽粒
imperfectible drainage　排水不完全
imperfectly drained　排水不完全
impervious soil　不透水土壤；渗透性不良土壤
implant　移植；植入
implantation　植入（法）；移植
impoverished soil　贫瘠土壤
impoverishment　（土壤）贫瘠化
improved land　已改良土地
improved mulching　改良地膜覆盖
improved practice　改良措施
improved seed strains　改良种系；良种
improved variety　改良品种
improved variety of seeds　良种（子）
improvement degree of soil　土壤的熟化度

improvement of contaminated soil 污染土壤改良

improvement of medium and low-yielding land 改造中低产田

improvement of saline and alkali soil 盐碱土改良

improvement of soil fertility 提高土壤肥力

improvement of variety 品种改良

impure aroma 香气不纯

impurity of seeds 种子混杂度

in situ bioremediation 原位生物修复

in situ chemical remediation 原位化学修复

in situ sampling method 原位采样方法

in situ soil leaching 原位土壤淋洗法

in vitro culture method 离体培养法

in vitro pollination 试管授粉

in vitro propagation 离体繁殖

inactive porosity 无效孔隙度

inanition 营养不良(营养不足)

inavailable water 无效水

inaxial root 侧根

incandescent electric lamp 白炽灯

incidence of disease 发病率

incineration 焚烧

incinerator 焚烧炉

incipient wilting 初萎

inclined field 坡地

incomplete block design 不完全区组设计

incomplete fertilizer 不完全肥料

incomplete flower 不完全花

incompletely riped 成熟度不足;未成熟的(烟叶)

incorporation of organic residues 有机残体混入

increase in quality 提高质量;改善质量

incubation 培养

incubation method 培养法

incubator 恒温箱;培养箱;孵化箱;细菌培养器

incursion 侵入(过程)

indensity of photosynthesis 光合作用强度

independent sample T test 独立样本的检验

2 independent sample test 二个独立样本检验

index of aridity 干燥指数

index of irrigation need 灌溉需水指数

index of pollution 污染指数

indicator 指示剂

indicator organism 指示生物

indicator of pollution 污染指示物

indicator paper 试纸

indigenous microorganism 土著微生

物
indigenous species 本地品种(土种)
indigenous technology 当地技术
indigenous variety 本地品种
indirect effect 间接作用
indirect fertilizer 间接肥料
indirect ordination 间接排序
indiscriminate use of pesticides 农药的滥用
indissoluble 不溶解的
individual development 个体发育;个体发展
individual plants 个体植株;单株
individual pollution index of soil 土壤单项污染指数
individual seed 单粒种子;种子个体
individual selection 单株选择;个体选择
indole 吲哚
indoleacetic acid 吲哚乙酸(IAA)
indoor climate 室内气候
indoor growing 室内栽培
induced deficiency 诱导缺乏
induced effect 诱导效应
induced enzyme 诱导酶
induced susceptibility 诱发感病性
inductively coupled plasma 电感耦合等离子体
inductively coupled plasma atomic emission spectrometry 电感耦合等离子发射光谱仪(ICP-AES)
inductively coupled plasma-mass spectrometry 电感耦合等离子体质谱分析法
industrial agriculture 农业产业化
industrial crop 经济作物;工业原料作物
industrial pollution 工业污染
inert humus 惰性腐殖质
infectious disease 侵染性病害;传染病
infertile 贫瘠的
infertile land 瘠地
infertile soil 贫瘠地;瘠土
infertility 不育性;不育;不肥沃
infested with weeds 长满杂草的
infiltrability 入渗强度
infiltrating irrigation 渗灌
infiltration 入渗
infiltration capacity 入渗量
infiltration flux 入渗通量
infiltration irrigation through shallow furrow 浅沟灌溉
infiltration rate 入渗率
infiltration velocity 入渗速度
inflorescence 花序;开花期
information 信息
information agriculture 信息农业
information carrier 信息载体
information era 信息时代

information society 信息社会
information system 信息系统
information theory 信息理论
information transfer 信息传递
infrared 红外线
infrared absorption spectrum 红外吸收光谱
infrared baking 红外烘烤;红外焙烤
infrared spectrum 红外光谱
ingraft 嫁接
ingrowth bag 生长袋法
inhalable particulate 可吸入颗粒物
inherent fertility 潜在肥力
inheritance resistance 遗传抗性
inhibiting germination 抑制发芽
inhibition 抑制
inhibition of axillary bud 抑制腋芽
inhibition of root growth 根生长受抑制
inhibitor 抑制剂
inhibitor for nitrification 硝化抑制剂
inhibitory effect 抑制效应
initial fertility 原始肥力
initial infiltration rate 初始入渗率
initial instar caterpillars 初龄幼虫
initial soil 原始土壤
initial soil moisture content 土壤初始含水率
injurious weed 有害杂草
injury 破损

injury by continuous cropping 连作危害;重茬危害
injury by hail 雹害
injury by pollution 污染危害
innoxious 无害的
innutrition 营养缺乏;营养不良
inoculant 接种剂
inoculate 接种
inoculated flat test 平板接种试验
inoculation （人工)接种;芽接
inoculum size 接种量
inorganic acid 无机酸
inorganic agriculture 无机农业
inorganic anions 无机阴离子
inorganic chemical pollution 无机化学品污染
inorganic contamination 无机污染物
inorganic elements 无机元素
inorganic fertilizer 无机肥料
inorganic nitrogen 无机氮
inorganic nitrogen compound 无机氮化合物
inorganic nitrogenous fertilizer 无机氮肥
inorganic nutrient 无机养分
inorganic nutrition 无机营养
inorganic phosphorus 无机磷
inorganic pollutant 无机污染物
inorganic pollution 无机污染
inorganic pyrophosphatase 无机焦磷酸酶

inorganic salt 无机盐类
inorganic substance 无机物质
inorganic sulfur 无机态硫
input by fertilizer 肥料输入
input-output analysis 投入产出分析
input-output ratio 投入产出比率
insect 昆虫
insect bait 诱虫饵
insect enemy 昆虫天敌
insect infestation 遭受虫害
insect injury 虫害
insect resistance 抗虫性;昆虫抗药性
insecticidal efficiency 杀虫效果
insecticide 杀虫剂
insectifuge 驱虫剂
insolation action 暴晒作用
insolubilization 不溶化;不溶解
insoluble oxide 难溶性氧化物
instar （虫）龄;龄期
instrumental analysis 仪器分析
instrumental insemination 人工授精
instrumental method 仪器方法
insufficiency 不足
insufficiency of supplies 供给不足
insufficient supply 供应不足
integrated control of environmental pollution 环境污染综合防治
integrated nutrient management 综合养分管理
integrated nutrient resource management 养分资源综合管理
integrated pest control 病虫害综合防治
integrated soil fertility index 综合土壤肥力指数
integrated survey 综合考察
intelligent agent 人工智能
intense culture 集约栽培
intense fall 暴雨
intense odor 强烈香气;强烈气息
intensification of cropping 种植集约化
intensity factor 强度因子
intensity of illumination 光照强度
intensity of photosynthesis 光合作用强度
intensity of soil respiration 土壤呼吸强度
intensive agricultural region 集约化农区
intensive agriculture 集约农业
intensive and meticulous cultivation 精耕细作
intensive cultivation 集约栽培;精耕细作
intensive farming 集约农业
intensive pattern 集约型
interact soil(= suppressive soil) 抑制（菌）土壤
interaction between nutrients 养分互作

interaction of genes 基因互作
interactions among nutritional elements 各营养元素间的相互作用
interblock(= between block) 区组间
interbreed 品种间杂交;变种间杂交
interception 截获
intercrop 间作物
intercropping 间混作;间作
intercropping crop 间作作物
intercropping interplanting 间作套种
intercropping model 间作模式
intercrossing 互作
intercultivation 中耕
interculture 间作物
interdeme selection 群间选择
interface of root and soil 根－土界面
interference 干扰
interfering component 干扰组分
intergrade soil(= transitional soil) 过渡性土壤
interionic relationship 离子间关系
interlayer charge 层间电荷
interlayer fixation 层间固定
interlayer potassium 层间钾
interlayer surface 层间表面
interlinear hybrid 品系间杂交;单交杂种

intermediate culture 补植;补栽;间作
intermediate decomposed organic horizon 半分解有机层
intermediate oxide 两性氧化物
intermediate soil 过渡土壤;中间土壤
internal cycle 内循环
internal standard method 内标法
internal surface 内表面
internal surface area 内表面积
International Organization for Standardization 国际标准化组织(ISO)
International System of Units 国际单位制(SI)
internode 节间;节距
internode length 节间长度;节距
interplant 间作;套种
interplanting 间种;套作
interrow crop 间作物
inter-row tillage 行间中耕
interspecies crossing 种间杂交
interspecies interaction 种间关系
interspecific gene transfer 种间基因转移
interspecific relationship 种间关系
interstices of soil 土壤孔隙
intertill(age) 中耕
intertillage crop 中耕作物
intertilled 中耕
interval 间隔;间距

interval estimation 区间估计
interval of safe use of pesticide 农药安全使用间隔期
interval of spraying 喷洒间隔
interveinal chlorosis 脉间失绿
interveinal necrosis 脉间坏死
interveinal yellowing 脉间黄化
intraspecific competition 种内竞争
intraspecific relationship 种内关系
introducer 引种者
inundation 浸水(洪水;泛滥)
inundation method 漫灌法
invading water 侵蚀水
inverse correlation 逆相关(负相关)
inverse Fourier transform 傅里叶逆变换
inverse matrix 逆矩阵
invertase 蔗糖酶;转化酶
invigoration of seeds 提高种子活力
involution 退化
ion antagonism 离子拮抗作用
ion binding affinity 离子亲和力
ion channel 离子通道
ion chromatography 离子色谱法
ion concentration 离子浓度
ion diffusion 离子扩散
ion exchange 离子交换
ion exchange adsorption 离子交换吸附
ion exchange chromatography 离子交换色谱(法)

ion exchange column 离子交换柱
ion exchange electrophoresis 离子交换电泳(法)
ion exchange resin 离子交换树脂
ion exclusion 离子排斥
ion pump 离子泵
ion selective electrode 离子选择性电极法
ion selectivity 离子选择性
ion synergism 离子协同作用
ionic equilibrium 离子平衡
ionic exchange 离子交换
ionic nutrient 离子态养分
ionic radius 离子半径
ionic replacement 离子取代
ionic size 离子大小
ionic state 离子态
ionic strength 离子强度
ionic substitution 离子取代
ionic valence 离子价
ionization 电离
ionol 紫罗兰醇
ionone 紫罗兰酮
iron 铁
iron and aluminum oxide 铁、铝氧化物
iron bacteria 铁细菌
iron chelates 螯合铁
iron chlorosis 缺铁失绿症
iron clay 铁质黏土
iron deficiency 缺铁症;缺铁
iron deficiency chlorosis 缺铁褪绿病

iron disturbance 铁素营养失调
iron fertilizer 铁肥
iron oxide 氧化铁
iron starvation 缺铁症
iron sulfate 硫酸铁
irradiance 辐照度
irradiate 照射
irradiation of ion 离子辐射
irregular 不规则
irregular chlorosis 不规则的失绿
irregular growth 生长异常
irregularly distributed necrotic spots 不规则分布的坏死斑点
irreversible adsorption 不可逆吸附（作用）
irreversible mechanical compaction 不可逆机械压实
irrigate 灌溉
irrigated crop 灌溉作物
irrigated farming 灌溉农业
irrigated farmland 灌溉农田
irrigated land 灌溉地;水浇地
irrigated paddy fields 灌溉水田
irrigating net 灌溉网
irrigation 灌溉
irrigation amount 灌水量
irrigation before seeding 播前灌水
irrigation by gravity flow 自流灌溉
irrigation by infiltration 浸润灌溉
irrigation canal 灌溉渠
irrigation date 灌溉时间

irrigation efficiency 灌溉效率
irrigation equipment 灌溉设备
irrigation erosion 灌溉侵蚀
irrigation facility 灌溉设备
irrigation farming 灌溉农业
irrigation field 灌溉地;水浇地
irrigation frequency 灌水频率
irrigation in growing period 生育期灌水
irrigation in upper rhizosphere 根层上部灌溉
irrigation installation 灌溉设备
irrigation interval 灌溉间隔(距)
irrigation method 灌溉方式
irrigation norm 灌溉标准
irrigation plant 灌溉设备
irrigation principal 灌溉原则
irrigation procedure 灌溉制度
irrigation project 灌溉工程;灌溉计划
irrigation quota 灌溉定额
irrigation regime 灌溉制度
irrigation requirement 灌溉需水量;灌水定额
irrigation system 灌溉系统
irrigation technique 灌水技术
irrigation tool 灌水工具
irrigation water 灌溉(用)水
irritancy 刺激性
isobutyric acid 异丁酸
isolated culture 隔离栽培;隔离培养;离体培养

isolation 分离
isolation ditch 隔离沟
isolation field 隔离田
isolation of microbes 微生物分离
isolation strip 隔离行;保护行
isoleucine 异亮氨酸
isomer 同分异构体
isomerase 异构酶
isomorphous substitution 同晶置换
isophane 等物候线;等始花线
isophenological line 等物候线
isophorone 异佛尔酮
isophyte 等高植生带

isoprene 异戊二烯
isotherm 等温线
isothermal 等温的;等温线(的)
isothermal drying oven 恒温干燥箱
isothermal line 等温线
isotope 同位素
isotope abundance 同位素丰度
isotope effect 同位素效应
isotope tracer method 同位素示踪法
isotopic labeling 同位素标记
isotopic tracer technique 同位素示踪技术

J

jar fermentor 发酵罐
jarovization 春化作用;春化处理
join monitor (盘纸)接头检测器
joint 接头;黏合
joint pollution 复合污染
jointing compound 密封剂;密封胶;黏合剂

judicious mating 定向杂交;合理交配
juglone 胡桃醌;胡桃酮
juvenile resistance 早期抗性;幼龄抗病性
juvenile soil 幼年土(沉积物上发育的);原生土壤
juvenile stage 幼期;幼龄期

K

K(variegated) 杂色(烟叶分级);灰色

Kachinsky classification system of soil fraction 卡庆斯基土粒分级制

Kachinsky classification system of soil texture 卡庆斯基土壤质地分类制
Kachinsky pipet 卡庆斯基吸管
kainit 钾盐镁矾(肥料)
kali salt 钾盐
kalk 石灰
kaolin 高岭土
kaolinite 高岭石
kapillary(= capillary) 毛细管
karst 喀斯特;岩溶;水蚀石灰岩地区
karst landform 喀斯特地貌
karst region 喀斯特地区
katabolism(= catabolism) 分解代谢
katalase 过氧化氢酶
katalaze(= enzyme) 催化酶
K-Ca activity ratio 钾钙活比性
KDP (= potassium dihydrogen phosphate) 磷酸二氢钾
keep away from humidity 避免受潮
keep away from sunshine 避免日晒
keep suckers on stalk 留杈
keeping 保存;饲养;维护
keeping quality 耐藏性;保存性
kernel starchiness 种子淀粉含量;种子淀粉率
kernel texture 种子结构;种子质地
4-keto-α-ionone 4-氧代-α-紫罗兰酮
keto-acetic acid 丙酮酸
ketone 酮
ketonic link(= ketonic linkage) 羰基键
key horizon 标志土层(关键土层)
Key Laboratory for Tobacco Cultivation of Tobacco Industry 烟草行业烟草栽培重点实验室
key to varieties 品种检索表;品种鉴定说明
key water requirement period 需水关键期
KF(= variegated orange) 杂色橙色
killing out (调制过程中的)干筋
killing temperature 致死温度
kiln 烤房;烟炕
kiln room 干燥室
kinase 激酶;致活酶
kind of cake 饼肥种类
kind of disease 病害种类
kindling point 燃点;着火点
kinetic equilibrium 动态平衡
kinetin(e) 激动素;6-呋喃甲基腺嘌呤
kingdom 界
king-size 特大的;特长的
king-size cigarette 大号卷烟(84~85 mm长)
kinins 激肽;细胞分裂素
Kjeldahl flask 凯氏烧瓶;长颈烧瓶
Kjeldahl method 凯氏定氮法

klendusity 功能性避病
knife coverer 刀式覆土器
knife harrow 刀式耙
knife tooth harrow 刀齿耙
knife-type rotary harrow 刀式回转耙
knot root 根结(病)
knotter 打捆机
knowledge base 知识库
Kolle flask 克氏瓶
kraal manure 厩肥
krasnozem 红壤
Krebs cycle 三羧酸循环;克雷伯氏循环
Kriging 克立格法(空间局部插值法)
Kriging map 克立格法制图
krilium 土壤改良剂
kurtosis 峰度(陡度)
Kyoto protocol 京都议定书;气候变化公约
kypman <商>代森锰
kypzin <商>代森锌

L

L 长度
L(= lemon) 柠檬色;淡黄色(分级)
labdan 赖百当
labdanoids 赖百当类化合物
label 标签
labeled atom 示踪原子;标记原子
labeled compound 标记化合物
labeling information 标签内容
labelled (同位素)标记的;示踪的
labelled isotope 标记同位素
labile humus 易分解腐殖质
labile organic nitrogen 易分解有机氮
labile organic phosphorus 活性有机磷
labile oxidization organic matter 易氧化有机质
labile phosphorus 活性磷
labor cost 人工成本;劳务费用
labor efficiency 劳动效率;人工效率
labor expense 人工费用
labor productivity 劳动生产率
labor rent 劳务用工
labor saving 人力节约;人工节约
laboratory certification 实验室认证
laboratory chemical 实验室化学制品
laboratory data base 实验室数据库
laboratory equipment 实验室设备
laboratory sample 实验样品
laboratory simulation 实验室模拟
labor-output ratio 劳动产出比率

lactate-soluble phosphorus 乳酸盐溶性磷
lactic acid 乳酸
lactonase 内酯酶
ladybird beetle(= ladybirds) 瓢虫
LAFC(= low alkaloid flue-cured tobacco) 低植物碱烤烟
lamella 薄层;薄板;薄片
lamina 叶片;片烟;去梗后叶片;薄片;薄板;层状体
lamina blending bulker 配叶柜
lamina blending regulator 配叶调节器
lamina bruising 叶片损伤
lamina casing cylinder 叶片加料机
lamina content in stem 梗中带叶率
lamina packer 叶片打包机
lamina pre-roaster 叶片预热机
lamina processing line 叶片加工线
lamina redryer 叶片复烤机
lamina redrying machine 叶片复烤机
lamina size 叶片尺寸
lamina size tester 叶片分选筛
laminar 薄片状的;层状的
laminar flow 层流
land 土地
land abandonment 土地撂荒
land application 土地利用
land area 土地面积
land aridization 土地干旱化
land arrangement 土地整理
land capacity 土地生产能力

land classification 土地分等
land classification map 土地分级图
land consolidation 土地整理
land contamination 土地污染
land cover 土地覆盖
land desertification 土地荒漠化
land development 土地开发
land drainage 地面排水;土地排水
land element 土地要素
land exhaustion 土地耗竭
land farming 土地耕作
land for growing field crops 大田
land grading 土地分等
land improvement 土地改良
land in good tilth 耕性良好的土地
land information 土地信息
land information system 土地信息系统
land leveler 平地机;压地机
land leveling 平整土地;填土
land management 土地管理
land map 土地图
land non-agriculturalization 土地非农化
land on fallow rotation 轮休地
land ownership 土地所有权;土地所有制
land plan 土地规划
land plat 土地图
land pollution 土地污染
land preparation 土地平整(整地)

land productivity 土地的生产力
land quality 土地质量
land quality rank 土地质量等级
land rearrangement 土地整理
land reclamation 土地开垦;土地改良
land resource exploitation 土地资源开发
land resources 土地资源
land restoration 土地恢复;土壤改良
land retirement 土地休耕
land sandification 土地沙化
land suitability 土地适宜性
land suitability evaluation 土地适宜性评价
land type 土地类型
land use 土地利用
land use efficiency 土地使用效率
land use map 土地利用图
land use type 土地利用类型
land utilization 土地利用
land utilization rate 土地利用率
land utilization type 土地利用类型
land valuation 土地评估
landform type map 地貌类型图
landrace 农家品种;地方品种
landslide 滑坡
lap code （卷烟）搭口印码
lap in-fold （卷烟）搭口内折
lap out-fold （卷烟）搭口外折
lap seal （卷烟）搭口粘封
lap width （卷烟）搭口宽度
large cigar 大雪茄
large continuous increase of crop 作物大幅度持续增产
large leaf 大型叶
large sample 大样本
large scale production 大规模生产
large scale survey 大比例尺调查
large seedlings 大苗
large yield 丰收;丰产
large-scale agriculture 大农业
large-scale soil map 大比例尺土壤图
largest leaf 最大叶
larval mortality 幼虫死亡率;幼虫致死率
Lasioderma serricorne Fabricius 烟草甲虫
last frost 终霜
late blight （马铃薯）晚疫病
late blooming 晚花的
late crop tobacco 晚烟
late fall ploughing 晚秋耕
late frost 晚霜
late grade 末级
late harvesting 晚期采收
late maturing 晚熟
late maturing variety 晚熟品种
late planting tobacco 晚烟
late ripening 晚熟
late sowing 晚播

late spring coldness　倒春寒
late transplanting　延迟移栽
late variety　晚熟品种
late-maturing cropland　晚茬地
late-maturity　晚熟性
lateness　晚熟性
latent bud　潜伏芽;休眠芽
latent deficiency　潜在缺乏
latent nitrogen　残留氮
latent parameter　特征参数
latent phase　潜育期
latent root　特征根
latent seed　休眠种子
later period resistance　后期抗病
later period susceptibility　后期感病
lateral　侧生的;侧面的;横向的
lateral branch　侧枝
lateral bud　侧芽(腋芽)
lateral irrigation　侧灌
lateral percolation　侧漏
lateral root　侧根;次生根
lateral vein　侧脉
laterite　砖红壤
lateritic soil　砖红壤性土
Latin square design　拉丁方试验设计
latitude　纬度
latosol(red soil)　砖红壤
lattice structure　晶格结构
lauric acid　月桂酸;十二酸
lauryl aldehyde　月桂醛
law of decreasing utility　效用递减规律
law of diminishing fertility of soil　土壤肥力递减定律
law of diminishing marginal productivity　边际生产率递减规律
law of diminishing returns　报酬递减规律;报酬递减法则
law of diminishing returns of land　土地报酬递减率
law of limiting factor　限制因子律
law of partition　分配(定)律
law of periodicity　周期率
law of plant mineral nutrition　植物矿物质营养律
law of the minimum　最小因子定律
law of the minimum nutrient　最小养分律
law of the optimum　最适因子律
law of zonality　地带性规律
law on tobacco monopoly　烟草专卖法
layer　(烟叶)层
layer by layer fertilization　分层施肥
layer manuring　分层施肥
layer mixing plow　混耕犁
layer numbers of rack　烟架层数
1:1 layer silicate　1:1型层状硅酸盐
1:1 layer structural aluminosilicate　1:1型层状结构的铝硅酸盐
2:1 layer silicate　2:1型层状硅酸

盐

2∶1 layer structured silicate　2∶1型层状结构硅酸盐

layer thickness　层厚度

layers of seed bed soil　苗床土层

layout　设计;布局;布置;陈设

LC(= lethal concentration)　致死浓度

LC(= liquid chromatography)　液相色谱(法)

LD(= lethal-dose)　致死剂量;致死药量

LD0(= lethal dose-0)　安全剂量

leached cinnamon soils　淋溶褐土

leached horizon　淋溶层

leaching　浸析(的);淋溶(的);沥滤;浸提;淋洗

leaching liquid　浸提液

leaching loss　淋失

leaching loss of nutrient　养分淋失

leaching solution　浸提液

lead　铅(Pb)

lead arsenate injury　砷酸铅害

lead diseases　主要病害

lead pollution　铅污染

lead toxicity　铅的毒性

leading crop　主要作物

leading variety　主要品种

leaf　烟叶;上部烟叶(代号B);上二棚(分级);叶片;上中部叶

leaf age　叶龄

leaf analysis　叶片分析

leaf apex　叶(尖)端

leaf area　叶面积

leaf area coefficient　叶面积系数

leaf area index　叶面积指数(LAI)

leaf area rate(= leaf area ratio)　叶面积比

leaf aroma and flavor　烟叶香气和香味

leaf arrangement　叶序(列)

leaf arranging in basket　烟叶装框

leaf axil　叶腋

leaf base　叶基

leaf blade　叶身;叶片

leaf blending　配叶

leaf body(= mesophyll)　叶肉

leaf bud　叶芽

leaf burn　叶焦病;灼糊叶

leaf butt　叶把

leaf change　烟叶变化

leaf characteristics and quality　烟叶特性与品质

leaf classifying by stalk position　分部位烟叶

leaf color　叶色

leaf color change　叶色变化

leaf color groups　烟叶颜色组

leaf colour diagnosis method　叶色诊断法

leaf composition and smoke delivery　烟叶成分与烟气释送

leaf condition 烟叶状态
leaf curing 烟叶调制
leaf curl 卷曲叶;扭叶
leaf curl disease 卷叶病
leaf curl of tobacco 烟草卷叶病
leaf damage 烟叶损伤
leaf density of baling 打包烟叶密度
leaf development 烟叶发育
leaf diagnosis 叶片诊断
leaf discolorations 叶变色
leaf drop disease 落叶病
leaf emergence date 出叶日期
leaf emergence rate 出叶速率
leaf expansion 展叶
leaf exudates 叶分泌物
leaf fall off 烟叶脱落
leaf grade limiting elements 烟叶等级控制因素
leaf grain 叶面颗粒
leaf group 叶组
leaf hair 茸毛
leaf hanging device 挂烟工具
leaf knitting 绑叶;编叶
leaf lamina 叶片;片烟
leaf length 烟叶长度;叶长
leaf lipids 烟叶脂肪类
leaf litter 枯枝落叶层
leaf loss 烟叶损耗
leaf margin 叶缘
leaf markings 叶片特征
leaf maturity 烟叶成熟度

leaf maturity stage 烟叶成熟期
leaf mesophyll 叶肉
leaf nature 叶质
leaf number 叶数
leaf nutrient analysis 叶片营养分析
leaf position 烟叶部位;叶位
leaf position angle 茎叶角度
leaf position group 烟叶部位组
leaf processing 烟叶加工
leaf processing plant 复烤厂
leaf quality 烟叶质量;烟叶品质
leaf quality and usability 烟叶品质与可用性
leaf respiration 叶片呼吸
leaf roll 卷叶
leaf scald 晒伤;日灼
leaf scars 叶痕
leaf scorch 灼叶;叶焦;叶腐烂病
leaf scrap 碎叶片
leaf segment 叶节;断叶片
leaf senescence 烟叶衰老;叶片老化
leaf shape 叶形
leaf shape and size 叶形和大小
leaf shape index 叶形指数
leaf shaped 叶状的
leaf sheath 叶鞘
leaf shoulder 叶基;叶肩
leaf size 叶片尺寸
leaf slipping 烟叶打滑
leaf sorting 选叶

leaf spot 叶斑(病)
leaf spreading 叶伸展
leaf stalk 叶柄
leaf stemming 抽梗;除梗
leaf storage 贮叶
leaf storing period 烟叶储存期
leaf structure 叶片结构;叶片组织
leaf surface 叶面;表面光滑程度(叶片)
leaf surface chemistry 叶片表面化学
leaf surface crimping 叶面皱缩
leaf surface evenly 叶面平坦
leaf surface form 叶面状态
leaf surface lipids 叶面脂质
leaf surface organism 叶面生物
leaf surface slightly folded 叶面稍皱缩
leaf temperature 叶温
leaf texture 烟叶组织;烟叶质地
leaf thickness 叶片厚度
leaf tip 叶尖
leaf tip necrosis 叶尖坏死
leaf tip yellowing 叶尖发黄
leaf tobacco 未加工的烟草原料;叶用烟草
leaf vein 叶脉
leaf vein form 叶脉状态
leaf weight 叶重
leaf width 叶宽
leaf with brown spots 褐斑叶;蒸片

leaf with brown streak around the stem <俗>泗片
leaf wrapped cigar 全叶卷雪茄
leaf-cast 落叶病
leaf-clipping (烟苗修剪时)掐叶;剪叶
leafit 小叶
leafless 无叶的
leaflet 小叶
leafy 多叶的
leakage 渗漏
lean 缺乏(油分)
lean soil 贫瘠土壤
least significant difference 最小显著差
least water-holding capacity 最小持水量
least-squares fit 最小二乘法拟合
left side(= underside) 烟叶的背面
leggy 徒长
leggy seedling 徒长苗
legume 豆科作物
legume bacterium 豆科植物根瘤菌
legume inoculation 豆科植物根瘤菌接种
leguminosae 豆科
leguminous crop 豆科作物
leguminous green manure 豆科绿肥
leguminous green-manure crop 豆科绿肥作物
lemon 柠檬黄色;柠檬

lemon leaf 柠檬叶
length 长度
length of corolla 花冠长度
length of day 日照长度
length of leaf 叶片长度;叶长
length/width ratio of leaf 叶片长宽比
less bodied 身份不足
less flavour （香气）淡薄
less harmful cigarette 低害卷烟
less hazardous cigarette 低害卷烟
less oily 油分稍有
less thin 稍薄
less-persistent pesticide 低残留农药
less-tillage system 少耕法
lethal dosage 致死剂量
lethal dose 致死剂量
lethal dose-o 安全剂量
Leu(=leucine) 白氨酸;亮氨酸
leucine 白氨酸;亮氨酸
level 水准仪
level of significance of difference 差异显著水平
level of the test 显著性水平
level terrace 水平梯田
LG(=light green) 淡青色(分级)
license for tobacco monopoly production enterprise 烟草专卖生产企业许可证
license for tobacco monopoly retail trade 烟草专卖零售许可证
Liebig's law of the minimum 李比希最低定律
Liebig-type limiting factors 李比希型限制因子
life cycle 使用周期;使用年限;生活周期;生活史
life history 生活史
ligase 连接酶
light and temperature potential productivity 光温生产潜力
light brown 浅褐色
light clay(soil) 轻黏土
light climate 光照气候
light color intensity 色度淡
light compensation point 光补偿点
light deficiency 轻度缺乏
light energy use efficiency 光能利用率
light fraction carbon 轻组碳
light fraction of humic substance 轻组腐殖质
light frost 轻霜
light germination 受光发芽
light green 淡青色(分级)
light inhibited seed 光抑制发芽种子;暗发芽种子;嫌光性种子
light injury 光害
light intensity 光强度
light intensity on leaf 叶面受光强度
light irrigation 轻灌(浅灌)

light lemon 淡黄(色,分级)
light loam 轻壤土
light loam soil 轻壤土
light mahogany 浅棕色(分级)
light mulches 淡色覆盖物
light orange 淡橙色;淡橘黄
light petroleum 石油醚
light promoted seed 光促进发芽的种子
light quality 照明质量
light rain 小雨
light ray 光线
light reaction 光反应
light requirement 需光量
light resistance 耐光性
light resources 光资源
light saturation point 光饱和点
light silt soil 轻粉沙土
light soil 轻质土
light stress 光胁迫
light texture soil 轻质壤土
light tillage 浅耕
light tobacco 淡色烟(烤烟)
light toxicity 轻度毒性
light transmission rate 透光率
light trap 诱虫灯
light treatment 光照处理
light up 点燃
light-favored seed 好光性种子
light-germination 光发芽;光萌发
light-requiring seed 需光种子

light-temperature index 光温度指数
ligneous odour 木质气
lignification 木质化作用
lignin 木质素
lignoprotein theory of humus 腐殖质的木质素-蛋白质学说
lime 石灰
lime chlorosis 石灰性褪绿病;缺钙绿病
lime concretion black soils 砂姜黑土
lime for agricultural use 农用石灰
lime for farm 农用石灰
lime hydrate 消石灰;熟石灰;氢氧化钙
lime hydroxide 消石灰;熟石灰;氢氧化钙
lime hypochlorite 次氯酸钙
lime requirement 石灰需要量
lime rock 石灰岩
lime soil 钙质土
lime sower 石灰撒施机
lime superphosphate 过磷酸钙
lime-induced chlorosis 石灰诱导的失绿症
lime-requirement determination 石灰需要量测定
limestone 石灰岩
limestone soil 石灰岩土;石灰土
liming 施用石灰
liming effect 石灰效应

limiting factor 限制因子
limiting variables 限定变量
line breeding 品系繁育;系统育种;品系选育
line charts 折线图
line cross 品系间杂交
line isolation 株系隔离
line seeding 条播
line selection 纯系选择;品系选择
line separation 品系分离
line(= strain;pedigree) 品系;线;轮廓;家系;谱系
linear burning rate 线性燃烧速度
linear regression 线性回归
linear regression analysis 线性回归分析
linear regression equation 线性回归方程
linear statistical analysis 线性统计分析
linoleate 亚油酸;亚油酸酯(或盐)
linoleic acid(= linolic acid) 亚油酸;9,12-十八碳二烯酸
linolenic acid 亚麻酸
lipid metabolism 脂类代谢
lipid synthesis 脂类的合成
lipid(e) 类脂;脂类
lipin 类脂;脂类;脂质
lipoids 类脂
liquid ammonia 氨水;液氨
liquid ammonia applicator 液氨施肥机
liquid application 液施
liquid chromatograph 液相色谱仪
liquid chromatography 液相色谱法
liquid dung 粪水;厩液
liquid fertilizer 液体肥料
liquid film 液体地膜
liquid fumigant 液体熏蒸剂
liquid manure 液体厩肥;粪水
liquid manure spreader 厩液施肥机
liquid mixed fertilizer 液体复合肥料
liquid nitrogen 液态氮
liquid nitrogen cryopreservation 液氮保藏法
liquid nitrogen fertilizer 液体氮肥
liquid phase 液相
liquid water 液态水
lister 双壁开沟犁
lister cultivator 沟播地中耕机
lister planter 沟播机
lithosol 石质土
lithosphere 岩石圈
lithotrophy 无机营养
litter 凋落物(枯枝落叶)
litter decomposition 植物残体分解
litter layer 枯枝落叶层
little cigar 小雪茄
livestock 家畜
livestock dung 家畜粪
livestock manure 畜粪尿

livestock wastewater 牧畜废水
living insecticide 生物杀虫剂
living nature 生物界
living organism 生物
loading door 装烟门；装料门
loading in barn 装炕
loading material 填料
loam 壤土；沃土
loam clay 壤黏土
loam group 壤土组
loam soil 壤土
loamification 壤质化
loamy clay 壤质黏土
loamy coarse sand 壤质粗沙（土）
loamy fine sand 壤质细沙（土）
loamy fine soil 壤质细土
loamy sand 壤沙土
loamy soil 壤质土
loamy texture 壤质
local climate 地方气候；局部气候
local exhaust ventilation 局部通风；局部排气
local fertilizer 当地肥料
local infection 局部侵染
local lesion 局部枯斑
local necrosis 局部坏死
local race 地方种；地方种族
local symptom 局部症状
locality protection 区域保护
localized application 集中施肥
localized experiment （定）区域试验
localized placement 局部施用
lock ring 密封圈
locule 室（指子房、花药等）；子囊腔
lodging 倒伏
lodging resistance 抗倒伏性
loess 黄土
loess plateau 黄土高原
loessial soil 黄土性土；黄绵土
log residuals 对数残差
logarithmic model 对数模型
logarithmic normal distribution 对数正态分布
logistic growth model 逻辑斯谛增长模型
log-linear regression 对数线性回归
lognormal distribution 对数正态分布
Lolium multiflorum L. 黑麦草；多花黑麦草
long day crop 长日照作物
long day plants 长日照植物
long lasting fertilizer 长效肥料
long raining 连雨天气
long term site research 长期定位研究
long-cultivated land 久耕地；熟地
long-day 长日照
longday type 长日照型
long-distance transport 长距离运输
long-flowering 开花期长的

123

long-flowing period　长花期
longitude　经度
long-keeping　耐储藏的
long-lasting　长效的;(香气)持久
long-range forecast　长期预测;长期预报
long-range weather forecasting　长期天气预报
long-run test　长期试验
longs　加长卷烟
long-stalked variety　高秆品种
long-term experiment　长期试验
long-term fertilizer experiment　长期肥料试验
long-term field experiment　长期田间试验
long-term soil fertility experiment　长期土壤肥力试验
long-term stationary experiment　长期定位试验
long-term storage　长期储存;长期储藏
loop　循环
loose　松散
loose combined humus　松结态腐殖质
loose leaf　散叶(分级)
loose leaves　散烟叶
loose sand　松沙土
loose soil　松散土壤
loose texture　疏松质地

loosely bound　松把(分级)
loosen　放松;疏松(土壤)
loosener　松土机;松土犁
looseness of soil　土壤松散(性)
looseness of structure　结构松散
loosening　松动;松散
loosening　疏松
loss by leaching　淋溶损失
loss of soil nutrient　土壤养分流失
loss of water and soil　水土流失
lot　批量;批号;地段;烟堆
Lou soil　垆土
low　低的;贫乏的;营养不足的;下部的
low alkaloid flue-cured tobacco　低植物碱烤烟
low alkaloid variety　低生物碱品种
low analysis complex fertilizer　低纯度复合肥
low analysis fertilizer　低成分肥料
low combustibility　低燃烧性
low flat ground　低平地
low grade infection　轻度感染
low ground　低地;洼地
low latitude　低纬度
low level laterite　低地砖红壤
low mountain and hill district　低山丘陵区
low nicotine　低烟碱
low nicotine variety　低烟碱品种
low nutrient level　低养分水平

low productive soil 低产土壤
low productivity land 低产田
low profile 矮生株型;低矮植株;矮株型烟草
low rates of stem elongation 茎伸长缓慢
low residue 低残留
low tar 低焦油
low tar cigarette 低焦油卷烟
low temperature resistance 耐寒性;耐低温性
low-analysis fertilizer 低成分肥料;低浓度肥料
low-class cigarette 低档烟
low-down manure spreader 低架厩肥撒施机
low-energy agriculture 低能耗农业
lower leaf 下部叶
lower plastic limit 塑性下限
lower the temperature halfway 中途掉温(烘烤)
lower yield 产量低
low-grade 劣等的;低等的;低级的;低质量的
lowland 低地;洼地;盆地
lowly water-soluble phosphate fertilizer 难溶性磷肥
low-lying land 低地;洼地
low-profile tobacco 打顶早的矮株烟草

low-stemmed 矮秆的
low-temperature-induced dormancy 低温诱导休眠(再度休眠)
low-toxic pesticide 低毒农药
low-water season 枯水期
low-yield field 低产田
lucerne 紫花苜蓿
lugs 下二棚(分级);中下部叶:代号X
luminometer(=photometer) 光度计
luminous intensity 光照强度
lump lime 生石灰
lumps 结团烟丝
lumpy soil 结块(结构)土壤
lumpy structure 碎块结构
lush 生长茂盛的
luster 光泽(雪茄烟叶)
luvisol 淋溶土
luxmeter (光)照度计
luxury supply 奢侈供应
luxury uptake 奢侈吸收
lyase 裂解酶;裂合酶
lyophilization 冻干保藏法;冷冻干燥
lyosol 液溶胶
Lys(=Lysine) 赖氨酸
lysimeter 测渗仪
lysimeter systems 蒸渗系统
lysine 赖氨酸

M

machine transplanting 机械移栽
macro 常量;大量
macroaggregate 大团聚体
macroclimate 大气候
macroelement 大量元素(常量元素)
macroelement fertilizer 大量元素肥料
macroenvironment 大环境
macrominerals 常量矿物质
macronutrient 大量元素养分;大量营养物
macula 斑点
madefaction 浸泡(润湿作用)
maduro 深色雪茄;色深味浓的(雪茄)
magazine 贮仓;烟支库
magnesium 镁(Mg)
magnesium ammonium phosphate 磷酸铵镁
magnesium carbonate 碳酸镁
magnesium chloride 氯化镁
magnesium deficiency 镁缺乏
magnesium deficient 缺镁症
magnesium disturbance 镁素营养失调
magnesium fertilizer 镁肥
magnesium hydroxide 氢氧化镁
magnesium lime 镁石灰
magnesium oxide 氧化镁
magnesium oxide injury 氧化镁药害
magnesium phosphate 磷酸镁
magnesium sulfate 硫酸镁
magnetic stirrer 磁搅拌器
magnetic stirrer paddle 磁性搅拌棒
magnitude 大小;尺寸;量;数值
magnitude of soil property variability 土壤性质变异幅度
magnum-size cigarette 粗支烟
mahogany 红褐色;(烟叶颜色)赤黄
mahogany leaf 橘色烟叶
Maillard reaction 美拉德反应;棕色化反应
main axis 主茎;主轴
main bud 主芽
main clay mineral 主要黏土矿物
main crop 主要作物
main drain 主排水沟;总排水管
main nerve(= main vein) 主脉
main note 主香
main plot 主区;整区;主(试验)小区
main root 主根;直根
main shaft 主轴;总轴;主传动轴;主茎
main soil group 主要土类
main source of pollution 主要的污染源

main stalk 主茎
main stem 主脉
main treatment 主处理
main vein 主脉
mainstream cigarette smoke 卷烟主流烟气
mainstream smoke 主流烟气
maintenance (菌种)保藏
maintenance of fertility 肥力保持
maintenance of soil fertility 维持土壤肥力
maize 玉米;黄色(的)
major character 主要特征;主要性状
major constituent 主要成分;大量成分
major element 大量元素(常量元素)
major fertilizer component 肥料主要成分
major fertilizer element 肥料主要元素
major horizon 主要土层
major nutrient element 主要营养元素
major pests 主要害虫
major substrate 主要基质
major tobacco producing countries 主要产烟国
major tobacco types 主要烟草类型
makhorka 莫合烟
making decision for fertilization 施肥决策
making preparations for ploughing and sowing 备耕
maladaptation 适应不良;适应不全;失调
maladjustment 失调
malate 苹果酸盐(或酯)
malate synthase 苹果酸合酶
malate synthetase 苹果酸合成酶
maldevelopment 发育不良;畸形
maleic acid 马来酸;顺丁烯二酸
maleic hydrazide 马来酰肼;顺丁烯二酰亚胺(MH)
malformed leaf 畸形叶
malic acid 苹果酸;羟基丁二酸
malic dehydrogenase 苹果酸脱氢酶
malnutrition 营养不良;营养不足;营养失调
malnutritional diseases 营养失调症
malnutritional diseases of tobacco 烟草营养失调症
malonic acid 丙二酸
management of seed bed 苗床管理
management of water and fertilizer 水肥管理
mancozeb <商>代森锰锌
mandibulate mouthparts 咀嚼式口器(昆虫)
manganese 锰(Mn)
manganese chlorosis 缺锰褪绿症
manganese deficiency 锰缺乏症
manganese disturbance 锰营养失调
manganese fertilizer 锰肥

127

manganese oxide 氧化锰
manganese sulfate 硫酸锰
manganese toxicity 锰毒害
man-made erosion 人为侵蚀
man-made soil 人工(造)土壤
man-rating 人员评价
mantle of soil 土壤表层
manual decapitation 手工打顶
manual farm implements 人力农机具
manual kneading 手工捏合(雪茄)
manual simulated rainfall 人工模拟降雨
manual suckering 人工抹杈
manual topping 人工打顶
manual transplanting 人工移栽
manual weeding 人工除草
manually-fed type 手工送苗式(栽苗机)
manufacture department 卷烟车间;制造部
manufacturing technology 生产工艺;制造工艺
manurance 施肥
manure 肥料;粪便;施肥;厩肥
manure accumulation 积肥
manure deficiency 缺肥
manure distributor 厩肥撒施机
manure heap 粪堆
manure heating 堆肥发热
manure juice 厩肥液汁
manure loader 装肥机
manure of bark 茎皮堆肥
manure pile 粪堆
manure pit 粪池;粪坑
manure spreader 施肥机
manure storage 粪坑;积粪窖
manure trial 肥效试验
manure-drill 厩肥条施
manurial application 施肥
manurial effect 肥效;肥料效应
manurial value 肥料价值
manuring 施肥
manuring effect 肥料效应
manuring field trial 肥料田间试验
manuring in planting hole 移栽穴施底肥
manuring irrigation 肥水灌溉
manuring time 施肥时期
many-bottom plough 多铧犁
many-sided utilization 综合利用的
margin of leaf 叶缘
marginal analysis 边际分析
marginal chlorosis 边缘失绿
marginal contribution ratio 边际贡献率
marginal cost 边际成本
marginal crop rows 边行作物
marginal distribution 边缘分布
marginal effect 边际效应
marginal necrosis 边缘坏死
marginal plant chlorosis 叶缘失绿

marginal production 边际产量
marginal revenue 边际效应;边际收益
marginal rot 边缘腐烂
marly soil 泥灰质土壤;泥灰土
marsh 沼泽;沼地
marsh land 沼泽地
marsh podzol 沼泽灰壤
marshy soil 沼泽土
Maryland tobacco 马里兰烟
masked symptom 隐形症状
mass 质量
mass analysis 大批量分析
mass breeding 混合育种;群体育种;大群繁育
mass flow 质流
mass spectral analysis 质谱分析
mass spectrogram 质谱(图)
mass spectrography 质谱法
mass spectrometric analysis 质谱分析
mass spectrometry 质谱法;质谱分析法
mass spectroscopy 质谱学;质谱分析(法)
mass water content 质量含水量
master horizon 基本发生层(基本土层)
master sample 标准样品;标准试样
master sample of agriculture 农业的规范试样

material flow 物质流
material migration 物质迁移
mathematical model of ecology 生态学数学模式
mathematical simulation 数学模拟
matric potential 基质势(衬质势)
matrix 矩阵
matrix analysis 矩阵分析
matrix calculus 矩阵计算
matrix correlation 矩阵相关
matrix eigenvalues 矩阵特征值
matrix equation 矩阵方程
matrix method 矩阵方法
matted bed 苗床;畦
matting of roots 根系交织
maturation 成熟
maturation period 成熟期
maturation phase 成熟期
mature 成熟的;工艺成熟;尚熟
mature field 熟地
mature phase 成熟期;成熟阶段
mature plant resistance 成株抗性
mature soil 成熟土;熟土
mature stage 成熟期
mature tissue 成熟组织
mature tissue resistance 成熟组织抗性
matured compost 腐熟堆肥
maturing period of capsule 蒴果成熟期
maturing stage 成熟期

maturity 成熟;成熟度
maturity degree 成熟度
maturity index 成熟度指数
maturity period of capsule 蒴果成熟期
maturity stage 成熟期
maximum 最大值
maximum absorption ratio 最大吸收率
maximum absorption stage 最大吸收期
maximum adsorptive capacity 最大吸附量
maximum capillary capacity 最大毛管持水量
maximum demand 最大需要量
maximum density 最大密度
maximum economic yield 最高经济产量
maximum efficiency stage of nutrition 营养最大效率期
maximum evapotranspiration 最大蒸散量
maximum field capacity 最大田间持水量
maximum growth temperature 最高生长温度
maximum humidity 最大湿度
maximum hygroscopicity 最大吸湿度
maximum immobilization 最大固定作用
maximum moisture capacity 最大持水能力
maximum permissible 最大允许值
maximum precipitation 最大降水量;最大沉淀量
maximum residue limit 最高残留限制
maximum temperature 最高温度
maximum water-holding capacity 最大持水量
maximum yield application rate 最高产量施肥量
meadow cinnamon soil 潮褐土
meager 贫瘠的
mealy sand 粉沙(土)
mean 平均值(中数)
mean annual precipitation 年平均降雨量
mean annual temperature 年平均温度
mean deviation 平均偏差
mean output 平均产量
mean precipitation 平均降水量
mean residence time of soil organic matter 土壤有机质的平均残留期(MRT)
mean square error 均方误差
mean temperature difference 平均温差
mean value 平均值

mean yield 平均产量
measure for saving water 节水措施
measuring cylinder 量筒
measuring instrument 测量仪器
measuring pipet 吸管
mechanical composition （土壤）机械组成
mechanical composition analysis （土壤）机械组成分析
mechanical cultivation 机械耕作
mechanical curing barn 机械式烤房
mechanical damage 机械损伤(分级)
mechanical defence 机械防御
mechanical harvesting 机械采收
mechanical impedance 机械阻力（机械阻抗）
mechanical injury 机械损伤(分级)
mechanical inoculation 机械接种;摩擦接种
mechanical resistance 机械阻力（机械阻抗）
mechanical tissue 机械组织
mechanical ventilation drying 机械通风干燥
mechanical work 机械作业
mechanical-stable aggregate 力稳性团聚体
mechanism of plant adaptation to phosphorus deficiency 植物缺磷响应机制

mechanized agriculture 机械化农业
mechanized composting 机械化堆肥
mechanized farming 机械化栽培;机械化耕作
mechanized harvesting 机械化采收（机械化收获）
mechanized intensive agriculture 机械化集约农业
mechanized irrigation 机械化灌溉
mechanized production 机械化生产
median 中位数
median charts 中位数图
Medicago sativa 紫花苜蓿
medical cigarette 药用烟;药物型卷烟
mediterranean climate 地中海气候
medium analysis fertilizer 中成分（浓度）肥料
medium and low-yielding field 中低产田
medium clay 中黏土
medium clay soil 中黏土
medium composition 培养基成分
medium loam 中壤土
medium loam soil 中壤土
medium nutrient level 中等养分水平
medium scale survey 中等规模调查
medium tar 中焦油
medium texture 中等质地
medium-fleshy 中等－偏厚

· 131 ·

medium-growing 中长生育期
medium-toxic pesticide 中毒农药
megastigmatrienone 巨豆三烯酮
megatherm 高温
megathermal climate 高温气候
melioration （土壤）改良；改善
mellow 完熟
mellow 丰满；成熟度好；完熟度；熟透的；松软的；肥沃的（指土壤）；柔和
mellow consistency 松软结持（度）
mellow earth 熟土；松软海绵土
mellow effect 熟化作用
mellow loam 疏松壤土
mellow soil 松软土壤；熟土
mellow soil layer 熟土层
mellowed 醇和的
mellowing 醇化
mellowing process 醇化过程
mellowness 成熟度；松软性
mellowy （香气）醇厚
Meloidogyne 根结线虫属
Meloidogyne hapla Chitwood 北方根结线虫
Meloidogyne incognita 南方根结线虫
membrane permeability 质膜渗透性
mercaptan 硫醇
mercury 汞
mercury pollution 汞污染
mesic layer 半分解有机质层

mesofauna 中型动物区系
mesology 生态学
mesophilic phase 中温期
mesophyll 叶肉
metabolic disturbance 代谢失调
metabolic equilibrium 代谢平衡
metabolic path 代谢途径
metabolic process 代谢过程
metabolic product 代谢产物
metabolism 代谢；新陈代谢
metal-humic acid complex 金属－胡敏酸络合物
metallic odour 金属气
metallic pollutant 金属污染物
metallophyte 超积累植物
metatrophic bacteria 腐生细菌
metatrophy 腐生营养
meteoric water 大气水
meteorological drought 气象干旱
meteorological factor 气象因子；气象因素
meteorological station 气象站
methane fermentations waste 沼气发酵肥
methane(= methylene) 甲烷；沼气
methanogen 产甲烷菌
methanol 甲醇
methionine 蛋氨酸
method of application 施用方法
method of bio-assay 生物试验法
method of fertilization for target yield

目标产量施肥法
method of fertilizer application 施肥方法
method of least squares 最小二乘法
methods of pesticide effectiveness rest on tobacco disease 烟草病害药效试验方法
6-methyl-3,5-heptadiene-2-one 6-甲基-3,5-庚二烯-2-酮
2-methoxy-3-methylpyrazine 2-甲氧基-3-甲基吡嗪
methyl acid 甲酸
methyl orange 甲基橙
methyl red 甲基红
methylation 甲基化作用
methylbromide 溴甲烷
methylene blue 亚甲蓝
mica 云母
micro 微量
micro syringe 微量注射器
microaerobes 微好氧菌
microaerophilic bacteria 微好氧菌
microaggregate 微团聚体
microbe 微生物
microbe in rhizospheric soil 根际微生物
microbe in root 根内微生物
microbial activity 微生物活性
microbial biomass 微生物生物量
microbial biomass carbon 微生物生物量碳

microbial biomass nitrogen 微生物生物量氮
microbial breakdown of pesticide 农药的微生物降解
microbial contamination 微生物污染
microbial decomposition 微生物分解(作用)
microbial degradation 微生物降解(作用)
microbial diversity 微生物多样性
microbial ecology 微生物生态学
microbial fixation 微生物固定(作用)
microbial herbicide 微生物除草剂
microbial immobilization 微生物固持(作用)
microbial inoculants 微生物接种剂
microbial insect pathogens 微生物杀虫剂
microbial insecticide 微生物杀虫剂
microbial manure 微生物肥料
microbial metabolism 微生物代谢
microbial parameter 微生物参数
microbial pesticide 微生物农药
microbial population 微生物种群
microbial process 微生物过程
microbial synthesis 微生物合成(作用)
microbial(=microbian, microbic) 微生物的

· 133 ·

microbiodegradation 微生物降解
microbiological activity 微生物活性
microbiological analysis 微生物分析
microbiological assay 微生物鉴定
microbiological cultivation 微生物培养法
microbiological culture 微生物培养
microbiological degradation 微生物降解
microbiological factor 微生物因素
microbiological immobilization 微生物固定作用
microbiological specification 微生物属性
microbiologically degradable pollutant 可用微生物降解的污染物
microbion 微生物
microbioscope 微生物显微镜
microbiota 微生物区系
microclimate 小气候
microclimate differences 小气候差异
microclimate in the field 农田小气候
micro-ecological-environment 微生态环境
microecology 微生态学
microecosystem 微生态系统
microelement 微量元素
micro-environment 小环境;微域环境;微环境(叶围,根围)

microfauna 微型动物区系
microfertilizers 微量元素肥料
microflora 微生物区系
micro-habitat 微生境
micro-irrigation 微灌
micro-jet irrigation 微喷灌溉;微量注射灌溉
micronutrient 微量营养元素
micronutrient deficiency 微量元素缺乏症
micronutrient fertilizer 微量元素肥料
microorganic contaminant 微生物污染物
microorganism 微生物
microorganism decompose 微生物分解
microorganism-root association 微生物-根系联合体
micropedology 微土壤学
micro-plot experiment 微区试验
micropopulation 微生物群
microscopic analysis 显微分析法
microscopic imaging system 显微成像系统
micro-spray 微喷
microstructure 微观结构;显微结构
microtitration 微量滴定
microwave 微波
microwave digestion 微波消解
middle latitude 中纬度

middle leaf 中部叶
middle rate 中间价;平均价;中间汇率;平均汇率
middle ribs 叶子中脉;主脉
middle stress 中度胁迫
middle-class cigarette 中档烟
middle-early 中早熟的
middle-late 中晚熟的
middle-low leaf proportion 中下部叶比例
middle-low position yellow leaf group 中下部黄烟组
middle-tar cigarette 中焦油卷烟
midpoint 中值
midrib 主脉;中脉;烟筋;烟梗
midrib proportion 烟梗比例
midvein(= midrib) 中脉
migration 迁移
migration capacity 迁移能力
mild （烟味）和顺;醇和
mild cigarettes 柔和卷烟
mild flue-cured tobacco 醇和型烤烟
mild flue-cured variety 醇和型烤烟品种
mild mosaic 轻度花叶病
mild odor 温和香气
mild smoke 烟气柔和
mild symptoms 轻微症状
mildethane 托布津
mildewed leaf 霉烟
mildly alkaline soil 轻度碱性土壤

mildness 醇和型;柔和性;温（缓）和;(柔)软适度
mildothane 甲基托布津
milk vetch 紫云英
millipore filter paper 微孔过滤滤纸
miner 深耕犁;潜叶蝇
mineral 矿物;矿物的;无机的
mineral acid 无机酸
mineral associated organic matter 矿物结合态有机物
mineral constituent 矿物组成;矿物成分
mineral deficiency 矿物质缺乏
mineral deficiency symptoms 矿物质缺乏症状
mineral deprivation 矿物质流失
mineral element 矿质元素
mineral excess 矿物质过量
mineral fertilizer 矿质肥料;无机肥料
mineral horizon 矿质层
mineral iron 矿物态铁
mineral matter 无机质;无机物质;矿物质
mineral metabolism 矿质代谢
mineral nitrogen 无机氮;矿质氮
mineral nutrition 无机营养;矿质营养
mineral particle 矿质颗粒
mineral potassium 矿物钾
mineral salts 无机盐

mineral soil 矿质土壤
mineral weathering 矿物风化
mineral-free light fraction of soil organic matter 轻组土壤有机质
mineralizable carbon 可矿化碳
mineralizable nitrogen 可矿化氮
mineralizable nutrient 可矿化养分
mineralization 矿化作用
mineralization coefficient 矿化系数
mineralization of organic phosphorus 有机磷的矿化
mineralization potential 矿化势
mineralization process 矿化过程
mineralization rate 矿化速率
mineralization rate constant 矿化速率常数
mineralization rate of organic matter per year 有机质年矿化率
mineralization velocity 矿化速率
mineralized carbon 矿化碳
mineralized nitrogen 可矿化氮
miniature root observation box 小型根箱
mini-cycle 小循环
mini-fermentor 微型发酵罐;微型培养槽
minimum 最小值
minimum density 最小密度
minimum detectable quantity 最小检出量
minimum field capacity 田间最小持水量
minimum temperature 最低气温;最低温度
minimum tillage 少耕
minimum tillage system 少耕体系
minimum water-holding capacity 最小田间持水量
minimum-tillage seed placement 少耕播种
mining plough 深耕器
minirhizotrons 微根室法
mini-sprinkler irrigation 微喷灌
minor and rare elements 微量和稀有元素
minor constituent 微量成分
minor element 微量元素
minor element deficiency symptom 微量元素缺乏症
minor symptom 次要症状
minorelement fertilizer 微量元素肥料
minty 薄荷香的;晾香的
minty note 薄荷香;晾香
minute 分钟
miry soil 淤泥土
miscellaneous components 杂色成分
miscellaneous manure 土杂肥
miscellaneous types 杂色(或各种)类型
miscured leaf 调制不当(烟)叶
miss planted rate 缺株率;漏栽率

misshapen 畸形的
misshapen cigarette 畸形卷烟
missing carbon sink 碳失汇
missing element trial 缺素试验
missing filter 缺滤嘴;掉滤嘴
missing hill 缺穴
missing plant 缺株
missing plot 缺区
missing sink 未探明的汇
mist bed 喷雾式苗床
mist blower 喷雾器
mist sprayer 喷雾机;鼓风弥雾机
mistus 杂交;杂种;混种
mitigation 减排
mitigative capacity 减排能力
mixed 混合的;混杂;混级(烟叶)
mixed color 混色
mixed cropping 混作
mixed cultivation 混作
mixed erosion 混合侵蚀
mixed fertilizer 混合肥料
mixed fertilizer application 混合施肥
mixed group 混组(烟叶)
mixed heavy metal pollution 重金属复合污染
mixed indicator 混合指示剂
mixed infection 混合侵染
mixed intercropping 混作
mixing 混合
mixture cropping annually 多作一熟型
mixtus 杂交;混种
Mn deficiency 缺锰症
Mn toxicity 锰毒
Mo Fe protein 钼铁蛋白
Mo(= Molybdenum) 钼
mobile equilibrium 动态平衡
mobile hothouse 活动温室
mobile nutrient 活性养分
mobility of nitrogen 氮素流动
mobilization of phosphate 磷的活化
mode 众数
model for plant growth 植物生长模型
model for regression 回归模型
model test 模型试验
model verification 模型验证
moder 半腐殖质(半腐熟腐殖质;酸性腐殖质)
moderate 中(色度)
moderate alkalinity 中等碱度(中等碱化)
moderate availability 中度有效性
moderate deficiency 中度缺乏
moderate drainage 排水尚佳
moderate rain 中雨
moderate resistance 中度抗病;中抗
moderate susceptible 中度感病;中感
moderately alkaline 中等碱性的
moderately salinized 中盐渍化的

moderately well drained 排水尚佳
modern agriculture 现代农业
modern intensive sustainable agriculture 现代集约持续农业
modernization of farming 农业现代化
modernized agriculture 现代化农业
modification of tobacco cultivars 烟草品种改良
modified phosphate fertilizer 改性磷肥
modifier 改良剂（调节剂）
moist 湿润的
moist climate 湿润气候
moistening 回潮；加湿
moisture 水分；水汽；湿气；湿度
moisture capacity 持水量（容水量；保水量）
moisture circulation 水分循环
moisture content 含水率；含水量；水分
moisture content testing 水分检测
moisture determination 测定含水量；湿度测定
moisture equilibrium 平衡水分
moisture equivalent 持水当量
moisture flow 水流
moisture free weight 干重
moisture gradient 水分梯度
moisture holding capacity 保湿性能；保水能力

moisture index 水分指数；湿润指数
moisture injury 湿害
moisture percentage 湿度百分率（含水量百分数）
moisture pick-up 吸湿
moisture proof 防潮的；防水的；抗湿的
moisture regain 回潮
moisture regime 水分状况
moisture retaining quality 持水性
moisture retention curve 持水曲线
moisture run back 烟叶阴筋
moisture supply 水分供应
moisture uptake curve 吸湿曲线
moisture-holding capacity 持水量（容水量；保水量）
moisture-retaining power 持水能力
mold 霉；霉菌
moldboard plow 壁式铧犁耕作
molding 霉变
moldy 霉变
mole cricket 蝼蛄
molecular agriculture 分子农业
molecular farming 分子农业
molecular microbial ecology 分子微生物生态学
molecular shape 分子形状
molecular size 分子大小
molecular weight 分子量
mollisoil 松软土
Molto chiaro 色很浅的弗吉尼亚雪

茄
molybdenum 钼(Mo)
molybdenum cofactor 钼辅因子
molybdenum deficiency 缺钼症
molybdenum fertilizer 钼肥
molybdenum flavoprotein 钼黄素蛋白
mongrelism 品种间杂交现象
mongrelize 品种间杂交
monoammonium phosphate 磷酸一铵;磷酸二氢铵
monocalcium phosphate 磷酸一钙
monochromatic light 单色光
monoclinous 雌雄同花(的);两性花(的)
monocropping 单作
monocultural 单作
monocultural farming 单作农业(单一经营)
monocyclic aromatic hydrocarbon 单环芳烃
monoecious 雌雄同株(的);雌雄同体
monoliths 土柱法
monopotassium phosphate 磷酸二氢钾(磷酸一钾)
monosaccharide 单糖
monose 单糖
monotonous (香气)单薄
monthly maximum temperature 月最高温度
monthly mean temperature 月平均温度
monthly minimum temperature 月最低温度
montmorillonite 蒙脱石(微晶高岭石)
montmorillonoid 蒙脱矿
moptop 丛顶(症)
mor 粗腐殖质;酸性有机质
mor layer 粗腐殖质层
morbidity 病态;发病率
more green less yellow 青多黄少
more yellow less green 黄多青少
morphological change 形态变化
morphological characters of roots 根系形态学特征
morphological diagnosis 形态诊断
morphological diagnosis method 形态诊断法
mortar 研钵
mosaic 嵌合性;花叶病
most profitable rate 最高收益用量
mother material 母质
mother rock 母岩
motion state of aroma 香气状态
motor plough 自走犁;动力犁
motor roller 动力镇压器
mottle leaf 花斑叶
mottled rot 斑驳腐烂
mottling 杂色;斑点;花斑;斑驳
mould control technique 防霉技术

mould damage protection in storing 储存防霉
mould smell(＝moldy) 霉味
mould tobacco handling 霉烟处理
mould(＝mold) 霉;霉菌;松软土地
mouldboard plow 翻转犁;铧式犁
moulded leaf 霉烟
mouldy 霉的;霉变
mound 岗地
mound planting 垄作;墩植
mountain land 山地
mountain region farming 山区农业
mountain soil 山地土壤
mountain yellow-brown earth 山地黄棕壤
mountainous soil 山地土壤
mounted implement 悬挂式农具
mouth spear 口针(昆虫)
mouthpiece 烟嘴;切丝机刀门
movement 迁移
moving range 移动极差
mowing machine 割草机
muck 粪;粪肥;腐泥土;腐殖土
muck heap 粪堆
muck soil 腐殖质土壤
muckland 沼泽地
mud field 烂泥田
muddy soil 淤泥土
muffle furnace 马弗炉
mulch 覆盖层;覆盖物

mulch cultivation 覆盖栽培
mulcher 覆盖机;覆盖层
mulching 地面覆盖;覆盖
mulching film 覆盖薄膜
mulching material 覆盖物
mull 腐熟腐殖质;细腐殖质;混土腐殖质
multi-bottoms plough 多铧犁
multi-components fertilizer 多成分肥料
multicourse rotation 多区轮作
multi-cropping index 复种指数
multidimensional scaling 多维尺度分析
multi-factorial experiment 多因子试验
multifactorial fertilizer experiment 多因素肥料试验
multi-functional fertilizer 多功能性肥料
multinomial distribution 多项分布
multi-nutrient fertilizer 多种营养元素肥料
multiple comparison 多重比较
multiple correlation coefficient 多重相关系数
multiple crop and multiple cropping 多作多熟型
multiple crop index 复种指数
multiple cropping 多熟;复种;多茬;种植

multiple cropping pattern 复种模式
multiple cropping acreage 复种面积
multiple cropping rotation 复种轮作
multiple cropping system 复种制
multiple disease resistance 多抗（病）性
multiple factor experiment 多因子试验
multiple harvesting 多次收获
multiple linear regression 多元线性回归
multiple nutrients compound fertilizer 多元（复合）肥料
multiple range test 复极差检测法
multiple regression 多元回归
multiple regression analysis 多元回归分析
multiple sampling 多次抽样
multiple superphosphate 重过磷酸钙
multiple-cropping area 复种面积
multiple-furrow plough 多铧犁
multiple-row cultivator 多行中耕机
multiplication nursery 繁殖苗圃
multispectral scanning system 多光谱扫描系统
multi-spot experiment 多点试验
multistage sampling 多级抽样
multistage type 多段式
multi-storied agriculture 立体农业

multivariate 多变量
multivariate distribution 多维分布
multivariate statistics 多元分析（多变量分析）
muriate of potash 氯化钾
muscovite 白云母
musty tobacco 发霉烟叶
mutation breeding 突变育种
mutual unit of cooperation tobacco farmer 烟农互助组
mutualism 互惠共生；共栖；共生性
mutualistic symbiosis 互惠共生；共生
mycelioid 菌丝状的
mycelium 菌丝体
mycoflora 真菌区系；真菌群落
mycology 真菌学
mycorrhiza 菌根
mycorrhizal formation 菌根形成
mycorrhizal fungi 菌根真菌
mycorrhizal fungus 菌根真菌
mycorrhizosphere 菌根根际（菌根圈）
mycorrhyzal association 菌根群丛
mycorrhyzal symbiosis 菌根共生
mycotrophic 菌根营养的
mycotrophy 菌根营养
myristic acid（＝tetradecanoic acid） 肉豆蔻酸；十四酸
Myzus persicae Sulzer 烟蚜；桃蚜

N

N. rustica L. 黄花烟草
N. sylvestris Spegazzini & Comes 美花烟草;林烟草
N. tobacum L. 普通烟草
N. tomentosa Ruiz & Pavon 绒毛烟草
NAA① (= nicotinic acid amide) 烟酰胺
NAA② (= naphthalene acetic acid) 萘乙酸
naked sowing 无覆盖播种
naphthoic acid 萘甲酸
naphthylacetic acid 萘乙酸
narrow 变窄
narrow gutted leaf 叶梗比很低的窄薄烟叶
narrow leaf type 窄叶型
narrow spectrum insecticide 窄谱性农药(专效性农药)
narrow spectrum pesticide 窄谱性农药(专效性农药)
narrow-leaved 狭叶的
narrow-row culture 窄行栽培
national standard 国家标准
national standard of flue-cured tobacco 烤烟国家标准
National Tobacco Cultivation and Physiology and Biochemistry Research Centre 国家烟草栽培生理生化研究基地
native breed 当地品种
native organic matter 固有有机质
native soil 残积土
native soil organic matter 土壤原有机质
native species 本地种;土种
native variety 地方品种
natural aggregate 自然团聚体
natural aging 自然醇化;自然陈化
natural attenuation 自然衰减
natural balance 自然平衡
natural base 生物碱
natural bulk density 自然状况容重
natural circulation 自然循环
natural cross 自然杂交
natural curing 自然条件下调制
natural disturbance 自然干扰
natural enemy 天敌
natural erosion 自然侵蚀
natural fermentation 自然条件下发酵
natural fertility 自然肥力;天然肥力
natural moisture 自然含水量
natural purification 自然净化作用
natural purification characteristics 自

然净化特性
natural rainfall 自然降雨
natural resource ecosystem 自然资源生态系统
natural resources information system 自然资源信息系统
natural sand rate 自然沙土率
natural smouldering rate 自由阴燃速度
natural soil 自然土壤
natural soil drainage 土壤自然排水
natural ventilation 自然通风
near infrared 近红外
near infrared light 近红外线
near infrared reflectance spectroscopy 近红外分光光谱仪(NIRS)
near ultraviolet 近紫外
near-infrared diffuse reflectance spectrophotometer 近红外反射光谱仪
necrosis 坏死
necrotic depression 坏死陷斑
necrotic mosaic virus 坏死花叶病
necrotic spot 坏死斑(枯斑)
negative charge 负电荷
negative charged ion 阴离子(负离子)
negative correlation 负相关
negative priming effect 负激发效应
nematicide 杀线虫剂
nematode 线虫

nematode disease of tobacco 烟草线虫病害
nematoded disease 线虫病害
Neo-alluvial soils 新积土
neophytadiene 新植二烯
neoxanthin 新黄质;新叶黄素
nervure 主脉;叶脉
nested 掺假(烟叶)
nested tobacco (美俗)优质烟叶里夹藏劣质烟叶
net assimilation rate 净同化率
net blotch 网斑病
net consumptive requirement 净耗水量
net consumptive use 净耗水量
net cost 净成本;净价
net earning 纯收入;净利润
net immobilization 净固定作用
net irrigation requirement 净灌溉需水量
net maximum immobilization 最大净固定作用
net mineralization 净矿化作用
net photosynthesis 净光能合成;净光合作用
net production 净生产量
net production rate 净生产率
net productivity 净生产力
net tobacco weight 烟叶净重
net weight of fertilizer 肥料净重
network of root 根系;根丛

neutral 无性的;中性的;中和的
neutral component 中性成分
neutral compound 中性化合物
neutral flavoured cigarette 中性香味卷烟;中度加香的卷烟
Neutral purplish soils 中性紫色土
neutral soil 中性土壤
neutral solution 中性溶液
neutralization 中和作用
neutralization titration 中和滴定法
neutralizing value 中和值
neutron method 中子法
neutron moisture meter 中子测水仪
neutron probe 中子仪
new growth of soil 土壤新生体
new soil 生荒地;处女地
new variety 新品种
newly plowed field 新耕地
newly-broken virgin soil 新垦荒地
new-ploughed field 新翻地
N-fertilizer 氮肥
nickel 镍
nickel crucible 镍坩埚
nickel pollution 镍污染
Nicotiana 烟草属
nicotiana alkaloid 烟草生物碱
Nicotiana rustica L. 黄花烟草
Nicotiana tobacco L. 红花烟草
Nicotiana tabacum L. 普通烟草
Nicotiana virus I 烟草普通花叶病毒 I

nicotine 烟碱;尼古丁
nicotine content 烟碱含量;烟碱量（烟气）
nicotine degrading bacteria 烟碱分解细菌
nicotine filtration efficiency 烟碱过滤效率
nicotine synthesis and breakdown 烟碱合成与降解
nicotine/cig. 每支烟的烟碱
night soil 人粪尿
nightshade family 茄科
ninhydrin reaction 茚三酮反应
nip 掐;摘;摘心;霜害
nipping 摘心
niter 硝石;硝酸钾
nitramine 硝胺
nitrate 硝酸盐
nitrate bacteria 硝化细菌
nitrate fertilizer 硝态氮肥
nitrate leaching 硝酸盐淋失
nitrate nitrogen 硝态氮;硝酸盐氮
nitrate of lime 硝酸钙
nitrate of potash 硝酸钾
nitrate pollution 硝酸盐污染
nitrate pollution in groundwater 地下水硝酸盐污染
nitrate reductase 硝酸还原酶
nitrate reduction 硝酸盐的还原(作用)
nitrate supply 硝酸盐供应

nitrate toxicity 硝酸盐毒害
nitration 硝化作用
nitre(= niter) 硝石;硝酸钾
nitric acid 硝酸
nitric oxide 一氧化氮
nitric oxides 氮氧化合物
nitric phosphate 硝酸磷肥
nitrification inhibitor 硝化抑制剂
nitrifier 硝化细菌
nitrifying 硝化的
nitrite 亚硝酸盐
nitrite nitrogen 亚硝态氮
nitrite oxidase 亚硝酸盐氧化酶
nitrite reductase 亚硝酸盐还原酶
nitroamine 硝胺
nitrobacteria 硝化细菌
nitrogen absorption 氮(素)吸收
nitrogen activity index 氮素活度指标
nitrogen assimilation 氮(素)同化作用
nitrogen balance 氮(素)平衡
nitrogen compound 含氮化合物
nitrogen content 氮(素)含量
nitrogen cycle 氮(素)循环
nitrogen cycling in agricultural land 农田氮素循环
nitrogen cycling in agroecosystem 农田生态系统中的氮素循环
nitrogen deficiency 缺氮;氮缺乏
nitrogen deposition 氮沉降

nitrogen dioxide 二氧化氮
nitrogen disturbance 氮素营养失调
nitrogen excess 氮(素)过量
nitrogen excess disease 氮(素)过量症
nitrogen fertilization 施氮肥
nitrogen fixation 氮固定;固氮作用
nitrogen fixation activity 固氮活力
nitrogen fixation by rhizobia 根瘤固氮法
nitrogen fixation rate 固氮速率
nitrogen fixer 固氮者
nitrogen fixing microorganism 固氮微生物
nitrogen fixing symbiosis 固氮共生现象
nitrogen hunger 缺氮;氮饥饿
nitrogen immobilization 氮固定
nitrogen isotope ratio analysis 氮同位素比值分析
nitrogen leaching loss 氮的淋洗损失
nitrogen manure 氮素肥料;(有机)氮肥
nitrogen metabolism 氮代谢作用
nitrogen mineralization 氮素矿物化
nitrogen mineralization potential 氮矿化势
nitrogen mineralization-immobilization process 氮素矿化固持过程
nitrogen oxides air pollution 氮氧化

物大气污染
nitrogen productivity 氮素生产力
nitrogen recovery 氮回收
nitrogen runoff loss 氮径流损失
nitrogen source 氮源
nitrogen supplying capacity of soil 土壤氮素供应能力
nitrogen surplus 氮素过量;氮盈余
nitrogen transformation in soil 土壤氮素转化
nitrogen turnover 氮周转
nitrogen, phosphorus and potassium 氮磷钾肥
nitrogenase 固氮酶
nitrogen-fixing algae 固氮藻类
nitrogen-fixing bacteria 固氮细菌
nitrogen-fixing capacity 固氮能力
nitrogen-fixing plant 固氮植物
nitrogen-fixing symbiosome 固氮共生体
nitrogen-loving plant 喜氮植物
nitrogen-potassium ratio 氮钾比
nitrogen-starved plot 土壤缺氮区
nitrogen-sulphur ratio 氮硫比
(straight) nitrogenous fertilizer 氮肥
(straight) phosphatic fertilizer 磷肥
nitrophosphate 硝酸磷肥
nitrosonornicotine 亚硝基去甲基烟碱(NNN)
nitrous acid 亚硝酸
nitrous oxide 氧化亚氮(N_2O)

NK compound fertilizer 氮钾复混肥料
N-nitrosation N-亚硝化(作用)
no tillage mulching 免耕覆盖
noctuids 夜蛾
nodule 根瘤
nodule bacteria 根瘤菌
nodule formation 根瘤形成(作用)
NOG(= no grade) 不列级烟叶;级外烟
nonadecanoic acid 十九(烷)酸
non-agricultural land 非农用地
non-agriculturalization 耕地非农化
non-available nutrient 无效养分
nonavailable potassium 非有效性钾
nonbiodegradable 不能生物降解的
non-biological degradation 非生物降解
non-biological volatilization 非生物性挥发
non-capillary pore 非毛管孔隙
non-capillary porosity 非毛管孔隙度
noncoherent soil 非黏结性土壤
noncompetitive inhibition 非竞争性抑制作用
non-complexed organic material 非复合的有机物质
non-contamination pesticide 无污染农药
non-cropping 休耕

nondeep dormancy 浅度休眠；非熟休眠
nondegradable pollutant 非降解性污染物
nondescript group 级外烟组(N)
non-enzymatic browning reaction 非酶促棕色化反应
non-enzymatic decomposition 非酶催化分解
non-essential amino acid 非必需氨基酸
non-essential element 非必需元素
non-essential element of plant 植物非必需元素
non-exchangeable cation 非交换性阳离子
non-exchangeable ion 非交换性离子
non-exchangeable nutrient 非交换性养分
non-exchangeable potassium 非交换性钾
non-free sampling 非自由取样
non-germinable 不发芽的；无发育的
non-germinable seed 不能发芽的种子
nonhazardous 无害的
nonhumic substance 非腐殖物质
non-humified material 非腐殖化物质

nonhumus material 非腐殖物质
non-hydrolyzable humic acid 非水解性胡敏酸
non-hydrolyzable nitrogen 非水解性氮
non-infectious disease 非侵染性病害
non-leguminous green manure 非豆科绿肥
nonlinear regression 非线性回归
non-mulching 不覆盖
nonparametric test 非参数检验
nonpersistent pesticide 无残留性农药；非持久性农药
non-point source pollution 面源污染
non-radioactive carbon 非放射性碳
non-residual taste 余味干净；(烟味)干净；无残留味
non-reversible 不可逆的
non-salinized 非盐渍化的
non-soluble 不溶(解)的
non-specific 非专化的；非专性的
non-specific adsorption 非专性吸附
non-structural soil 无结构土壤
non-structure soil 无结构土壤
non-symbiotic nitrogen fixation 非共生固氮作用
non-tillage 免耕(法)
non-tillage seeding 不整地播种
non-topping 不打顶
nonvolatile 非挥发性的

non-volatile acids 非挥发性酸
nonvolatile component 非挥发性成分
nonvolatile organic acids 非挥发性有机酸
non-volatile terpenoids 非挥发性类萜
normal 正常
normal crop 常年收成
normal deviate 正态离差
normal distribution 正态分布
normal illumination 正常光照;正常照明
normal irrigation with water layer 常规灌溉
normal leaf 正常叶
normal moisture capacity 正常持水量
normal profile of soil 正常土壤剖面
normal root 定根
normal seeding 正常苗;健苗
normal soil 正常土
normal superphosphate 过磷酸钙(普钙)
normal thermometer 标准温度计
normal tillage 常规耕作
normalization method 归一化法
normalized difference vegetation index 归一化植被指数
normalized processing 归一化处理
nornicotine 降烟碱;去甲基烟碱

northern latitude 北纬
north-facing slope land 阴坡地
not readily degradable 难降解的
note 香韵;香型
no-tillage 免耕
no-tillage culture 免耕栽培
no-tillage farming 免耕农业
no-tillage system 免耕法
noxious substance 有毒物质
NP compound fertilizer 氮磷复混肥料
NPK compound fertilizer 氮磷钾复混肥料
NPK (= nitrogen, phosphorus and potassium) 氮磷钾肥
NR (= nitrate reductase) 硝酸还原酶
N-rates 氮用量;含氮率
nuclear magnetic resonance 核磁共振
nuclear magnetic resonance spectrometry 核磁共振波谱法
nucleic acid 核酸
nucleotide 核苷酸
nugget 块金
nugget effect 块金效应
nugget variance 块金方差
null hypothesis 零假设
number of apparent leaves 实际留叶数
number of hanging sticks 挂烟杆数

number of irrigation 灌水次数
number of pollinated flowers 授粉花数
number of replication 重复次数
nunja 湿地
N-uptake 氮素吸收
nurse crop 覆盖作物;保护作物
nursery bed 苗床
nursery bed planter 苗床播种机
nursery infection 苗床感染
nutmeg 肉豆蔻
nutriculture 营养(液)栽培;水培
nutrient 养分;养料;营养品;营养的
nutrient adsorption 养分吸附
nutrient analysis of complex fertilizer 复混肥料养分分析
nutrient analysis of organic fertilizer 有机肥料养分分析
nutrient availability 养分有效性
nutrient balance 养分平衡
nutrient bioavailability 养分生物有效性
nutrient budget 养分收支
nutrient cation 养分阳离子
nutrient combination 营养配比
nutrient concentration 养分浓度
nutrient concentration gradient 养分浓度梯度
nutrient content 养分含量
nutrient cycle(cycling) 养分循环

nutrient cycles 养分循环
nutrient cycling characteristics 养分循环特征
nutrient cycling in agroecosystem 农业生态系统中养分循环
nutrient deficiency 养分缺乏;营养不足
nutrient deficiency diagnosis 营养缺素诊断(缺素诊断)
nutrient deficiency disease 营养缺乏病
nutrient deficiency soil 瘦土
nutrient deficiency symptom 养分缺乏症状
nutrient deficiency symptom in plant 植物缺素症
nutrient deficiency zone 养分亏缺区
nutrient depletion 养分耗尽
nutrient depletion zone 养分损耗区
nutrient desorption 养分解吸
nutrient diagnosis 营养诊断
nutrient diagnosis and fertilization 营养诊断施肥
nutrient discharge 养分输出;养分流失
nutrient dynamics 养分动态
nutrient efficiency 营养效率
nutrient elements 营养元素
nutrient elements cycle 养分元素循环

nutrient enrichment 养分富集

nutrient equilibrium 养分平衡

nutrient excess 养分过剩

nutrient fixation 养分固定

nutrient flow 养分流

nutrient fluid 培养液

nutrient formula 营养配方

nutrient holding capacity 持肥力

nutrient income and expense 养分收支

nutrient input and output 养分收支

nutrient interaction 养分交互作用

nutrient leaching 养分淋失

nutrient limitation 营养限制

nutrient limiting factor 养分限制因子

nutrient loss 养分损失

nutrient management 养分管理

nutrient matter 营养物质

nutrient medium 营养培养基

nutrient mineralization 养分矿化

nutrient mobilization 养分活化

nutrient movement to root surface 养分向根表面的迁移

nutrient pollution 营养物质污染

nutrient pool 养分库

nutrient preserving capability 保肥性

nutrient preserving capacity 保肥性

nutrient recovery hypothesis 营养恢复学说

nutrient recycling 养分再循环

nutrient release 养分释放

nutrient release characteristics 养分释放特性

nutrient release curve 养分释放曲线

nutrient release summit 养分释放峰值

nutrient replenishment 养分补偿

nutrient return 养分归还

nutrient reutilization 养分再利用

nutrient sink 营养衰竭

nutrient solution 营养(溶)液

nutrient stratification phenomenon 养分分层现象

nutrient stress 养分胁迫

nutrient supplying capability 养分供应能力

nutrient supplying capacity 养分供应能力

nutrient transfer 养分迁移

nutrient transformation 养分转化

nutrient translocation 养分转移

nutrient transport 养分迁移

nutrient turnover rate 养分周转率

nutrient uptake 养分吸收

nutrient uptake efficiency 养分吸收效率

nutrient use efficiency 养分利用率

nutrient utilization efficiency 养分利用效率

nutrient utilization rate 养分利用效率
nutrient value 营养价值
nutrient-efficient genotype 养分高效基因型
nutrients transport in plant 植物体内养分运输
nutrient-supplying capacity 养分供应能力
nutrition 营养
nutrition deficiency 营养缺乏;营养不足
nutrition diagnosis 营养诊断
nutrition pot 营养钵
nutrition status 营养状况
nutritional bag 营养袋
nutritional balance index 养分平衡指数
nutritional disease 营养性病害
nutritional disorder 养分失调;营养失调
nutritional disturbance 营养不良（营养不足）
nutritional excess 养分过量
nutritional level 营养水平
nutritional stage 营养阶段
nutritional stress 营养胁迫
nutritional trait 植物营养性状
nutritive absorption 养分吸收
nutritive cube 营养钵
nutritive disturbance 营养失调
nutritive organ 营养器官
nutritive proportion 营养比例
nutritive root 营养根
nutty aroma 坚果香
nylon bag technique 尼龙袋法
nylon cloth bag 尼龙袋
nylon net 尼龙网
nymphosis 蛹化;成蛹
Nysius ericae Schilling 小长蝽象

O

O(= orange) 橙色的
object oriented database system 面向对象数据库系统
object resolution 地面分辨率
objectionable constituent 有害成分
obligate aerobes 专性好氧菌
obligate anaerobes 专性厌氧菌
obligate mutualism 专性互利
obligately thermophile 专性嗜热菌
observed value 观测值
occluded particulate organic matter 闭蓄态颗粒有机物
occluded water 封闭水
occlusion water 封闭水

occurrence prediction （病虫害）发生预测

odo(u)r characteristic 香气特征

odo(u)r classification 香气分类

odo(u)r concentration 气味浓度

odo(u)r intensity 香气强度

odo(u)r permanency 香气持久性

odo(u)r type 香型

odorant 香料；有气味的

odoriphor group 发香基团

odorous constituent 香气成分

odorous principle 香气主要成分

oecology 生态学

off flavor 臭味；杂气；杂味

off odor 臭气；臭味；杂气；异常气息

offal 下脚料

off-color 褪色（叶）；走色

off-color leaf 褪色烟叶

off-colored 颜色不合格

offensive odor 异味；杂气

offensive taste (= off-taste) 怪味；杂气

official variety test 法定品种鉴定试验

off-odor 异味

offshore tobacco 引种烟草

offspring 后代；结果；幼苗；后裔

offspring-parent regression 亲子间回归

offspring-parent relationship 亲子间关系

offtake 排水沟

oil 油分

oil cake 油渣饼；豆饼

oil-bath 油锅；油浴

oily 油分尚有

old arable land 久耕地；熟地

old crop 陈烟（叶）

old instar caterpillar 老龄幼虫

old land 久耕地；熟地

old period 衰老期

old seed 老种子；陈种子

old soil 老龄土

oleic acid 油酸

oligophyllous 少叶的

oligorganic layer 贫有机质层；矿质层

oligotrophic 贫瘠的

ombratropism 向雨性

ombrophyte 喜雨植物；适雨植物

ombrophyte plant 喜雨植物

on the spot investigation 实地调查

once single plant selection 一次单株选择

once-over harvesting 一次性收获

one course rotation 单作

one course system 单作制

one crop annually 一年一熟制

one crop per annum 一年一熟制

one crop succession 连作

one-crop agriculture 单一作物农业

one-crop system 连作制
one-lime phosphate 重过磷酸钙
one-man priming machine 单人采叶机;单人操作烟叶打顶机
one-off application 一次性施肥
one-sided confidence interval 单侧置信区间
one-sided test 单侧检验
one-tailed test 单尾检验
one-way plough 旋转犁;单向犁
ontogeny 个体发育;个体发生
open 稍疏松;(丝束)开松
open (ditch) drainage 明沟排水
open cultivation 露地栽培
open ditch 明沟
open grained 组织疏松烟草
open pollination 开放传粉;自然传粉
open weave 舒展(烟叶)
opener 开沟器;松土器
open-ground plant 露地植物
opening and closing of stomata 气孔的启闭
operating cost 经营成本
operators 运算
opportunity cost 机会成本
optical property 光学性质
optical spectrum 光谱
optimal dose 最适量
optimal root structure 理想根构型
optimal temperature 最适温度

optimal yield 最适产量
optimization scheme 优化配置
optimized fertilization 优化施肥
optimum 最佳
optimum allocation of land use 土地利用优化配置
optimum application rate 最佳施肥量
optimum application rate of nitrogen 最佳施氮量
optimum compaction 最佳压实度
optimum crop yield 最适作物产量
optimum density 最适密度
optimum design 最优设计
optimum dose 最适量
optimum economic yield 最适经济产量
optimum fertilizer-N management 氮肥优化管理
optimum growth temperature 最佳生长温度
optimum leaf area 最适叶面积
optimum leaf number 最佳留叶数
optimum level 最佳水平(适宜量)
optimum mature (= fine mature) 适熟
optimum maturity 最适成熟度
optimum moisture 最佳含水量
optimum N/K ratio 最适 N/K 比值
optimum rate of application 最佳施肥量

optimum regression 最优回归
optimum supply 最适供应
optimum temperature 最适宜温度；最佳温度
optimum texture 最佳质地
optimum tillage 最适耕作
optimum transplant time 最适移栽期
optimum water content of soil 土壤最佳含水量
optimum yield 适宜产量
orange 橘黄色；橙黄色；橙色
orange red 橘红
orange-yellowish chlorosis 桔黄色褪绿
order （烟叶）回潮；品级；品味（指烟叶含水率）；秩序；级；次；次序；等级
order form agriculture 订单农业
order of soil 土纲
order production 订单生产；分批生产
order statistic 顺序统计量
ordering 回潮；加湿；烟叶湿润；排列次序；次序关系；调整
ordering house （烟叶）增湿房；回潮室
orderly improvement 定向改良
ordinary Kriging 普通克立格法
ordinary superphosphate 普通过磷酸钙；普钙

ordinary temperature 常温
ordure 粪
organic 有机的；有机质的；有机物的；组织的
organic acid 有机酸
organic acid metabolism 有机酸新陈代谢
organic agricultural chemicals 有机农药
organic agriculture 有机农业
organic carbon 有机碳
organic carbon density 有机碳密度
organic cation 有机阳离子
organic chloride 有机氯化物
organic chlorine pesticide poisoning 有机氯农药中毒
organic colloid 有机胶体
organic combined form 有机结合态
organic fertilizer 有机肥料
organic foliage fungicide 有机叶面杀（真）菌剂
organic food 有机食品
organic form 有机结合态
organic herbicide 有机除草剂
organic horizon 有机层
organic insecticide 有机杀虫剂
organic manure 有机（粪）肥
organic mat 有机覆盖物
organic matter 有机质
organic matter degradation 有机物（质）降解

organic mineral soil 有机矿质土
organic nitrogen 有机氮;有机态氮
organic nitrogen compound 有机氮化合物
organic nitrogen fertilizer 有机氮肥
organic nitrogen pool 有机氮库
organic nitrogenous 有机氮
organic nitrogenous fertilizer 有机氮肥
organic nutrient 有机营养
organic nutrition 有机营养
organic nutrition fertilizer 有机营养肥料
organic particle 有机颗粒
organic pesticide 有机农药
organic phosphorus 有机磷
organic phosphorus compound 有机磷化合物
organic phosphorus insecticide 有机磷杀虫剂
organic phosphorus pesticide 有机磷农药
organic phosphorus pesticide poisoning 有机磷农药中毒
organic pollutant 有机污染物
organic pollutant degradation 有机污染物降解
organic pollution 有机污染
organic reducing substance 有机还原性物质
organic remain 有机残留物
organic soil 有机质土(有机土)
organic soil layer 有机土层次
organic soil material 有机土壤物质
organic soil phosphorus 土壤有机磷
organic solvent 有机溶剂
organic substance 有机物
organic sulfide 有机硫化物
organic sulfur 有机态硫
organic toxicant 有机毒物
organic waste fertilizer 有机废弃物肥料
organically complexed zinc 有机络合态锌
organichaloid compounds 有机卤代物
organic-inorganic compound fertilizer 有机-无机复混肥料
organic-mineral particle 有机-矿质颗粒
organic-mineral soil complex 土壤有机-无机复合体
organic-rich soil 富含有机质土壤
organoarsenic pesticide 有机砷农药
organoarsenic pesticide poisoning 有机砷农药中毒
organoarsenic residue 有机砷残留
organo-cation 有机阳离子
organochlorine insecticide 有机氯杀虫剂
organochlorine pesticide 有机氯杀虫剂

organochlorine pesticide residue 有机氯农药残留量

organocompound of arsenic 有机砷化合物

organofluorine insecticide 有机氟杀虫剂

organogenesis 器官发生;器官形成

organoleptic attribute 感官特征

organoleptic rating 感官评价

organoleptic test 感官检验;感官评定

organomercury compounds 有机汞化合物

organomercury pesticide 有机汞农药

organomercury pesticide poisoning 有机汞农药中毒

organo-mineral colloidal complex 有机-矿质胶体复合体

organo-mineral particle 有机-矿质颗粒

organonitrogen pesticide 有机氮农药

organonitrogenous insecticide 有机氮杀虫剂

organophosphorues pesticide 有机磷杀虫剂

organophosphorus compounds 有机磷化合物

organophosphorus fungicide 有机磷杀真菌剂

organophosphorus herbicide 有机磷除草剂

organophosphorus insecticide 有机磷杀虫剂

organophosphorus pesticide 有机磷农药

organophosphorus poison 有机磷毒物

organophosphorus residue 有机磷残留

organosulfur pesticide 有机硫农药

organotrophy 有机营养

organs of reproduction 生殖器官

organs of vegetation 营养器官

oriental aromatic 东方型香(烟叶);东方型香料

oriental cigarettes (= Turkish cigarettes) 香料型卷烟

oriental leaf curing 香料烟调制

oriental tobacco aroma type 香料烟香型

oriental tobacco budworm 烟青虫;烟草夜蛾

Oriental tobacco leaf 香料烟叶

oriental tobacco (= Turkish tobacco, aromatic tobacco) 东方型烟(香料烟,土耳其烟)

origin of pollution 污染源

original data 原始数据

original mineral 原生矿物

original record 原始记载;原始记录

original salinization 原生盐渍化
original seed 原种
original silt 淤积物
original soil 未垦地;成土母质
originator 培育者
ornithine 鸟氨酸
Orobanche 列当属
orogenic soil 山地土壤
orterde 硬磐
orthoclase(=orthose) 正长石
orthogonal design 正交设计
orthogonal experimental design 正交试验设计
orthogonal function 正交函数
orthogonal Latin square design 正交拉丁方设计
orthogonal regression 正交回归
ortho-phosphate （正）磷酸盐
ortho-phosphoric acid （正）磷酸
orthophyll 正常叶
osmoregulation 渗透调节
osmosis 渗透作用
osmotic potential 渗透势
osmotic pressure 渗透压
out of door drying 露天晒干
outch group 上部叶(香料烟)
outdoor culture 露地栽培
outdoor float bed 室外漂浮育苗床
outdoor planting 露地栽培
outdoor seed bed 露地苗床
outdoor sowing 露地播种

outfall ditch 排水沟
outflow of mineral substances from leaves 矿物质从叶片中的溢流
outlier 离群值(异常值)
output 产量;产品;排出物
output by leaching 淋失输出
output capacity 生产能力
output per hectare 每公顷产量
output quota 产量定额
output target yield index 产量指标
output value 产值
output-input ratio 产投比
outroot feeding 根外施肥
oven 烘箱
oven dry weight 烘干重
oven drying 烘干
oven drying method 烘干法
oven-dried sample 烘干样
oven-dry soil 烘干土
over clayed soil 过黏土壤
over cultivation 耕种过度;滥垦;过度开垦
over flavored 加香过量
over grown tobacco leaf 过大烟叶;憨烟
over heating 过热;加热过度
over matured wilt tobacco odor 过熟枯萎的烟叶气味
over nutrition 养分过量
over tillage 再耕
over wintering egg 越冬卵

overall application 全面施药;全面施用

overall cultivated land 耕地总量

overburden 覆盖层

overburden layer 覆盖层

overcropping 耕地使用过度

over-fertilization 施肥过量

overflow with heaviness 沉溢

overflow with lightness 飘逸

overflow with smoothness 悬浮

overgrowth 徒长

overharvesting 过度收获

overhead irrigation 喷灌;人工降雨

overhead sprinkler 顶喷式喷灌机

overhead sprinkling system 高架喷灌系统

overhead watering(= irrigation) 喷灌

overirrigation 灌溉过度

overland flow 地表径流

overland runoff 地表径流

overlay 覆盖;遮掩;覆盖物;覆面

overlayer 覆盖层

overlaying horizon 覆盖层

overlying strata 覆层;表层

over-manured 施肥过量

overmature(= overripe) 过熟

overnutrition 富营养化

overripe 过熟

overripeness 过熟

overripe 过熟

overshadow 遮蔽

overstocked 过密的

overwash 洪积土壤

overwinter control 越冬防治

overwinter(ing) 越冬

overwintering crop 越冬作物

overwintering of generations 越冬世代

overyearing 越冬;越年份

oxalic acid 草酸;乙二酸

oxaloacetate 草酰乙酸

oxialf 氧化淋溶土

oxidation 氧化作用

oxidation potential 氧化电位

oxidation reduction potential 氧化还原电位

oxidation reduction reaction 氧化还原反应

oxidation titration 氧化还原滴定法

oxidation-reduction 氧化还原(作用)

oxidation-reduction enzyme system 氧化还原酶系

oxidation-reduction indicating agent 氧化还原指示剂

oxidation-reduction indicator 氧化还原指示剂

oxidation-reduction potential 氧化还原电位

oxidation-reduction process 氧化还原过程

oxidation-reduction reaction 氧化还原反应
oxidation-reduction system 氧化还原体系
oxidative phosphorylation 氧化磷酸化作用
oxide 氧化物
oxidimetry 氧化还原滴定
oxidizable nutrient 可氧化养分
oxidizing action 氧化作用
oxido-reductase 氧化还原酶
oxoionone 氧化紫罗兰酮
oxygen 氧
oxygen-containing functional group 含氧官能团
oxyhydroxide 氢氧化物
oxyphile 适酸性植物;喜酸性植物
oxyphilous 喜酸性的;适酸性的
oxyradical 氧自由基
oxytolerant 耐氧的
oxytropic 亲氧的
oxytropism 向氧性
ozone 臭氧(O_3)

P

P(= primings) 脚叶组
pachynsis 肥厚
pachyphyllous 厚叶的
pack barn 堆(烟)房
packed fertilizer 包装肥料
packer 打包机;包装机
packet (烟)盒
paddle culture 窄垄(行)栽培
paddle seeding 窄垄(行)播种
paddy field 稻田
paddy soil 水稻土
paddy-upland rotation 水旱轮作
paedogenesis 早熟
pair of bamboo grates 烟折
paired sample t test 配对样本的 t 检验
pale (烟叶)暗色;淡色
pale back 浅色叶背;带灰白色(分级)
pale-yellow (烟叶)灰黄(色,分级)
palisade tissue 栅栏组织
palmitic acid(= hexadecanoic acid) 棕榈酸;十六酸;软脂酸
palmitic aldehyde(= hexadecanal) 棕榈醛;十六醛
palmitoleic acid 棕榈油酸
paludification 泥炭形成过程
paludization 泥炭形成(作用);泥炭化过程
pan irrigation 根圈灌溉

panel 评吸组;专门小组
panel test 小组评吸
pantothenic acid 泛酸
paper chromatography 纸色谱法
parallels 平行试验
parameter 参数
parasitic wasp of aphid 蚜虫寄生蜂
parasitism 寄生现象
parathion 对硫磷
paratrophy 营养不良(营养不足)
parent material 母质
parent material horizon C 母质层(C层)
partial autocorrelation analysis 偏自相关分析
partial correlation coefficient 偏相关系数
partial factor productivity 偏生产力
partial regression 偏回归
partial sill 局部基台
particle size fractionation method 粒度分级法
particle-size classification 粒径分级
particle-size composition 颗粒大小组成
particle-size distribution 颗粒大小分布
particulate 颗粒状物
pascal 帕斯卡
passivating agent 钝化剂
passivation 钝化(作用)

passive absorption 被动吸收
passive uptake 被动吸收
pastal (香料烟)烟把
pasteurization 巴氏消毒法
pasteurized soil 消毒土壤
path coefficient 通径系数
pathogen(e) 病原;病原体;病菌
pathogenic bacteria 致病细菌;病原细菌
pathogenic bacterium 病原性细菌
pathogenic microorganism 病原微生物
pathogenic organism 病原物
pathogen-suppressive soil 病原物抑制性土壤
pattern analysis 模式分析
pattern gene 模式基因
pattern of rotation 轮作方式
PCR(= polymerase chain reaction) 聚合酶链式反应
peak of growing season 生长盛期
peak temperature 最高温度
peak value 峰值
peak-heating room 高温灭菌室
peakness 峰度
peanut cake 花生饼
peat 泥炭;泥煤
peat cube 营养块;泥炭(腐殖质)块
peat fertilizer 泥炭肥料
peat formation 泥炭形成(作用)

peat resources 泥炭资源
peat soil 泥炭土
pebbles(=gravel) 沙砾
pectic substance 果胶物质
pectin 果胶
ped 土壤自然结构体
pedicel 花梗
pedigree breeding 系谱育种;系统育种
pedobiomes 土壤生物群落
pedocompaction 土壤压实作用
pedodiversity 土壤多样性
pedogenesis 土壤发生
pedogenetic weathering process 土壤发生风化过程
pedogenic process 成土过程(土壤发生过程)
pedogeochemistry 土壤地球化学
pedological drought 土壤干旱
pedological features 土壤形成物
pedological map 土壤图
pedosphere 土壤圈
pedotheque 土壤样品
pedoturbation 土壤扰动作用
PEE(=petroleum ether extract) 石油醚提取物
pellet (种子)丸粒
pelleted seed 丸化种子;丸衣种子;包衣种子
pelosol 重黏土
penetrability 渗透性
penetrate 穿透
penetration 渗透
penetration resistance 穿透阻力
penetration test 渗透试验(穿透试验)
Penicillium 青霉属
Penman formula 彭曼公式
pentose phosphate pathway 磷酸戊糖途径
peptone 蛋白胨
per unit area yield 单位面积产量
percentage base saturation 盐基饱和度
percentage humidity 湿度百分数
percentage of moisture 含水率
percentage of retention 残留率
percentage of saturation 饱和率
percolate 渗出液;滤出液
percolation of soil water 土壤水渗漏
perennial grass 多年生牧草
perennial weed 多年生杂草
perfect combustion 完全燃烧
period of color stabilization 定色期
period of crop rotation 轮作周期
period of dormancy 休眠期
period of duration 生育期
period of elongation 伸长期
period of maturation 成熟期
period of stem drying 干筋期
period of yellowing 变黄期
periodic drought 季节性干旱

periodic systematic sampling　周期系统抽样
periodicity of germination　发芽周期
perish　枯萎
perlite　珍珠岩
permafrost horizon　永冻层
permafrost layer　永冻层
permanent agriculture　不休闲耕作
permanent charge　永久电荷
permanent charge surface　永久电荷表面
permanent experiment　长期试验;永久性试验
permanent farming　连年耕作制(常年耕作制)
permanent humus　永久性腐殖质
permanent perched water　悬着水
permanent wilting　永久萎蔫
permanent wilting coefficient　永久萎蔫系数
permanent wilting percentage　永久萎蔫含水率
permanent wilting point　永久萎蔫点
permeameter　渗透仪
permeation　渗透作用
permiable soil　通透性土壤
peroxidase　过氧化物酶
perpetual　连续开花的;永久的;永恒的;不间断的;多年生植物
perphosphoric acid　过磷酸
persistence of pesticide　农药持久性

persistent development　持续发展
persistent organic pollutant　持久性有机污染物
perudic moisture regime　常湿润土壤水分状况
pervaporation　全蒸发(过程);蒸发作用
pervious soil　通透性土壤
pest and disease monitoring　病虫害监测
pest avoidance　病虫驱避性
pest breakout　害虫爆发
pest damage　虫害
pest resistance　抗虫性
pesticide　杀虫(菌)剂;农药
pesticide contamination　农药污染
pesticide degradation　农药降解
pesticide effect　药效
pesticide formulation　农药剂型
pesticide incorporated fertilizer　(含)农药肥料;农药化肥复合剂
pesticide injury　药害
pesticide metabolism　农药代谢
pesticide phytotoxicity　农药药害
pesticide pollution　农药污染
pesticide residual　农药残留物
pesticide residue　农药残留量
pesticide residue analysis　农药残留量分析
pesticide residue hazard　农药残留的危害

pesticide residue tolerance 农药残留限量;农药允许残留量
pesticide resistance 抗药性
pesticide tolerance 耐药性;农药允许量
pesticide toxicity 农药毒性
pesticide-added fertilizer 农药肥料
pesticides of fossil origin 矿物源农药
pests control of seed bed 苗床害虫防治
pests of stored tobacco 烟草仓储害虫
petiole 叶柄
petri dish 培养皿
petroleum ether 石油醚
petroleum ether extract 石油醚提取;石油醚提取物
pH meter 酸碱度计;pH 计
pH meter method pH 计法
pH of rhizosphere 根际 pH
pH paper pH 试纸
pH scale pH 标度
pH value pH 值;酸碱度
phaeozem 黑土
phenethyl alcohol 苯乙醇
phenogram 物候图
phenolphthalein 酚酞
phenotype 表现型
phenylacetic acid 苯乙酸
phenylalanine 苯丙氨酸

phenylglycine 苯甘氨酸
pheromone 信息激素;性引诱激素;味诱激素;性外激素
pheromone trap 激素诱捕剂;激素诱捕器
phloem 韧皮部
phloem transport 韧皮部运输
pH-meter pH 计
phosphatase 磷酸酶
phosphate 磷酸盐
phosphate adsorption 磷酸盐吸附
phosphate bacteria fertilizer 磷细菌肥料
phosphate buffer 磷酸盐缓冲液;磷酸缓冲剂
phosphate desorption 磷解吸
phosphate dissolving action 解磷作用
phosphate fertilization 施磷肥
phosphate fertilizer 磷酸盐肥料
phosphate fixation 磷固定
phosphate potential 磷位
phosphate retention 磷酸吸持
phosphate solubilization 溶磷作用
phosphate-dissolving microorganism 解磷微生物
phosphate-solubilizing bacteria 解磷细菌
phosphate-solubilizing fungi 溶磷真菌
phosphatic fertilizer 磷肥

163

phosphatide 磷脂
phosphatidylcholine 磷脂酰胆碱（卵磷脂）
phosphatidylethanolamine 磷脂酰乙醇胺
phosphine fumigation 磷化氢熏蒸
phosphobacteria 磷细菌
phosphoenolpyruvate 磷酸烯醇式丙酮酸
phosphogluconate pathway 磷酸葡萄糖酸途径
phosphoglycerides 甘油磷脂
phosphohumate 腐殖酸磷肥
phospholipid fatty acids analytic method 磷脂脂肪酸分析
phospho-nitrogen system 氮磷施肥制
phospho-potash system 磷钾施肥制
phosphor-calcic soils 磷质石灰土
phosphoric acid 磷酸
phosphorous deficiency symptom 缺磷症
phosphorus cycle 磷素循环
phosphorus fertilizer 磷肥
phosphorus fixation 磷固定
phosphorus pentoxide 五氧化二磷
phosphorylation 磷酸化作用
photoassimilation 光同化
photocatalysis degradation 光催化降解
photochemical degradation 光化学降解
photochemical systems of photosynthesis 光合作用的光化系统
photochrome 光敏色素;植物色素
photodecomposition 光解作用;感光分解作用
photodegradable film 光降解膜
photoelectric spectrophotometer 光电分光光度计
photoinduction 光诱导;光感应
photoinductive cycle 光诱导周期
photoinhibition 光抑制
photolysis 水的光解;水裂解;光解;光(分)解作用
photometer 光度计
photoperiod 光(周)期;光照期;日长;日照长度
photoperiodic sensitivity 日长敏感性;光(周)期敏感性
photoperiodical reaction 光周期反应
photophase(= photostage) 感光期;光照阶段;光阶段;(植物发育)光期
photophil 喜光的
photophosphorylation 光合磷酸化作用
photorespiration 光呼吸作用
photosynthesis 光合作用
photosynthesis inhibitors 光合作用抑制因子

photosynthesizer 光合作用系统
photosynthetic 光合的;光合作用的
photosynthetic active radiation 光合有效辐射
photosynthetic activity 光合活性
photosynthetic autotroph 光合自养生物
photosynthetic bacteria 光合细菌
photosynthetic efficiency 光合效率
photosynthetic function 光合功能
photosynthetic intensity 光合强度
photosynthetic process 光合作用过程
photosynthetic productivity 光合生产率
photosynthetic rate 光合作用率;光合速率
photosynthetic system 光合作用系统
photosynthetically active radiation 光合有效辐射
phototroph 光能自养生物
phoxime <商>辛硫磷
phreatic water 地下水
phthalic acid 邻苯二甲酸
Phthorimaea heliopa Loew 烟蛀茎蛾
Phthorimaea operculella Zeller 烟潜叶蛾
pH-value pH值(酸碱度值)
phyla 门;类群
phyllogen 顶芽;叶原

phyllogenous 叶上着生的
phyllome 叶性器官;叶原体
phyllosphere 叶围
phyllotaxy (= leaf arrangement) 叶序
physical absorption 物理吸附
physical absorption water 物理吸附水
physical adsorption and deposition of pollutant 污染物的物理吸附与物理沉淀
physical clay 物理性黏粒
physical degradation 物理降解
physical drought 物理干旱
physical fertility 物理肥力
physical remediation 物理修复
physical remediation of contaminated soil 污染土壤的物理修复
physical sand 物理性沙粒
physical signal 物理信号
physical weathering 物理风化
physiological acidic fertilizer 生理酸性肥料
physiological alkaline fertilizer 生理碱性肥料
physiological available water 生理有效水
physiological characteristic 生理特征
physiological disease 生理性病害(生理病)

physiological disorder 生理失调
physiological disorder-mineral nutrition 矿物营养生理失调
physiological drought 生理干旱
physiological ecology 生理生态学
physiological function 生理机能
physiological mature 生理成熟
physiological mechanism 生理机理
physiological neutral fertilizer 生理中性肥料
physiological process 生理过程
physiological property 生理特征
physiological reaction 生理反应
physiological ripe stage 生理成熟(生理成熟期)
physiological trait 生理特征
phytate 植酸盐(植素;肌醇六磷酸)
phytic acid 植酸
phytoactin 植物激素;植物活菌素
phytoalexin 植保素;植物抗毒素
phytoavailability 植物有效性
phytochelation 植物螯合
phytochrome (植物)光敏素;(植物)光敏色素
phytodegradation 植物降解
phytoecology 植物生态学
phytoedaphon 土壤微生物(群落)
phytoextraction 植物提取
phytokinin 细胞分裂素;植物胞裂素;植物胞激素

phytol 植(物)醇;叶绿醇
phytopesticide 植物性农药
phytophysiology 植物生理学
phytoremediation 植物修复
phytoremediation of contaminated soil 污染土壤的植物修复
phytoremediation of organic pollutant 有机污染物的植物修复
phytoremediation of soil contaminated with heavy metals 重金属污染土壤的植物修复
phyto-rhizospheric remediation 植物根际修复法
phytostabilization 植物稳定化
phytotoxicity 生理毒害
phytotron 人工气候室;育苗室;环境实验室
phytovolatilization 植物挥发
phytoxanthin 叶黄素;类胡萝卜醇
picking machine 烟叶分拣机
picking maturity 采摘成熟度
picking table 拣叶台
picking time 采收期
pie charts 饼图;圆形图
piece 支(卷烟)
pig excreta 猪粪
pig manure 猪粪
pig slurry 猪粪水
piggery wastewater 猪圈废水
pile turning 翻堆
pile yellowing 堆积变黄

pileorhiza　根冠
piliferous layer　根毛层
piling way covered stacks　塑料薄膜覆盖垛
pilling way　堆垛方法
pilot farm　试验农场
pilot investigation　试点调查
pilot study　试验性研究；探索研究；中间试验
pilot test　小规模试验
pimelic acid　庚二酸
pink　（烟叶）浅红色；粉红色；石竹
pinkish　（烟叶）微红色
pipe　烟斗；斗烟；管（子）
pipe draining　管式排水
pipe frame curing barn　管架式调制棚
pipe heater　管式加热器
pipe mouthpiece　烟嘴
pipe tobacco　斗烟丝；烟斗用的烟丝
pipe-heated hotbed　水管加热温床
pipet　吸管
pipetman　＜商＞自动移液器
pipette method　吸管法
pistil　雌蕊
pit planting　穴（坑）植
pit transplanting　穴移植；穴移栽
pitting　凹斑；陷斑
PK compound fertilizer　磷钾复混肥料

place of origin　原产地
placement of fertilizer　施肥位置
plagioclase　斜长石
plain　平原
plain cigarette（＝straight cigarette）　不带滤嘴的卷烟
plain filter　普通滤嘴
plane table　平板仪
plane transmitted light　单偏光
plant　植株；工厂
plant absorption　植物吸收
plant alkaloid　植物碱
plant analysis　植株分析
plant and animal residues　动植物残体
plant and leaf population　植株与叶片总数（密度）
plant ash　草木灰
plant bed　苗床；植床
plant chemical ecology　植物化学生态学
plant chemistry diagnosis method　植株化学诊断法
plant cultivation　植物栽培；植物栽培学
plant defence　植物防御
plant density　植株密度
plant development　植株发育
plant disease　植物病害
plant division　分株
plant endogenous hormones　植物内

源激素
plant exudates remediation 植物分泌物修复法
plant growth hormone 植物生长激素
plant growth regulator 植物生长调节剂
plant growth retardant 植物生长抑制剂
plant growth substance 植物生长物质
plant height 株高
plant hormones 植物激素
plant in rows 条栽;行栽
plant indicator 指示植物
plant juice analysis 植物液汁分析
plant lice 蚜虫
plant metabolism 植株新陈代谢
plant mineral nutrition 植物矿质营养
plant moisture stress 植物水分亏缺 (PMS)
plant monitoring 植物监测
plant morphology 植物形态学
plant nutrient 植物养分
plant nutrient deficiency disease 植物营养缺乏症
plant nutrient ratio 植物养分比例
plant nutrition status 植物营养状况
plant nutritional genetics 植物营养遗传学
plant nutritional physiology 植物营养生理学
plant organic nutrition 植物有机营养
plant organism 植株有机体
plant pathology 植物病理学
plant percent 成苗率
plant population 植株群体;种植密度
plant position 植株部位
plant potential 成苗数(潜力);植株潜力
plant pressure potential 植物(水分)压力势
plant residues 植株残体
plant resource 植物资源
plant root 植株根系
plant sample collecting and preparation 植物样品的采集与制备
plant setter 种植机
plant setting 植物栽植
plant spacing 株距
plant tissue culture 植物组织培养
plant trash 植株残体
plant vigor 烟株长势
plant virus 植物病毒
plantation land 种植业用地
planted area 种植区
planter 种植机;播种机
planting bed 种植苗床
planting belt 栽植带
planting date 栽种时期
planting density 种植密度

planting depth 种植深度
planting distance 种植距离
planting hole 栽植穴
planting hole machine 穴播机;穴植机
planting in line(= planting in rows) 条栽;列栽
planting nursery 种植苗圃
planting on mounds 垄作
planting pit 定植穴
planting plan 种植计划;种植设计
planting season 种植季节;种植期
planting stock 定植苗;栽植材料
planting time 播种期
plant-insect interaction 植物与害虫间相互影响
plant-soil interface 植物-土壤界面
plastic film covered stacks 塑料薄膜覆盖垛
plastic greenhouse 塑料温室
plastic property 塑性
plastic tunnel 塑料大棚
plastic waste 塑料废料
plasticity 可塑性
plate count 平板测数
plate count method 平板计数法
plate culture 平板培养;平皿培养物
plate culture method 平板培养法
plate dilution 平板稀释法
plateau climate 高原气候

plating procedure 平板计数;平皿法
pleasant note 愉快香韵
pleiophyllous 多叶的
pleiotomy 多分枝式
plenum 强制通风
Pleonomus canaliculatus Faldermann 沟金针虫
plot 小区
plot arrangement 田间排列;小区排列
plot assignment 小区划分;土壤小区划分
plot design 试验区设计
plot experiment 小区试验
plot layout 试验区配置
plot plan 区域设计
plot runoff 小区径流
plot sowing 块状播种
plough 耕地;犁;投资
plough horizon 耕作层
plough land 耕地;可耕地
plough layer 耕层;耕作层
plough under 翻入;耕翻
plough up 耕地
ploughing in green manure 翻埋绿肥
ploughing in winter 冬耕
ploughing plowing 耕地;犁地
ploughing season 耕期
ploughing with disk 圆盘耙耕作
ploughland 耕地;农田

169

plow 耕作;犁地
plow depth 耕深
plow layer 耕作层
plow pan 犁底层
plow sole 犁底层
plowed plot 犁耕区
plowing 翻耕
plowing depth 耕深
plowing horizon 耕作层
plowing width 犁幅宽度
plowpan 犁底层;犁磐
plow-sole 犁底层
plucking 采摘
plug tobacco 嚼烟;烟饼
plumula 胚芽;胚芽的
plumular 胚芽的
plumular axis 胚轴
plumule 胚茎;幼芽;胚芽
plurifoliate 多叶的
plurifolious 多叶的
pluvial 多雨的;洪水的;雨成的;雨期
pluvial region 多雨地区
pluviometer 雨量计;雨量器
pneumathodium 呼吸根;气孔
pneumatic 气动的;空气的
pneumatic classification 风选
pneumatic drying 通风干燥
pneumatic seeder 吸气式播种机
pneumatic selecting 风选
pneumatic separating 风分

pneumatic separating chamber 风分室;风分箱
pocket planting 袋装种植
podzol 灰化土;灰壤
podzolic 灰壤的
podzolic brown earth 灰化棕壤
podzolic horizon 灰化层
podzolic soil 灰化土
podzolized horizon 灰化层
podzolized laterite 灰化砖红壤
point estimation 点估计
point frame method 样点法
point observation method 样点观察法
point of zero charge 电荷零点
point source 点源
point source pollution 点源污染
pointed tip 锐形(叶)尖
poisoned by pesticide 农药中毒
poisoning symptom 中毒症状
poisonous element pollution 有毒元素污染
poisonous substance 有毒物质
Poisson distribution 泊松分布
polar organic molecular adsorption method 极性有机分子吸附法(测定比表面)
pole-burn 杆腐病;梗腐病
pollen 花粉
pollen abortion 花粉败育
pollen brush 花粉刷

pollen culture 花粉培养法;花粉培养
pollen odour 花粉气
pollen trap 花粉采集器
pollenize 传粉;授粉
pollinate 传粉;授粉
pollination medium 传粉媒介
pollutant accumulation in the soil 土壤中污染物积累
pollutant avoidance technology 减少污染物的技术
pollutant concentration 污染物浓度
pollutant control 污染物控制
pollutant degradation 污染物降解
pollutant index 污染物指数
pollutant threshold 污染物阈值
polluted soil 污染土壤
polluted water 污水
pollution by carbon oxide 碳氧化物污染
pollution by domestic animal manure 畜禽粪污染
pollution by fluorine 氟污染
pollution by livestock effluent 牲畜排放物污染
pollution by organochlorine pesticide 有机氯农药污染
pollution by organonitrogen pesticide 有机氮农药污染
pollution by organophosphorus pesticide 有机磷农药污染
pollution concentration 污染浓度
pollution control function 污染控制作用
pollution criterion 污染标准
pollution haven 污染"避难所"
pollution hazard 污染危害
pollution index 污染指数
pollution of environment 环境污染
pollution prevention technique 污染防治技术
pollution reducing measure 降低污染的措施
pollution source 污染源
pollution toxicity 污染毒性
pollution type 污染类型
pollutional equivalent 污染当量
pollution-free agriculture 无公害农业
pollution-free energy 清洁能源
pollution-free pesticide 无污染农药
polyacrylamide 聚丙烯酰胺
polyamine(PA) 聚胺(PA)
polyauxotroph 多重营养缺陷型
polychlorinated biphenyl 多氯联苯
polychlorobiphenyl pollutant 多氯联苯污染物
polycross 多系杂交
polycross method 多系杂交法
polycyclic aromatic hydrocarbon 稠环芳烃;多环芳烃
polyethylene blown film for agriculture

农业用聚乙烯吹塑薄膜
polyethylene film 聚乙烯薄膜
polyethylene(PE) 聚乙烯(PE)
polyetic disease 积年流行病害
polyhydric 多羟(基)的
polyhydric acid 多元酸
polymerization 聚合(作用)
polymorphic organic matter 无定形有机物质
polynomial 多项式
polyose 多糖类;聚糖
Polyoxins 多抗霉素
polyphenolic compound 多酚类化合物
polyphosphate 多磷酸盐
polypropylene 聚丙烯
polypropylene film 聚丙烯薄膜
polysaccharide 多糖
polyterpene 多萜(烯)
poor (烟叶)劣级;(烟叶)次级
poor class leaf 下等烟
poor drainage 排水不良
poor growth 生长缓慢
poor harvest 低产;歉收
poor land 薄地
poor quality seed 劣质种子
poor seed germination 种子萌芽率低
poor soil 贫瘠土壤
poor stand 出苗情况不好
poor tightening 紧固不良
poor-drained soil 排水不良土壤

poorly drained 排水不良
poor-structured soil 结构差的土壤
population 种群
population density 种群密度
population density of insect 虫口密度
population distribution 总体分布
population improvement 群体改良
population mean 总体平均值
population of glandular hairs 腺毛密度
population parameter 总体参数
population quality of insect 昆虫种群质量
population quantity 种群数量
population standard deviation 总体标准差
population variance 总体方差
porcelain crucible 瓷坩埚
porcelain evaporating dish 瓷蒸发皿
pore 孔;毛孔;气孔
pore diameter 孔径
pore size 孔径;孔径大小
pore space 孔隙空间
pore space distribution 孔隙分布
pore volume (烟支中)非填充空间;孔隙体积
pore water 孔隙水
poriferous 多孔的
poriness 多孔性;疏松性
porosimeter 孔度计;孔隙度测定仪;透气度测定仪

porosity 多孔性;孔隙度;透气度;孔性
porosity meter 空隙度测定仪
porous aggregate 多孔团聚体
position （烟叶）部位（分级）;位置
positive charge 正电荷（阳电荷）
positive ion 阳离子
positive priming effect 正激发效应
post plant 移栽后
post ripening 后熟
post-emergence 萌发后;出土后
post-emergence treatment 萌发后处理
post-harvest changes 收获后变化
post-harvest disease 贮藏病害;收获后病害
postponement of germination 延期发芽
postponing germination 延迟发芽
post-transplanting application 移栽后施用
pot culture 盆栽
pot culture experiment 盆栽试验
pot culture method 盆栽法
pot experiment 盆栽试验
pot incubation test 盆栽培养试验
pot plant 盆栽植物
pot test 盆栽试验
pot transplanting 盆钵移栽
potash 钾肥;草碱;钾碱;碳酸钾
potash deficiency 缺钾症

potash fertilization 施钾
potash fertilizer 钾肥
potash hunger 缺钾
potash status 钾素状况
potassic fertilizer 钾肥
potassium 钾
potassium acetate 乙酸钾
potassium balance 钾平衡
potassium bicarbonate 碳酸氢钾
potassium bichromate 重铬酸钾
potassium buffer capacity 钾缓冲容量
potassium carbonate 碳酸钾
potassium channel 钾离子通道
potassium chloride 氯化钾
potassium chromate 铬酸钾
potassium citrate 柠檬酸钾
potassium cyanide 氰化钾
potassium cycle 钾循环
potassium deficiency 缺钾
potassium dichromate 重铬酸钾
potassium dichromate oxidation 重铬酸钾氧化法
potassium dihydrogen phosphate 磷酸二氢钾（磷酸一钾）
potassium disturbance 钾素营养失调
potassium fertilizer 钾肥
potassium fixation 钾的固定;钾固定
potassium fixation capacity 钾固定

能力
potassium humate　腐殖酸钾
potassium hydrogen phosphate　磷酸氢二钾
potassium hydroxide　氢氧化钾
potassium in soil solution　土壤溶液钾
potassium liberation　钾的释放
potassium magnesium sulfate　硫酸钾镁
potassium nitrate　硝酸钾
potassium permanganate　高锰酸钾
potassium phosphate dibasic　磷酸钾
potassium potential　钾位
potassium release　钾的释放
potassium salts　钾盐
potassium sodium tartrate　酒石酸钾钠
potassium sulfate　硫酸钾
potassium sulphate　硫酸钾
potassium supplying capacity of soil　土壤钾素供应能力
potassium supplying intensity　钾供应强度
potassium transport system　钾吸收转运系统
potassium-bearing mineral　含钾矿物
potassium-enriching plant　富钾植物
potato tuber moth　马铃薯块茎蛾
Potato virus X　烟草马铃薯 X 病毒病；马铃薯 X（型）病毒
Potato virus Y　烟草马铃薯 Y 病毒病；马铃薯 Y（型）病毒
potential acidity　潜性酸度
potential available potassium　潜在有效钾
potential buffering capacity　潜在的缓冲容量
potential deficiency　潜在缺乏
potential evapotranspiration　潜在蒸散
potential fertility　潜在肥力
potential soil productivity　土壤潜在生产力
potential transpiration　潜在蒸腾
potential yield　潜在产量
potentiality of axillary buds　腋芽生长势
potentiality of field tobacco　田间烟草长势
potentiality of seed　种子发芽势
potentially available nutrient　潜在有效养分
potentially available potassium　潜在有效钾
potting　盆栽
potting compost　盆栽营养土
potting earth　盆栽土
potting mixture　盆栽营养土
potting shed　盆栽棚
potting soil　盆栽土
Potyvirus Y　马铃薯 Y（型）病毒属；

马铃薯 Y 病毒属
poultry dung　禽肥
poultry manure　禽粪肥
povamycin　多效霉素
powder fertilizer　粉状肥料
powdered rock phosphate　磷矿粉
powdery mildew　白粉病
power model　幂函数模型
power of the test　检测能力
power tiller　动力耕耘机
practical germination percentage　实际发芽率
practice ground　试验用地
precautionary principle　预防为主原则
preceding crop　前作(物)
precipitate　沉淀物
precipitation　降水;降水量;降雨量;沉淀(作用)
precipitation efficiency　降水效率
precipitation evaporation ratio　降水蒸发比
precipitation index　降水指数
precipitation intensity　降水强度
precipitation reaction　沉降反应
precipitation resources　降水资源
precision agriculture　精准农业(精确农业)
precision farming　精准农业(精确农业)
precision fertilization　精准施肥
precision fertilizing　精准施肥

precision seeding　精确播种
precision tillage　精耕细作
precocious　过早开花的;过早熟的
precocious germination　过早萌发;提前萌发
predatory exploitation　掠夺性开发
predicted yield　预测产量
prediction　预测
predictive models　预测模型
pre-emergence damping-off　出苗前猝倒
pre-emergence treatment　萌发前处理
preference temperature　适宜温度
preferred temperature　适宜温度
pregerminated seed　萌发前的种子;未萌发的种子
pregermination　催芽
pregermination treatment　发芽前处理
preharvest　采收前的;收获前的
preharvest period　采收前期;收获前期
preharvest spray　采前喷药
preinfective period　侵染前期
preliminary examination　初步调查
premature　假熟;欠成熟;未成熟的;早熟的
premature flowering　早花
premature harvesting　提前收获
premature seed　不成熟的种子
premature senescence　早衰

premature shedding of leaves　成熟前落叶
premature yellowing of lugs　（烟叶）底烘
premature　假熟
preparation of land　土地平整；整地
preparation of seeds　种子准备
preparation of soil　播前整地；整地
pre-planting soil fumigant　移栽前土壤熏蒸剂
presenility　早衰
preservation　（菌种）保藏
preservation of fertility　地力保持
pre-sowing application　播前施用
presowing treatment　播前处理
presprouting of seed　种子催芽
press　新闻报道；印刷；镇压（指土壤）
press drill　带有压土轮的条播机
press operation　镇压作业
pressed tobacco　压实的烟草
pressure potential　压力势
pressurized irrigation　加压灌溉系统
pre-transplanting application　移栽前施用
pretreatment　预处理
pretreatment procedure　预处理方法
prevention and control of soil pollution　土壤污染防治
prevention and treatment of soil pollution　土壤污染防治
prevention of crop nutrient deficiency　作物缺素症防治
prevention of land degradation　土壤退化防治
previous crop　前茬作物；前作
previous crop with its stubble field　茬口
previously cultivated land　熟地；已耕地
price of the output　产品价格
pricking out　选苗移植
primary causal factor　主要致病因素
primary element　主成分；主要元素
primary leaves　初生叶
primary mineral　原生矿物
primary nutrient　大量元素；主要营养
primary root　初生根；主根
primary vein　中脉；主脉
prime　采摘；初期；优质的；原始的；最初
prime agricultural land　上等农地
prime cropland　基本农田
prime farmland　基本农田
prime farmland protection　基本农田保护
primed seed　催芽种子
priming　采叶；逐叶采摘
priming effect　激发效应
priming machine　采收机
priming method　逐叶采摘法
primings　烟株下面第一片烟叶；脚叶

primitive soil 原始土壤
principal component 主要成分
principal component analysis 主成分分析
principal crop 主要作物
principal element 主要元素
principal leaf 主叶
principal of combining use and maintenance of cultivated land 耕地用养结合原则
principal root 主根
principal vein 主脉
principal yield 主产量;主收获
principle factor analysis 主因子分析
principle horizon 基本发生层(基本土层)
Pro(＝proline) 脯氨酸
probability 概率
probability analysis 概率分析
probability density function 概率密度函数
probability distribution 概率分布
probability map 概率图
probability sampling 概率抽样
probable error 概率误差
problem of resources 资源安全;资源问题
procedural regulations regarding the environment quality monitoring of air in agricultural regions 农区环境空气质量监测技术规范
procedural regulations regarding the environment quality monitoring of soil 农田土壤环境质量监测技术规范
process of ecological degradation 生态退化过程
process of nitrogen fixation 固氮过程
processing property 加工性能
processing technology 加工工艺
producing horizon 生产层(土壤)
product of nitro humic acids 硝基腐殖酸制品
production capacity 生产能力
production efficiency 生产效率
production output 产量
production potential 生产潜力
productive agriculture 高产农业
productive arable land 肥沃的可耕地
productive soil 肥沃土壤
productivity 生产率;生产力
profile form 剖面构造类型(剖面构型)
profile pattern 剖面构型
profit increment 利润增量
profit per unit area 单位面积利润
profuse flowering 盛开花的
progressive sampling 先进采样法
prokaryote 原核生物
prokaryotic microorganism 原核微生物

prolonged drought 持续性干旱
proluvial deposit 洪积物
proluvium 洪积物
promoting germination 促进发芽
promotion on animal health （土壤）对动物健康的促进
promotion on human beings health （土壤）对人类健康的促进
promotion on plant health （土壤）对植物健康的促进
promotive effect 促进作用
propagation farm 采种田;种子繁殖场;种子繁殖田;种子田
propane diacid 丙二酸
proper rotation 合理轮作
proper spacing 适宜株行距
proper use of fertilizer 合理施肥
proportion of NPK 氮磷钾比例
proportional sampling 比例抽样
protected agriculture 设施农业
protected cultivation 保护地栽培
protected culture 保护地栽培
protection of biological resources 生物资源保护
protective exploitation 保护性开发
protective measure 保护措施
protein 蛋白质
protein N 蛋白质氮
protein synthesis 蛋白质合成
proteinase 蛋白酶
proteoid root 排根

pruner 修剪机
pruning 修建;抑制
psychrometer 湿度计
psychrometric room 人工气候室
psychrophile 低温菌;嗜冷细菌
psychrophilic microorganisms 嗜冷微生物
psychrotolerant 耐冷的
p-Tolualdehyde 对甲基苯甲醛
public disaster 公害
public hazard 公害
puckering of leaves 叶片起皱
puff count （对烟支限定长度）抽吸次数
puff frequency 抽吸频率
puff interval 抽吸间隔;抽吸间隔时间
puff length （卷烟）抽吸长度
puff number 抽吸口数
puff profile 抽吸曲线
puff resistance 吸阻
puff volume 抽吸容量;每口抽吸烟量
puffed-cut tobacco 膨胀烟丝
pullulation 发芽;萌发;增殖
pulse crop 豆类作物
pulse family 豆科
pulverator plough 松土犁;碎土犁
pulverescent 粉块(粉状)
pulverization 粉碎(作用);研磨(作用)
pulverize 粉碎;磨成粉状;喷雾;松土

pulverizer 粉碎机;喷雾器
pulverizing harrow 浅耕松土耙
pump 泵
pump drain 抽水;排水沟
pumping irrigation 提水灌溉
Punctual Kriging 点克立格法
pungency 辛辣味
pungent 刺激;辛辣(烟味)
pungent odor 刺激性气味;辛辣味
pungent taste 刺激味;尖刺;辛辣味
purchase number 收烟编号
purchasing 收购
purchasing inspection 购进验查
pure breed 纯种
pure line cultivar 纯系品种
pure nugget effect model 纯块金效应模型
pureness 纯度;纯洁
purer flavor 香气纯正
pure-sweetness aroma 正甜香
pure-sweetness type 中间香型
purification 纯化
purification ability 净化能力
purification capacity 净化能力
purification efficiency 净化效率
purify 精制;净化;提纯
purifying effect 净化作用
purity 纯度;纯净

purity and cleanness (烟气)纯净
purity check 纯度检验
purity of seeds 种子纯度
purple alfalfa 紫花苜蓿
purple soil 紫色土壤
purplish brown tip 紫褐色叶尖
Purplish soil 紫色土
putrefaction 腐败;腐烂
putrefaction ferment 腐败发酵
putrid odor 腐败气味
putrid taste 腐烂味
pycnometer 比重计;比重瓶
pyrethrin 除虫菊酯
pyridine 吡啶;氮(杂)苯
pyrogenic decomposition 高温分解;热解
pyrogenic distillation 干馏;高温蒸馏
pyrogenous 干馏的;高温蒸馏的
pyrolysis temperature 热解温度
pyrophilous 喜高温的
pyrophosphatase 焦磷酸酶
pyrophosphate 焦磷酸盐
pyroracemic acid 丙酮酸
pyrrole 吡咯
pyruvate kinase 丙酮酸激酶
pyruvic acid 丙酮酸
Pythium 腐霉属

Q

quadruple cropping 一年四作(四茬复种)
qualitat(= quality index) 品质指数
qualitative analysis 定性分析
qualitative approach 定性方法
qualitative character 质量分析;质量性状
qualitative filter paper 定性滤纸
qualitative method 定性方法
qualitative technique 定性方法
quality parameter 质量参数
quality appraisal 品质鉴定;品质评价
quality assurance system 质量保证体系
quality certification 质量认证制度
quality character 品质特征
quality characteristic 品质特征
quality class 品位级;质量等级
quality comprehensive judgement 质量综合判定
quality control 质量管理;质量控制;品质监督
quality deterioration 品质下降;品质劣变
quality evaluation 品质评价
quality factors 质量因素
quality grade 品质等级(质量分等);质量等级
quality improvement 质量改进
quality index 品质指数;质量指数
quality inspection 质量检测
quality loss 品质损失
quality management system 质量管理体系
quality manual 质量手册
quality norm 质量标准
quality objective 质量目标
quality of aroma 香气品质;香气质量
quality of ecological environment 生态环境质量
quality of flavor 香味质量
quality of products 产品品质
quality of requirements 品质要求;质量要求
quality of seedlings 幼苗质量
quality rating 品质等级(质量分等)
quality seed 优质种子
quality sorting 品质分级
quality specification 技术规格;质量指标
quality standard 品质标准;质量标准
quality standard for ground water 地

下水质量标准
quality standard of soil environment 土壤环境质量标准
quality supervision 质量监督
quality supervision and inspection 质量监督检验
quality system 质量体系
quality test 质量检测
quantitative analysis 数量分析;定量分析
quantitative approach 定量方法
quantitative character 数量性状
quantitative determination 定量测定
quantitative estimation 定量测定
quantitative filter paper 定量滤纸
quantitative measurement 定量测定
quantitative method 定量方法
quantitative technique 定量方法
quantity factor 数量因素
quantity of aroma 香气量
quantity of cultivated land 耕地数量
quantity of nutrient 养分含量
quantity of seeding 播种量
quantity of soil organisms 土壤生物数量
quarantine seed health testing 种子检疫检验
quartering 四分法
quartile deviation 四分位差
quartz sand 石英砂
quick cluster 快速聚类
quick lime 生石灰
quick soil 暖土;早发土
quick soil analysis 快速土壤分析;土壤速测
quick test 速测
quick-acting fertilizer 速效肥料
quiescence 被遏制状态;静止;休眠
quinone 苯醌;醌

R

R. H. (= relative humidity) 相对湿度
R/S ratio （微生物）根土比
race （生理）小种;亚种;族
rack 夹烟架;烟架
radial transport 径向运输
radiant emittance 辐射率(辐射度)
radiant heat 辐射热
radiated heat 辐射热
radiating heat 辐射热
radiation 辐射
radiation breeding 辐射育种
radiation heat 辐射热
radical biology 根系生物学

·181·

radicant 从茎部生根的
radication 根系
radicel 胚根
radiciform 根状;根形
radicle 胚根
radiocarbon 放射性碳同位素
radioisotope 放射性同位素
rag 碎片;烟丝
ragged leaf spot 破烂叶斑病
rain duration 降雨持续时间
rain fall amount 雨量
rain wash 雨水冲刷
rain water 雨水
rainfall 降雨;降雨量
rainfall amount 降雨量
rainfall catchment and supplementary irrigation 集雨补灌
rainfall distribution 雨量分布
rainfall duration 降雨持续时间
rainfall erosion 雨蚀(作用);降雨侵蚀
rainfall erosivity 降雨侵蚀程度
rainfall frequency 降雨频率;降雨强度
rainfall infiltration 降雨入渗
rainfall resource 降水资源
rainfall runoff 降雨径流(地面径流)
rainfall simulation 模拟降雨
rainfall use efficiency 降水使用效率

rainfed farming 雨养农业
rainmaking 人工降雨
rainstorm 暴雨
rainwash 雨水冲刷
rainwater harvesting 集雨;雨水集蓄
rainwater harvesting agriculture 集水农业
rainwater resource 雨水资源
rainwater use 雨水利用
rainy climate 多雨气候
rainy season 雨季
rainy year 丰水年
raise seedling 育苗
raised field 台田
raising 培育;饲养;提高;栽培
random distribution pattern 随机分布型式;无序分布型式
random effect 随机效应
random error 随机误差
random model 随机模型
random pollination 自然授粉
random sample 随机样本
random sampling 随机取样
random sowing 撒播
random variable 随机变量;随机向量
randomisation 随机化
randomized arrangement 随机排列
randomized block design 随机区组设计

randomness 随机性
range 极差;变程(范围)
range method 极差法
range of nicotine content 烟碱含量范围
rank 等级;繁茂的;腐臭(烟叶);过于肥沃的;列;排
rankness 繁茂;肥沃
rape cake 菜籽饼
rape seed cake 菜籽饼
rapeseed cake fermentation 菜籽饼发酵
rapid analysis method 快速分析方法
rapid release fertilizer 速效肥
rapid test 速测
rare earth element 稀土元素;稀土
rareripe 早熟的
raster data 栅格数据
rate of absorption 吸收速率
rate of accumulation 积累率
rate of adsorption 吸附速率
rate of biodegradation 生物降解速度
rate of broken bundles 烟把破损率
rate of budding 萌芽率
rate of burn 燃烧速率
rate of diffusion 扩散速率
rate of evaporation 蒸发量
rate of fertilizer application 施肥量
rate of growth 生长速率;增长率
rate of net photosynthesis 净光合作用速率
rate of transpiration 蒸腾率
ratio of cut tobacco 出丝率
ratio of lamina yield 出叶率
ratio of sugar and nicotine 糖碱比
rational application of fertilizer 合理施肥
rational close planting 合理密植
rational development and utilization 合理开发利用
rational fertilization 合理施肥
rational manuring 合理施肥
rational rotation 合理轮作
rational use of cultivated land 耕地合理利用
raw data 原始数据;原始资料
raw green odour 生青气
raw humus 粗腐殖质
raw soil 生土
raw soil layer 生土层
raw tobacco 原烟
readily available fertilizer 速效肥料;速效性肥料
readily available moisture 速效含水量
readily available nutrient 速效性养分
readily available phosphorus 速效磷
readily available potassium 速效钾;速溶性钾

· 183 ·

readily fixed nutrient 易被固定养分
readily oxidation carbon 易氧化碳
reagent 试剂
real specific gravity 真比重
real tillage 翻耕土地
realized production 净生产量
real-time N management 实时氮素管理
real-time nutrient management 实时养分管理
reap 收割;收获
rearing of larvae 幼虫养育
recalcitrant pollutant 难降解污染物
receiving inspection 购进验查
reciprocal averaging 相互平均法
reciprocating shaker 往复式振荡机
recirculation （再）循环
reclaim 开垦荒地;土壤改良;驯养
reclaimed soil 已改良土壤
reclamation of saline-alkali soil 盐碱土改良
reconnaissance 踏勘
reconnaissance survey 概查
reconstituted sheet 再造烟叶
reconstituted tobacco 再造烟叶
reconstituted tobacco sheet 再造烟叶
recovery 回收率
recovery of fertilizer 肥料利用率
recovery of nutrient 养分利用率;养分的回收

recovery of surrogate standard 加标样回收率
rectification 校正
recurrent parent 回交(轮回)亲本
recycle 再循环
recycling 再循环
recycling enforcement 强制性再循环
recycling of waste 废物再利用
red （烟叶)红棕色
red earth 红壤;红土
red loam 红壤
red podzolic laterite 灰化砖红壤
red primitive soils 红黏土
red rust 红锈(缺铜,烟叶斑点病)
red soil 红壤
reddish-brown 红棕色
redox 氧化还原(作用)
redox equilibrium 氧化还原平衡
redox process 氧化还原过程
redox reaction 氧化还原反应
redox regime 氧化还原状况
redox status 氧化还原状况
redox system 氧化还原体系
redried tobacco 复烤烟叶
redry 复烤
reduce 减量化;减量
reduced tillage 少耕
reduced-tillage system 少耕法
reducing humidity 去湿
reducing sugar 还原糖

reduction in grow rate 生长速率减慢
reduction of carbon dioxide emissions 碳减排
reductive TCA cycle 还原性三羧酸循环
red 红棕色
reference 坐标;参考文献
reference material 标准物质;标准样品;参考物质(标准物质)
refermentation 二次发酵;后发酵
reflectance 反射率
reflectance factor 反射因子
refractory organic pollutant 难降解有机污染物
refuse composting 垃圾堆肥
refuse-to-energy 废物变能源
regarding 验级;再鉴定
regeneration 再生作用(更新作用)
regenerative growth 再生生长
regenerative technology of soil 土壤修复技术
regional development 区域发展;区域开发
regional eco-environmental issues 区域生态环境问题
regional innovation capability 区域创新能力
registered pesticides 注册农药
registered variety 注册品种
regression 回归;退化

regression analysis 回归分析
regression coefficient 回归系数
regression design 回归设计
regression equation 回归方程
regression function 回归函数
regression-orthogonal design 回归正交设计
regular cultivation 常规耕作
regular fertilizing 常规施肥
regular planting 常规播种(移栽);常规栽培
regularity of fertilizer requirement 需肥规律
regulated deficit irrigation 调亏灌溉
regulation for safe use of pesticides 农药安全使用规定
regulation of nicotine 烟碱调整
reject 残烟
rejectment 排泄物
2 related sample test 二个相关样本检验
relational database management system 关系式数据库管理系统
relative air humidity 相对空气湿度
relative density 相对密度
relative effect of control 相对防治效果
relative error 相对误差
relative frequency 相对频数
relative humidity 相对湿度
relative moisture 相对湿度

· 185 ·

relative moisture content 相对含水量

relative moisture of the soil 土壤相对湿度

relative saturation 相对饱和度

relative temperature 相对温度

relative transpiration 相对蒸腾

relative water content 相对含水量

relative yield 相对产量

relay cropping 套作

relay intercropping 套作

release 释放

releasing rate 释放速率

reliability 可靠性

reliability analysis 可靠性分析

relief map 地形图;地势图

remaining stubble 留茬

remediation goal 修复目标

remediation technique 修复技术

remote measurement 遥测

remote sensing 遥感

remote sensing diagnosis method 遥感诊断法

remote sensing monitoring 遥感监测

remote sensing technique 遥感技术

remote sensor imagery 遥感成像(遥感图像)

removal 去除

renovation of variety 品种复壮;品种更新

repeatability 重复性

repeated cultivation 连作

repeated plots 重复小区

repeated test 重复试验

repeated tillage 多次耕翻;重复耕作

replaceable ion 可置换离子

replaced pesticide 替代农药

replant obstacle 连作障碍

replanting 补栽;补植

replicate determination 重复测定;平行测定

replication 重复

replication number 重复次数

representative sample 代表性样品

reproducibility 再现性

reproductive bud 繁殖芽;花芽

reproductive growth 生殖生长

reseeding 补播;重播;追播

reservoir 储存库

residual acidity 残留酸度

residual deposit 残积物

residual dose 残留量

residual effect 后效

residual effect of fertilizer 肥料残效

residual effect of pesticide 农药残效

residual effect of phosphate fertilizer 磷肥残效

residual error 残差

residual form 残留态

residual life 残留期

residual pesticide 农药残留;残留

性农药
residual rate 残留率
residual ratio 残留率
residual soil 残积土
residual taste 后味;余味
residuals 残差
residue plastic film 残膜
residue-prone agricultural chemical 残留性农药
resin aroma 树脂香
resistance 抗性
resistance to adversity 抗逆性
resistance to insect 抗虫性
resistance to penetration 穿透阻力
resistance to pesticide 抗药性
resistant nutrient 难分解养分
resistant strain 抗性品系
resistant variety 抗病品种
resolution 分辨率
resource 资源
resource assessment 资源评价
resource availability 资源可用性
resource circulation 资源循环
resource conservation 资源保护
resource depletion 资源枯竭
resource development 资源开发
resource diversity 资源多样性
resource elements 资源要素
resource management 资源管理
resource protection 资源保护
resource recovery 资源化;资源回收
resource recovery technique of solid waste 废物资源化技术
resource scarcity 资源缺乏
resource use 资源利用
resource use efficiency 资源利用效率
resource utilization 资源利用
resource utilization efficiency 资源利用效率
resources and environment 资源环境
resources and environment information system 资源环境信息系统
resources degeneration 资源退化
resources ecosystems 资源生态系统
resources environment 资源环境
resources exhaustion 资源枯竭
resources exploitation 资源开发
resources remote sensing 资源遥感
resources stress 资源胁迫
resource-saving economy 资源节约型经济
resource-saving technology 节约资源型技术
resources-saving agriculture 资源节约型农业
respiration 呼吸作用
respiration quotient 呼吸商
respiratory exchange 气体交换
respiratory intensity 呼吸强度

respiratory metabolism 呼吸代谢（作用）
respiratory quotient 呼吸商
response 感应;响应;反应
response time 响应时间
response to fertilizer 肥料效应;施肥反应
restbalk 畦;垄
resting 静止的;休眠的
resting bud 休眠芽;潜伏芽
restoration of soil productivity 土壤生产力恢复
retardation of growth 生长阻滞;延缓生长;生长减缓
retarded growth of apical growing point 顶部生长点生长停滞
retarding effect 抑制作用
retention 保持(固持;吸持)
retention of electric charge balance 电荷平衡的维持
retranslocation 再运转
retrogradation 退化
retrogressive 退化
return 归还
return farmland to forestland or grassland 退耕还林还草
return of farm wastes to the land 农业废物还田
returns from land 土壤报酬
reuse 再利用
reversible 可逆的

reversible plough 翻转犁;转向犁
reversible plow 翻转犁;转向犁
revitalization of variety 品种复壮
revolute leaf 外卷叶
rhizobia 根瘤菌
rhizobium 根瘤菌
rhizobium fertilizer 根瘤菌肥料
rhizocoenosis 根瘤联合共生(根际联合作用)
rhizodeposition 根际淀积
rhizogenic layer 生根层
rhizome 根茎;地下茎
rhizoplane 根面;根表
rhizoremediation 根际修复
rhizosphere 根际
rhizosphere acidification 根际酸化（作用）
rhizosphere bacteria 根际细菌
rhizosphere dynamic 根际动态
rhizosphere effect 根际效应
rhizosphere flora 根际微生物区系;根围区系
rhizosphere management 根际调控
rhizosphere microecosystem 根际微生态系统
rhizosphere microflora 根际微生物区系
rhizosphere microorganism 根际微生物
rhizosphere nutrition 根际营养;根圈营养

rhizosphere technology　根际技术
rhizosphere-to-soil ratio　根土比率
rhizospheric microorganism　根际微生物
rhizotrons　根室法
rice field　稻田
rice following wheat　麦茬稻
rice soil　水稻土
rice straw　稻草
rice-wheat rotation　稻麦轮作
rich　丰富;充足(指香气);多(油分)
rich body(＝fleshy)　丰满;身份充实
rich clay　肥沃黏土
rich flavor　浓香;香味足
rich in aroma　香气足
rich soil　肥沃土壤
richness　(香气)充足
riddling of remedying plants　修复植物的筛选
ridge　垄;田埂
ridge building　作垄
ridge culture　垄栽;垄作
ridge drill　垄作播种机
ridge furrow　垄沟
ridge irrigation　畦灌;垄灌
ridge making　作垄
ridge planting　垄作
ridge plowing　垄耕;培土
ridge site　缓岗地
ridge sowing　垄播
ridge tillage　垄作(垄畦耕作)
ridge tillage and no-tillage　垄作免耕
ridge-and-furrow irrigation　垄沟灌溉
ridger　培土器;起垄机;松土器
ridging　培土;起垄;作垄
ridging by cultivating　中耕培土
ridging plough　培土犁;起垄犁
riding primers　乘坐式采烟机
right side　(烟叶的)正面
ring spot virus　环斑病毒
ringing　环割
ripe　成熟;充分成熟;工艺成熟
ripen　使成熟;催熟
ripen in advance　提前成熟
ripened seed capsule　成熟蒴果
ripener　催熟剂
ripeness　(烟叶)成熟;成熟度
ripening　成熟;熟化;陈化
ripening agent　成熟催化剂;促熟剂
ripening process　成熟过程
rising gradient of temperature　升温梯度
river erosion　河流侵蚀
river valley　河谷
river wash　河床冲积物;河滩荒地
rock phosphate　磷矿石
roll in　辗平;镇压
rolled stem shreds　梗丝
roller　辊子;卷筒;卷(雪茄)外包皮工;滚压器

roller cutting machine 滚刀式切丝机
roll-over plough 翻转犁
room germinator 发芽室
room temperature 常温;室温
root 根
root absorbing area 根系吸收面积
root absorbing power 根的吸收力
root activity 根系活力
root and stalk disease 根茎病害
root apex 根端
root biology 根系生物学
root cap 根冠
root collar 根颈
root culture 根培养
root density 根密度
root development 根系发育
root diameter 根径
root digger 拔根机
root dip 蘸根
root distribution 根系分布
root division 分根(分株)
root dry weight 根系干重
root elongation 根系伸展
root extension 根系伸展
root exudate 根系分泌物
root exudation 根系分泌物
root fibril 须根
root field 根域
root fresh weight 根鲜重
root growing space 根系生长空间
root growth 根系生长

root hair 根毛
root hair density 根毛密度
root hair region 根毛区
root hair zone 根毛区
root knot 根结病;根瘤病
root knot nematode 根结线虫病
root length 根长
root length density 根长密度
root mass 根群
root medium 根围
root neck 根颈
root nodule 根瘤
root nodule bacteria 根瘤菌
root pressure 根压
root products 根产物
root region 根区
root respiration 根系呼吸
root rot 根腐
root rot disease 根腐病
root spreading stage 伸根期
root stalk 根茎;初生主根
root stretching stage 伸根期
root surface area 根表面积
root system 根系
root system configuration 根系构型
root system length 根系长度
root tip 根尖
root tubercle 根瘤
root volume 根体积
root volume weight 根容重
root zone 根区

root/shoot ratio 根冠比
root/soil ratio 根土比
rooting ability 生根能力
rooting capacity 生根能力
rooting powder 生根粉
rooting stage 烟苗生根期
root-inhabiting fungi 栖根真菌
root-knot 根结
root-knot nematode 根结线虫
rootless 无根的
root-promoting hormone 生根激素
root-rot disease 根腐病
root-shoot ratio 根冠比
root-soil interface 根土界面
root-top ratio 地上部与地下部比率
rosette 簇叶症(缺锌或缺硼)
rosette stage 团棵期
rotary tillage 旋耕;旋涡式耕作
rotary tobacco cutter 旋转式切丝机
rotary type shaker 旋转式振荡机
rotating system 轮作制
rotation 轮作;旋转
rotation cropping 轮作栽培
rotation cropping system 轮作制
rotation cycle 轮作周期
rotation design 旋转设计
rotation flow method 轮灌方式
rotation flow system 轮灌制度
rotation plan 轮作规划
rotation planting 轮作栽培;轮作种植

rotation plough land 轮耕地(轮作田)
rotation sequence 轮作顺序
rotational application 轮换施药
rotational crossing 轮回杂交
rotational pasture 轮牧草地
Rothamsted Experimental Station 洛桑试验站
rotovation 翻土
rotted manure 腐熟厩肥
rotten root 烂根
rough soil 未耕耘土壤
route map of soil 土壤路线图
routine analysis 常规分析;例行分析
rove beetle 隐翅虫
row 行;排
row application 行施(肥料);条施
row crop 行间作物;中耕作物
row crop tractor 中耕拖拉机
row culture 条作;条播栽培
row drill 条播机
row fertilization 条施
row humus 粗腐殖质(地面枯叶层)
row intercropping 间作
row sowing 条播
row spacing 行距
row-crop cultivator 行间中耕机
RS(= remote sensing) 遥感
rubber plug 橡胶塞

rubber stopper 橡胶塞
rubber tube 橡皮管
rule for name of tobacco cultivar 烟草品种命名原则
rule for rational fertilization 肥料合理使用准则
rules for basic curing technique of flue-cured tobacco 烤烟基本烘烤技术规程
rules for original and produced tobacco seed in producing technique 烟草原种、良种生产技术规程
rules for tobacco introduction technique from abroad 烟草国外引种技术规程
run length 运转周期
runnel erosion 地下水侵蚀
running survey 勘测(查勘;考察)
runoff 径流
runoff amount 径流量
runoff coefficient 径流系数
runoff erosion 径流侵蚀
runoff erosivity 径流侵蚀力
runoff gathering 径流汇集
runoff generation 产流
runoff loss of nitrogen 氮的径流损失
runoff modulus 径流模数
runoff plot 径流小区;径流场
runoff ratio 径流率
runoff volume 产流量
rural domestic waste 农村生活污水
rural energy 农村能源
rust disease 锈病
Rustica 黄花烟
rustica tobacco 黄花烟草
rusty brown colour 锈褐色
rusty brown elongated streak 锈褐色细长条斑
ryegrass 黑麦草

S

S(= scrap) 碎叶组
sabulous 沙质的;多沙的
sabulous clay 沙黏土
sabulous loam 沙壤土
saccharase(= sucrase) 蔗糖酶
saccharide 糖类
saccharose 蔗糖
safe dose 安全剂量
sailing-ship tobacco 肯塔基深色明火烤烟
sajong soil 砂姜土
sal 硅铝层
saliferous 含盐的
salination 盐渍化

saline 盐水;含盐的
saline and alkaline land 盐碱地
saline environment 盐性环境
saline soil 盐性土
saline tobacco 盐碱地烟草
saline tolerance 耐盐性
saline-alkali soil 盐碱土
salineness 含盐度
salinity 盐浓度;含盐量
salinity adaptation 盐度适应
salinity and waterlogging stresses 盐分和水淹胁迫
salinity stress 盐分胁迫
salinity tolerance 耐盐性
salinization 盐渍化
salinization of soil 土壤盐渍化
salinized soil(=salty soil) 盐渍土
salinizing 盐化
salt 盐
salt concentration obstacle on protected land 保护地盐害
salt damage(=salt injury) 盐害
salt endurance 耐盐性
salt injury 盐害
salt leaching 洗盐
salt resistance 抗盐性
salt stresses 盐胁迫
salt tolerance 耐盐性
salt toxicity 盐害
salt water resistance 耐盐水性;耐海水性
salt weathering 盐成风化
salting out 盐析
salt-loving plant 喜盐植物
saltpeter 硝酸钾
salt-tolerance plant 耐盐植物
salt-tolerant 耐盐的
salty 高氯烟叶
sample 样本
sample average 样本均值
sample bag 样品袋;样本袋
sample cabinet 样品柜
sample cell 样品室
sample collection 样品收集
sample collector 采样器
sample conservation 样品保存
sample correlation coefficient 样本相关系数
sample covariance 样本协方差
sample mean 样本均值
sample median 样本中位数
sample number 样本数
sample pouch 标本袋
sample preparation 样品制备
sample pretreatment 样品前处理;样品预处理
sample size 样本量;样本大小
sample standard deviation 样本标准差
sample storage 样品贮藏
sample taking 抽样;取样
sample unit 抽样单位

sample variable coefficient 样本变异系数
sample variance 样本方差
samplers 采样器
sampling 抽样
sampling apparatus 取样设备
sampling bias 取样偏差
sampling bottle 采样瓶
sampling density 取样密度
sampling device 取样器;采样设备
sampling distribution 抽样分布
sampling error 取样误差;抽样误差
sampling fraction 抽样比例
sampling interval 采样间隔;抽样区间
sampling location 采样点
sampling method for commodity pesticides 商品农药采样方法
sampling period 采样周期
sampling plan 抽样方案
sampling point 采样点
sampling process 采样过程
sampling rate 采样率
sampling site 采样点
sampling system 采样系统
sampling technique 采样技术
sampling time 采样时间
sampling tool 取样工具
sampling unit 取样单元
Samsun 沙姆逊(香料烟品种)
sand 沙粒

sand clay 沙质黏土
sand content 沙土率
sand culture 沙培
sand culture experiment 沙培试验
sand grain 沙粒
sand leaf 底脚叶;沙土叶
sand lug 脚叶(分级)
sand on leaf testing 烟叶沙土检验
sand soil 沙土
sandstone 沙岩
sandy 沙质的;多沙的
sandy clay 沙黏土
sandy clay loam 沙质黏壤土
sandy clay loam soil 沙质黏壤土
sandy loam 沙壤土
sandy loam soil 沙壤土
sandy silt loam 沙质粉沙壤土
sandy silt soil 沙质粉沙土
sandy soil 沙土
sanmate 多菌灵
sapric soil material 高腐有机土壤物质
saprophitic nematode 腐生线虫
satellite photography 卫星摄影
saturated acid 饱和酸
saturated fatty acid 饱和脂肪酸
saturated humidity 饱和湿度
saturated hydraulic conductivity 饱和导水率
saturated soil 饱和土壤
saturated soil water flow 土壤饱和

水流
saturated water content 饱和含水量
saturation （烟叶）色度；饱和（分级）
saturation curve 饱和曲线
saturation deficit 饱和亏缺
saturation moisture capacity 饱和持水量
saturation moisture content 饱和持水量
saturation point 饱和点
saturation rate of soil 土壤水分饱和率
scab 叶部斑点病
scalding 挂灰（叶）
scalding leaf 挂灰烟叶
scale 尺度
scale axis 尺度数据
scale of pollution 污染标度
scanning electron microscope 扫描电子显微镜(SEM)
scanning transmission electron microscope 扫描透射电子显微镜
scarifier 松土机；刨土机；种子破皮机
scarify 松土
scatter diagram 散点图
scatter plots 散点图
scattergram 散点图
scent 香气；气味；香水；嗅；闻
schizomycetotrophy 细菌营养

Schmuck value 施木克值
scorched 烫片（叶）；焦片（叶）；糊片
scorched appearance 枯萎状
scorched leaf 烤焦叶；烫伤烟叶；高温损害叶；烤红
scorched odour 枯焦气
scorched smell 烤焦味
scorching 高温干燥；烤焦
scorching smell 焦糊味
score 评分；记分；标记；划痕
score card 记分表
scoring technique 打分技术
scouring erosion 冲蚀
scrap tobacco 碎叶片
scrapping 造碎
scratchy 辣口的
screen 筛子
screenhouse research 网室研究
Scrobipalpa heliopa Lower 烟草蛀茎蛾
sea foam 海泡石
sealed tube method 闭管法
searching table 选叶台
seasonal cycle (of disease) （病害的）季节循环
seasonal frozen soil 季节性冻土
seasonal variation 季节性变化
seasoning 调味；醇化；调味品；佐料；干燥；变干
second 秒

second leaf 上二棚叶(分级)
secondary alkalinity 次生碱度
secondary alkalization 次生碱化(作用)
secondary amine 仲胺
secondary buds 副芽;次生芽
secondary clayified horizon 次生黏化层
secondary element 中量元素
secondary element nutrient 中量元素养分
secondary gleyization 次生潜育作用
secondary infection 二次侵(感)染;再度感染
secondary metabolite 次生代谢产物
secondary mineral 次生矿物
secondary nutrient 次要营养
secondary nutrient fertilizer 中量元素肥料
secondary ploughing 复耕
secondary pollution effect 次生污染影响;二次污染效应
secondary root 次生根
secondary salinization 次生盐渍化
secondary soil 次生土壤
secondary suckers 腋芽
secondary tillage 复耕;再耕
secreta 分泌物(渗出物)
sediment 沉淀物;沉积
sediment concentration 含沙量
sedimentary clay 沉积黏土

sedimentation 沉积(作用);沉降
sedimentation analysis 沉降分析
sedimentation coefficient 沉降系数
sedimentation cylinder 沉降筒
sedimentation equilibrium 沉降平衡
sedimentation parameter 沉降参数
sedimentation soil 沉积土壤
sedimentation time 沉降时间
sedimentation velocity 沉降速度
sedimentation volume 沉降容积
seed 种子;籽;结籽;播种
seed agitator 种子搅动器;种子搅拌器
seed amelioration 种子改良
seed bag 种子袋
seed bank 种子库
seed bed 苗床;下种床;播种床
seed bed covers 苗床覆盖物
seed bed stage 苗床期
seed behavior 种子习性;种子特性
seed board 播种板
seed cake 种子饼;饼肥
seed cake fertilizer 饼肥
seed certification 种子鉴定;种子证书
seed coat 种皮
seed coat treatment 种皮处理
seed coating 种子包衣
seed coating machine 种子包衣机
seed covering 种子覆土;埋种;盖种
seed culture 种子培养

seed dressing 拌种
seed dusting 药剂拌种
seed fertilizer 种肥
seed germinability 种子发芽力
seed germination 种子发芽;种子萌发
seed grade 种子等级
seed grading 种子分级
seed grower 种子生产者;制种者
seed leaf 子叶
seed level 播种深度
seed lobe 子叶
seed longevity 种子寿命
seed manipulation 种子处理
seed manure 种肥
seed maturation 种子成熟
seed method 播种方法
seed packaging 种子包装
seed pelleting 种子包衣
seed quantity 播种量;种子量
seed respiration 种子呼吸
seed scarification 种子破皮(处理)
seed screening 筛选
seed selection 选种
seed setting percentage 结实率
seed soaking 浸种
seed standardization 种子标准化
seed sterilization 种子消毒
seed sunning 晒种
seed time 播种时期;播种期;播种时间

seed treatment 种子处理(拌种)
seed vitality 种子活力
seed with poor quality 种子品质差
seed yield 种子产量
seedage 播种法;播种技术
seedbed diseases 苗床病害
seedbed fertilization 苗床施肥
seed-borne pathogenic fungi 种子携带的病菌
seeder 播种机;播种者
seeding 播种
seeding date 播种期
seeding density 播种密度
seeding depth 播种深度;种子覆盖深度
seeding in hole 点播;穴播
seeding in line 条播;沟播
seeding in strip 带播
seeding machine 播种机
seeding quality 播种质量
seeding rate 播种速率;发芽率
seeding test 幼苗试验
seeding time 播种时间;播种期
seedling 幼苗
seedling disease 苗期病害;幼苗病害
seedling emergence 出苗
seedling establishment 出苗;成苗
seedling evaluation 幼苗评价
seedling for transplant 移栽苗
seedling growth rate 幼苗生长速率

（活力）
seedling hardening 锻苗
seedling raising and transplanting 育苗移栽
seedling resistance 幼苗抗病性;苗期抗性
seedling restitution stage 返苗期;还苗期
seedling root with earth 苗根带土
seedling stage 苗床期
seedling stand percent 成苗率
seedling test 幼苗试验
seedling thinning 疏苗;间苗
seedling vigo(u)r 幼苗活力
seedlings hardening 锻苗
seed-soak 浸种
seep irrigation 渗灌
seepage 渗漏
seepage prevention 防渗
segetal 农田杂草
segregation 分离
selection of strain 品系选育
selective absorption 选择性吸收
selective feeding 重点施肥
selective sampling 选择取样
selective uptake 选择性吸收
selenium 硒
selenium pollution 硒污染
self-cleaning 自净作用
self-controlled greenhouse 自控温室
self-purification 自净作用

self-recording apparatus 自动记录仪
self-regulation 自我调控
SEM(= scanning electron microscope) 扫描电子显微镜
semi-air-cured and semi-flue-cued 半晾半烤(烟叶)
semi-arid 半干旱
semi-arid climate 半干旱气候
semi-arid land 半干旱地
semi-arid land ecosystem 半干旱地区生态系统
semi-arid region 半干旱(地)区
semi-aromatic tobacco 半香料烟(草)
semicellulose 半纤维素
semi-leaf wrapped cigar 半叶卷雪茄
semimicro 半微量
semimicro-method 半微量法
seminal leaf 子叶
semi-oriental tobacco 半香料烟
semiquantitative 半定量的
semivariance 半方差
semivariogram 半方差图
semivariogram modeling 半方差函数建模
semivariogram value 半方差函数值
semivolatile 半挥发(性)的
semivolatile component 半挥发性成分;半挥发性物质
semi-volatile organic compounds 半

挥发性有机化合物
senescence 衰老
senescence period 衰老期
sensation 感觉
sensitive plant 敏感植物
sensory analysis 感官分析
sensory evaluation 感官评定
sensory quality 感官品质
separated layer fertilization 分层施肥
separation 分离
separation coefficient 分离系数
separation of lamina and stem 叶梗分离
separator 分离器
separatory funnel 分液漏斗
sequence chart 序列图
sequential chemical extraction 连续化学提取
sequential cropping 连作
sequential sampling 顺序抽样
sequential test 顺序检验
serial root-washing technique 根连续洗涤法
serine 丝氨酸
serious scalded 严重挂灰
sesame cake 芝麻饼
sessile leaf 无柄叶
setted plant 定苗
setting depth 种植深度
setting of tobacco plant 烟株定苗

setting space of seedlings 定苗距离;定苗间距
severe deficiency 严重缺乏
severe salinized 重盐渍化的
severity of infection 病害严重度
sewage 污水
sewage irrigation 污水灌溉
sewing 绑烟
sexual attracting method 性引诱法
shade plant 阴地植物;耐阴植物
shade tobacco 荫植烟草
shade tolerant plants 耐阴植物
shade-bearing 耐阴的
shadegrown 遮阴栽培
shade-grown tobacco 荫栽烟草
shade-tolerant 耐阴的
shading 遮光;(雪茄烟支)分色;遮阴的
shading film 遮光膜
shading-growth tobacco 荫生烟叶(高档雪茄烟叶)
Shajiang black soil 砂姜黑土
shaker 振荡器
shale 页岩
shallow application 浅施
shallow cultivation 浅耕;浅松土
shallow groundwater 浅层地下水
shallow placement 浅施
shallow ploughing 浅耕
shallow roots 浅根
shallow soil 薄土;浅层土

shallow tillage 浅耕

shallow transplanting 浅植

shallow winter breaking 浅耕灭茬；冬浅耕

sharp 鲜明

sharp flavor 辛辣味；辛辣调料

sharp odor 尖锐气息；辛辣气息

sharpening image 锐化图像

sharpness 辛辣味；尖锐度；清晰度；尖刺

shatter index （烟叶）破碎指数[分级]

shatter resistance 抗破碎性；耐碎性

shatter resistance property 抗破碎性（分级）

shear 剪切

shear strength 剪切强度

shear stress 剪切应力

shed leaf 落叶

sheep manure 羊粪

shifting balance 动态平衡

shifting cultivation 撂荒

shoot initiation 腋芽生长处

short day crop 短日照作物

short day plants 短日照植物

short rotation 短期轮作

shortage （香气）短少

short-day 短日照

short-distance transport 短距离运输

shortening of internodes 节间缩短

shortening of vegetative growth stage 营养生长期缩短

short-lived pesticide 短效农药

shovel plough 中耕培土犁

shred 烟丝

shredded tobacco 烟丝

shrink 萎缩；收缩；皱缩

shrinkage （烟叶）皱缩（分级）；因水分和造碎引起的烟叶重量损失

shrivel 枯萎

sial 硅铝层

siccation 干燥

side dressing 侧施肥料

side-dressed 侧施

sidestream smoke 侧流烟气

sierozem 灰漠钙土

sieve mesh 筛孔

significance level 显著水平

significance of difference 差异显著性

significance of regression coefficient 回归系数的显著性

significance testing 显著性检验

significant difference 显著差异

significant digits 有效数字

significant figure 有效数字

significant test 显著性检验

silica 氧化硅

silica crucible 石英坩埚

silica gel 硅胶

silicate bacteria 硅酸盐细菌（钾细菌）

silicate bacteria fertilizer 硅酸盐细菌肥料
silicate-dissolving bacteria 硅酸盐分解细菌
silicon 硅
silicon fertilizer 硅肥
silt 粉(沙)粒
silt clay 粉沙黏土
silt clay loam 粉沙黏壤土
silt content of runoff 径流含沙量
silt loam 粉沙壤土
silt particle 粉沙颗粒
silt soil 粉沙土;粉粒土
siltation 淤积(作用)
silting 淤积(作用)
silty clay 粉沙黏土
silty loam 粉沙壤土
silty sand 粉(沙)质沙土
silver 银
silver sand 细沙
similarity 相似性
simple correlation 简单相关
simple hydrometer method 简易比重计法
simple hypothesis 简单假设
simple interest disease 单循环病害;单一性病害
simple random sampling 简单随机抽样
simulated rainfall 模拟降雨
simulation in ecology 生态模拟

simultaneity of ripening 同时成熟
single cropping 一年一熟制;单作
single culture 单作
single element fertilizer 单元素肥料
single element manure 单质肥料
single factor experiment 单因子试验
single fertilizer 单一肥料
single infection 专一浸染
single sampling inspection 一次抽样验查
single superphosphate 普通过磷酸钙(普钙;过磷酸一钙)
single-dose application 一次性施肥
single-season experiment 单季试验
sink 库(汇)
sintered glass crucible 微孔玻璃坩埚
sintering method 烧结法
Sistan 威百亩
site condition 立地条件
site-specific farming 定位农业(定点农业)
site-specific fertilization 定位施肥
site-specific fertilizer application 定位施肥
site-specific fertilizer management 定位施肥管理
site-specific field management 田间定位精确管理
site-specific management 精确定位

·201·

管理

site-specific nitrogen management　定位氮素管理

site-specific nutrient management　定位养分管理

size distribution　尺寸分布;(颗粒)大小分布;粒径分布

size fractionation　大小分级

size of mesh　筛孔尺寸

size of soil grain　土壤粒径

size test　粒度分析

skeleton soil　粗骨土

skewness　偏度

skipped-row culture　隔行栽培

slack farming season　农闲季节

slaked lime　熟石灰;消石灰

sleep　烟叶陈储期

slick　光滑(叶组);光滑叶

slick leaf　光滑烟叶

slick　光滑(叶):代号S

slight alkalinity　微碱性

slight immature leaf　轻度未熟叶

slight mold　轻度霉变

slight scalded　轻度挂灰

slightly acid　微酸性

slightly contaminated　轻度污染的

slightly decomposed organic horizon　弱分解有机层

slightly hard　略硬实

slightly salinized　轻盐渍化的

slip　滑坡

slope　坡地

slope aspect　坡向

slope exposure　坡向

slope gradient　坡度

slope inclination　坡度

slope land　坡地

slope land-runoff　坡面径流

slope management for erosion control　坡面治理工程

slope runoff　坡面径流

sloping cropland　坡耕地

sloping field　坡地

slow available nitrogen fertilizer　缓效氮肥

slow release　缓释

slow release fertilizer　缓释肥料;缓效肥料

slow release method　缓释方法

slow release pesticide　缓释型农药

slow release technique　缓释技术

slow-acting fertilizer　缓释肥(长效肥)

slow-acting manure　缓效农肥

slow-acting nitrogen fertilizer　缓效氮肥

slowly available nutrient　缓效性养分

slowly available potassium　缓效钾

slow-release fertilizer　缓释肥(长效肥)

slow-release nitrogen fertilizer　缓释

氮肥
slow-release organic nitrogen fertilizer 缓释有机氮肥
sludge and refuse 废弃物肥料
slug 蛞蝓;蛞蝓型幼虫
slurry 泥杂肥
slush paddy field 烂泥田
small carrier for inter-ridge work 垄间作业机
small sample 小样本
small scale survey 小比例尺调查（小比例尺测图）
small stiff leaves 小而僵直的叶片
small valley treatment 小流域治理
smaller leaf 叶片较小
smear 涂片
smoke analysis 烟气分析
smoke characteristics 烟气特征
smoke component 烟气成分
smoke tar 烟气焦油
smoke trap 烟气捕集器
smokeless products 无烟气（烟草）制品
smoking characteristic 抽吸（烟）特性
smoking compartments 吸烟室
smoking habit 吸烟习惯
smoking leaf 完熟(叶);代号 H
smoking machine 吸烟机
smoking panel 评吸小组
smoking samples 样品烟

smoking without inhaling 抽烟时不把烟气咽下
smooth （烟味)细腻;(烟叶)平滑;（烟气)和顺;柔和
smooth note 圆和香韵
smooth odor 圆和香韵;柔和香气
smoothing （烟气)和顺;(烟气)柔和;光滑化;平整
smoothing harrow 细平耙
smoothly tight 致密
smoothness 烟叶表面光滑程度;柔和性;平滑度
smoothy （烟气）细腻
smouldering 阴燃
smouldering zone 阴燃区
snail 蜗牛
snapping beetle 叩头虫
snout beetles 象甲;象鼻虫
snuff 鼻烟
social benefit 社会效益
soda 苏打
sodium 钠
sodium bicarbonate 碳酸氢钠;小苏打
sodium chloride 氯化钠;食盐
sodium dihydrogen phosphate 磷酸二氢钠
sodium hexametaphosphate 偏磷酸钠
sodium hydrogen carbonate 碳酸氢钠

203

sodium hydroxide 氢氧化钠

sodium molybdate 钼酸钠

sodium nitrate 硝酸钠

sodium oxalate 草酸钠

sodium pyrophosphate 焦磷酸钠

sodium salt 钠盐

soft 松软

soft leaf 水分大的软叶

soft rot 软腐病

soft soil 松软土壤

softness 柔和

soggy land 湿地

soil 土壤

soil ^{14}C age 土壤^{14}C年龄

soil absolute water content 土壤绝对含水量

soil acid-base equilibrium 土壤酸碱平衡

soil acidification 土壤酸化

soil acidity 土壤酸度

soil active acidity 土壤活性酸度

soil adhesion 土壤黏附力(土壤黏着力)

soil adsorption 土壤吸附

soil aeration 土壤通气;土壤通气性

soil age 土壤年龄

soil aggregate 土壤团聚体

soil agitator 碎土器

soil air 土壤空气

soil air capacity 土壤空气容量

soil air composition 土壤空气组成

soil air diffusion 土壤空气扩散

soil air exchange 土壤空气交换

soil air permeability 土壤通气性

soil air regime 土壤空气状况

soil alkalinity 土壤碱度

soil alkalinization 土壤碱化作用

soil altitudinal zonality 土壤垂直地带性

soil amelioration 土壤改良

soil amelioration material 土壤改良物质

soil amendment 土壤改良剂

soil analysis 土壤分析

soil and terrain digital data base 土壤与地形数字化数据库

soil and water conservation 水土保持

soil and water conservation tillage measures 水土保持耕作措施

soil and water conservation vegetation measures 水土保持植物措施

soil and water loss 水土流失

soil animal 土壤动物

soil application 土壤施肥

soil assessment 土壤评价

soil auger 土钻;土壤采样钻

soil available nutrient 土壤有效养分

soil available water 土壤有效水

soil background value 土壤本底值(土壤背景值)

soil bacteria 土壤细菌

soil bearing capacity 土壤承载力
soil biochemistry 土壤生物化学
soil biodiversity 土壤生物多样性
soil biodynamics 土壤生物动力学
soil biological activity 土壤生物活性
soil biological pollution 土壤生物污染
soil biology 土壤生物学
soil biomass 土壤生物量
soil biota community 土壤生物群落
soil black carbon 土壤黑炭
soil borne disease 土传病害
soil buffer action 土壤缓冲作用
soil buffering 土壤缓冲性
soil buffering ability 土壤缓冲能力
soil bulk density 土壤容重
soil capability 土壤生产力
soil capillarity 土壤毛细管作用
soil carbohydrate 土壤碳水化合物
soil carbon 土壤碳
soil carbon cycle 土壤碳循环
soil carbon flux 土壤碳通量
soil charcoal 土壤黑炭
soil chemical fixation 土壤化学固定
soil chemistry 土壤化学
soil chemistry diagnosis method 土壤化学诊断法
soil classification 土壤分类
soil classification system 土壤分类制

soil clay 土壤黏粒
soil colloid 土壤胶体
soil color 土壤颜色
soil column 土柱
soil compaction 土壤压实
soil compaction index 土壤压实度指标
soil conditioner 土壤改良剂;土壤调理剂
soil conservation 土壤保持
soil constraint factor 土壤障碍因子
soil contamination 土壤污染
soil cultivation 土壤耕作
soil culture 土壤培养
soil culture experiment 土培试验
soil decontamination 土壤净化
soil degradation 土壤退化
soil density 土壤密度
soil denudation 土壤侵蚀(作用)
soil depth 土壤深度
soil description 土壤描述
soil deterioration 土壤退化
soil development 土壤发育
soil disease 土壤病害
soil disinfection 土壤消毒
soil dispersing agent 土壤分散剂
soil dispersion 土壤分散
soil distribution 土壤分布
soil drainage 土壤排水
soil drifting 土壤风蚀
soil drill 土钻

soil drought 土壤干旱
soil dynamics 土壤动态变化
soil ecological amelioration 土壤生态改良
soil ecological environment 土壤生态环境
soil ecological environment effect 土壤生态环境效应
soil ecological system 土壤生态系统
soil ecology 土壤生态学
soil ecosystem 土壤生态系统
soil environment 土壤环境
soil environment capacity 土壤环境容量
soil environment monitoring 土壤环境监测
soil environment protection 土壤环境保护
soil environment quality 土壤环境质量
soil environmental factor 土壤环境因子
soil environmental quality assessment 土壤环境质量评价
soil environmental quality index 土壤环境质量指数
soil environmental quality standard 土壤环境质量标准
soil enzyme 土壤酶
soil enzyme activity 土壤酶活性

soil erosion 土壤侵蚀(作用)
soil erosion agency 土壤侵蚀营力
soil erosion classification 土壤侵蚀分类
soil erosion degradation 土壤侵蚀退化
soil erosion degree 土壤侵蚀程度
soil erosion factor 土壤侵蚀因素
soil erosion intensity 土壤侵蚀强度
soil erosion modulus 土壤侵蚀模数
soil erosion pollution 土壤侵蚀污染
soil erosion process 土壤侵蚀过程
soil erosion rate 土壤侵蚀速率
soil erosion regular 土壤侵蚀规律
soil erosion type 土壤侵蚀类型
soil evaluation 土壤评价
soil exhaustion 土壤耗竭
soil expansion 土壤膨胀
soil extract 土壤浸出液
soil failure 土壤破坏
soil failure diminution 土壤肥力减退
soil family 土族
soil fauna 土壤动物区系
soil fertility 土壤肥力
soil fertility evaluation 土壤肥力评价
soil fertility factor 土壤肥力因子
soil fertility grade 土壤肥力等级
soil fertility grading 土壤肥力分级
soil fertility index 土壤肥力指标

soil fertility indexing system 土壤肥力指标体系

soil fertility maintenance 土壤肥力保持

soil fertility management 土壤肥力管理

soil fertility map 土壤肥力图

soil fertility monitoring 土壤肥力监测

soil fertility quality 土壤肥力质量

soil fixing 土壤固定

soil flushing 土壤淋洗

soil formation 土壤形成

soil former 土壤形成因素(简称成土因素)

soil forming process 成土作用;成土过程

soil free water 土壤自由水

soil fumigant 土壤熏蒸剂

soil fumigation 土壤熏蒸

soil fungi 土壤真菌

soil fungicide 土壤杀菌剂

soil genesis 土壤发生

soil genetic classification 土壤发生分布

soil genetic horizon 土壤发生层

soil genus 土属

soil geochemistry 土壤地球化学

soil granulation 土壤团粒作用

soil gross nitrogen mineralization 土壤氮素总矿化量

soil group 土类

soil hardness 土壤坚实度(土壤硬度)

soil hardness meter 土壤硬度计

soil health 土壤健康

soil health quality 土壤健康质量

soil heat flux 土壤热通量

soil heat regime 土壤热状况

soil horizon 土壤层次

soil horizontal distribution 土壤水平分布

soil horizontal zonality 土壤水平地带性

soil humification 土壤腐质化(作用)

soil humus 土壤腐殖质

soil hydraulic conductivity 土壤导水率

soil improvement 土壤改良

soil improvement agent 土壤改良剂

soil improving crop 养地作物

soil inavailable water 土壤无效水

soil individual 土壤个体

soil infection 土壤感染

soil infestation 土壤侵染

soil information 土壤信息

soil information system 土壤信息系统

soil inhabitant 土壤习居菌

soil inoculation 土壤接种

soil inorganic carbon 土壤无机碳

·207·

soil inorganic nitrogen 土壤无机氮
soil insect 地下害虫
soil insecticide 土壤杀虫剂
soil invader 土壤寄居菌
soil invertase 土壤蔗糖酶
soil invertebrate 土壤无脊椎动物
soil layer 土壤层次(土层)
soil lime potential 土壤石灰位
soil local type 土种
soil loss 土壤流失
soil loss amount 土壤流失量
soil loss tolerance 允许土壤流失量
soil lump 土块
soil macrofauna 土壤大型动物区系
soil macroorganism 土壤大生物
soil management 土壤管理
soil mantle 土壤覆盖;土被
soil map 土壤图
soil mass application 土壤深层施药
soil material 土壤物质
soil matrix 土壤基质
soil mechanics 土壤力学
soil meiofauna 土壤小型动物区系
soil microbe 土壤微生物
soil microbial biomass 土壤微生物生物量
soil microbial diversity 土壤微生物多样性
soil microbial population 土壤微生物群体
soil microbiology 土壤微生物学

soil micro-ecology 土壤微生态
soil microfauna 土壤小型动物区系
soil microflora 土壤微生物区系
soil microorganism 土壤微生物
soil microscopy 土壤显微镜学
soil mineral 土壤矿物
soil mineral chemistry 土壤矿物化学
soil moisture 土壤湿度;土壤水分;土壤水
soil moisture content 土壤含水量
soil moisture deficiency 土壤水分亏缺
soil moisture regime 土壤水分状况
soil moisture stress 土壤水分胁迫
soil moisture tensiometer 土壤水分张力计
soil moisture tension 土壤水分张力
soil monolith 土壤整段标本
soil mulch 土壤覆盖
soil nematode 土壤线虫
soil net nitrogen mineralization 土壤氮素净矿化量
soil nitrogen apparent balance 土壤氮素表观平衡
soil nitrogen supplying capacity 土壤氮素供应能力
soil nomenclature 土壤命名
soil nutrient 土壤养分
soil nutrient availability 土壤养分有效性

soil nutrient budget 土壤养分收支
soil nutrient chemistry 土壤养分化学
soil nutrient status 土壤养分状况
soil organic carbon 土壤有机碳（SOC）
soil organic carbon pool 土壤有机碳库
soil organic carbon storage 土壤有机碳储量
soil organic matter 土壤有机质（SOM）
soil organic matter balance 土壤有机质平衡
soil organic nitrogen 土壤有机氮
soil organism 土壤生物
soil organo-mineral complex 土壤有机-无机复合体
soil oxidation-reduction reaction 土壤氧化还原反应
soil parent material 土壤母质
soil particle 土壤颗粒
soil particle cementation 土粒胶结（作用）
soil particle charge 土粒电荷
soil particle coagulation 土粒凝聚（作用）
soil particle composition 土壤颗粒组成
soil particle density 土粒密度（曾称土壤比重）

soil particle size fractionation 土壤颗粒分级
soil particle-size analysis 土壤颗粒大小分析
soil penetration resistance 土壤穿入阻力
soil permeability 土壤通透性
soil persistent pesticide 土壤残留性农药
soil pH 土壤 pH
soil pH map 土壤 pH 图
soil pH(value) 土壤 pH 值
soil phase 土相
soil phosphate adsorption 土壤磷吸附
soil physics 土壤物理学
soil plasticity 土壤塑性
soil pollutant 土壤污染物
soil pollution 土壤污染
soil pollution by heavy metal 土壤重金属污染
soil pollution by organic matter 土壤有机污染
soil pollution chemistry 土壤污染化学
soil pollution control 土壤污染控制
soil pollution critical value 土壤污染临界值
soil pollution index 土壤污染指数
soil pollution source 土壤污染源
soil pool 土壤库

soil pore space 土壤孔隙
soil porosity 土壤孔隙度(土壤孔隙率)
soil potassium supplying capacity 土壤供钾能力
soil potential acidity 土壤潜性酸度
soil potential evaporation 土壤潜在蒸发
soil preparation 整地
soil productivity 土壤生产力
soil productivity grading 土壤生产力分级
soil profile 土壤剖面
soil profile sample 土壤剖面样品
soil property 土壤性质
soil protection 土壤保护
soil purification 土壤净化
soil quality 土壤质量
soil quality assessment 土壤质量评价
soil quality criteria 土壤质量标准
soil quality enhancement 提高土壤质量
soil quality evaluation 土壤质量评价
soil quality index 土壤质量指数
soil quality indicators 土壤质量指标
soil quality indices 土壤质量指标
soil quality standard 土壤质量标准
soil reaction 土壤反应

soil reclamation 土壤改良;复垦
soil redox reaction 土壤氧化还原反应
soil remediation 土壤修复
soil remote sensing 土壤遥感
soil replacement 客土法
soil residue 土壤残留
soil resource 土壤资源
soil resource information system 土壤资源信息系统
soil respiration 土壤呼吸
soil respirator 土壤呼吸仪
soil run-off 土壤流失
soil saccharase 土壤蔗糖酶
soil salinity 土壤盐渍度
soil salinity sensor 土壤盐分传感器
soil salinization 土壤盐渍化
soil salinization-alkalization 土壤盐碱化
soil salt content 土壤含盐量
soil sample 土样
soil sample container 土壤样品储存容器
soil sample drying 土壤样品干燥
soil sample preparation 土壤样品制备
soil sample quartation 四分法(土壤取样)
soil sample sieve 土壤样品筛
soil sample storage 土壤样品储存
soil sampler 土壤取样器

soil sampling 土壤样品采集
soil sanitation 土壤净化
soil science 土壤学
soil self-purification 土壤自净作用
soil self-purification function 土壤自净化功能
soil series 土系
soil shear 土壤剪切
soil solidification 土壤固化
soil solution 土壤溶液
soil solution boron 土壤溶液硼
soil solution copper 土壤溶液铜
soil solution potassium 土壤溶液钾
soil species 土种
soil specific gravity 土壤比重(曾用名)
soil specific surface area 土壤比表面积
soil specific volume 土壤比容
soil specific weight 土壤比重
soil split preparation 土壤切片制备
soil sterilant 土壤灭菌剂
soil sterilization 土壤杀菌;土壤消毒
soil strength 土壤强度
soil structure 土壤结构
soil structure classification 土壤结构分类
soil structure type 土壤结构类型
soil submicromorphology 土壤超微形态学

soil suitability 土壤适宜性
soil surface 土表
soil survey 土壤调查
soil survey manual 土壤调查手册
soil survey program 土壤调查规划
soil swelling 土壤膨胀
soil taxon 土壤类别
soil temperature 土壤温度
soil temperature regime 土壤温度状况
soil testing 土壤测试
soil testing and fertilizer recommendation 测土推荐施肥
soil texture 土壤质地
soil texture classification 土壤质地分类
soil texture map 土壤质地图
soil texture profile 土壤质地剖面
soil thermal regime 土壤热状况
soil thermometer 土壤温度计
soil tillage 土壤耕作
soil tilth 土壤耕性
soil total nitrogen 土壤全氮
soil type 土壤类型
soil urease 土壤脲酶
soil utilization 土壤利用
soil utilization map 土壤利用图
soil vapor diffusion 土壤水汽扩散
soil vapor extraction 土壤蒸汽提取
soil vertical distribution 土壤垂直分布

soil vertical zonality 土壤垂直地带性
soil viscosity 土壤黏度
soil volume 土壤容积
soil warming 土壤增温
soil water 土壤水分
soil water and salt movement 土壤水盐运动
soil water characteristic curve 土壤水分特征曲线
soil water classification 土壤水分类
soil water constant 土壤水（分）常数
soil water content 土壤含水量
soil water evaporation 土壤水（分）蒸发
soil water flow 土壤水流
soil water form 土壤水形态
soil water hysteresis 土壤水（分）滞后现象
soil water in liquid phase 土壤液态水
soil water in solid phase 土壤固态水
soil water in vapor phase 土壤气态水
soil water potential 土水势
soil water regime 土壤水分状况
soil water resource 土壤水资源
soil water storage 土壤贮水量
soil weathering 土壤风化
soil wetness 土壤湿度
soil wind erosion 土壤风蚀
soil zonality 土壤地带性

soil zoology 土壤动物学
soil-conserving crop 保土作物
soild state fermentation 固态发酵
soil-forming condition 土壤形成条件
soil-forming environment 成土环境
soil-forming factor 土壤形成因素（简称成土因素）
soil-forming process 土壤形成过程
soil-forming rock 成土母岩
soil-free culture 无土栽培
soilless agriculture 无土农业
soilless culture 无土栽培
soilless culture mixture 无土栽培基质
soil-plant-atmosphere continuum 土壤-植物-大气连续体
soil-protective cover 土壤保持覆盖
soil-ripening 土壤熟化
soiltex 土壤酸度试剂
Soil-Water-Atmosphere-Plant Model 土壤-水分-大气-植物模型（SWAP）
sol 溶胶
solanesol 茄呢醇
solanum tuberosum 马铃薯
solar greenhouse 日光温室
solar radiation 日照量；太阳辐射
solar-heating barn 太阳能加热烤房
solarium 网室
sole and multiple cropping 单作多

212

熟型
sole cropping 单作
sole cropping annually 单作一熟型
solid fertilizer 固体肥料
solid manure 厩肥
solid phase 固相
solid phase volume 固相容积
solid portion 固体部分
solid residue 固体残渣
solid waste 固体废物
solid waste treatment 固体废弃物处理
solodi 脱碱土
solodization 脱碱作用
solodized soil 脱碱化土壤
solonchak 盐土
solonetz 碱土
solonization 碱化(作用)
solotization 脱碱作用
solubility of fertilizer 肥料溶解度
soluble ash 可溶性灰分
soluble carbohydrate 可溶性碳水化合物
soluble humus 可溶性腐殖质
soluble nitrogen 可溶性氮
soluble nutrient 可溶性养分
soluble organic carbon 可溶性有机碳
soluble organic nitrogen 可溶性有机氮
soluble salt 可溶性盐(溶性盐)

soluble sugar 可溶性糖
solum 土体层
solute diffusion 溶质扩散
solute migration 溶质运移
solute potential 溶质势
solute transfer 溶质运移
solute transport 溶质运移
solution culture experiment 溶液培养试验;水培试验
solution culture method 溶液培养法
solution fertilizer 溶液肥料
solution phosphorus 溶液中磷
sombric horizon (暗色)腐殖质淀积层
sooty mold (= fumago vagans Pers.) 烟草煤污病
sorption 吸附作用;吸持作用
sorption and desorption of ammonium 铵的吸附和解吸
sort 选叶
sound 无沙土(烟叶);健全(烟叶);健康叶
sound seedling 壮苗
sound soil health quality 安全的土壤健康质量
sour humus 酸性腐殖质
source 源
source of pollution 污染源
source-sink hypothesis 源库假说
south-facing slope land 阳坡地
sowing 播种

· 213 ·

sowing area 播种面积
sowing depth 播种深度
sowing in furrows 沟播
sowing in hole 穴播
sowing in line 行播;条播
sowing machine 播种机
sowing manure 种肥
sowing period 播种期
sowing season 播种季节
sowing-drill 播种机
sowing time 播种期
soybean cake 大豆饼
space fumigant 空间熏蒸剂
space inter-plants 株距
space inter-rows 行距
space remote sensing 航天遥感
spaced cropping 宽行栽培
spaced sowing 点播
spacing 间隔;间距;株行距
spacing in rows 行距
spacing of plants 株距
spade 铁锹
sparsely soluble nutrient 难溶性养分
spatial 空间的
spatial analysis 空间分析
spatial autocorrelation index 空间自相关指数
spatial characteristics 空间特征
spatial comparison 空间比较
spatial continuity 空间连续性

spatial scale 空间尺度
spatial variability 空间变异性
spatial variability of soil properties 土壤性质空间变异性
spearing 烟株挂起调制法
special fertilizer 专用肥料
special grading 特殊分级
special odor 特殊气味
species 种
species diversity 物种多样性
species equilibrium 物种平衡
species richness 物种丰富度
specific adsorption 专性吸附
specific fertilizer 专用肥料
specific gravity 比重
specific heat 比热
specific pore volume 比孔容
specific retention of soil 土壤持水率
specific surface 比表面
specific surface area 比表面积
specific weight 比重
specifically adsorbed potassium 专性吸附钾
specification 技术规范;说明书
specification of insecticide 杀虫药剂规格(说明书)
specimen 样本;样品
specimen chamber 样品室
specimen preparation 样品制备
spectral absorption 光谱吸收

spectral analysis 波谱分析
spectral analysis method 光谱分析法
spectral characteristics of soil 土壤波谱特性
spectral curve 光谱曲线
spectral data 光谱数据
spectral library 光谱数据库
spectral property 光谱特性
spectral radiance 光谱辐射率;光谱辐射强度
spectral range 光谱范围
spectral reflectance 光谱反射
spectral response 光谱响应
spectrogram 光谱图
spectrograph 摄谱仪;光谱仪
spectrophotometry 分光光度法
spectrum of carbon sources 碳源谱
spherical model 球状模型
spice 香料
spicy 辛香;辛辣的
spicy taste 辛辣味
spindling 徒长;徒长的
spindly growth 徒长
spiral phyllotaxy 螺旋形上升叶序
spiral root 螺旋根;气生根
spirilla 螺旋菌
spirillum 螺旋菌
split application 分次施用
split plot 裂区
split plot experiment 裂区试验
split root culture 分根培养
split root culture experiment 分根培养试验
split root technique 分根技术
split-block design 裂区组设计
split-plot design 裂区设计
split-plot experiment 裂区试验
split-root equipment 分根装置
split-root experiment 分根试验
split-root method 分根法
splitting 分裂;劈(烟茎);劈(叶);丝束分裂
splitting-root culture 分根培养
sponge function of forests 森林的水源涵养功能
sponged leaf 潮红烟叶
sponging 潮红(专指烤烟调制变黄期过长所产生的红褐色斑)
spongy 柔软
spongy tissue 海绵组织
spore 分生孢子
spore-bearing mycelium 孢子丝
spot application 穴施
spot fumigant 局部熏蒸剂
spot irrigation 点灌
spot sowing 点播
spotted necrosis 斑点状坏死
spotted wilt 斑萎病
spray 喷雾
spray agent 喷雾剂
spray application 喷施

spray injury 喷药伤害;喷雾药害
spray irrigation 喷灌
spraying 喷施;喷雾
spread plate method 涂布培养法
spring crop 春播作物
spring cultivation 春耕;春季栽培
spring frost 晚霜
spring ploughing 春耕
spring tobacco 春烟
spring vetch (*Vicia sativa* L.) 箭舌豌豆
springy land 湿润土地
sprinkler 喷洒器;喷灌器;人工降雨器
sprinkler application 喷灌
sprinkler irrigation 喷灌
sprinkler irrigation system 喷灌系统
sprinkling equipments 喷淋设备
sprinkling irrigation 喷灌;洒灌
sprinkling irrigation system 喷灌系统
sprinkling norm 浇水量
sprout 幼苗;发芽
sprout inhibition 抑制发芽
sprout seeding 催芽播种
sprouting 发芽
square grid 方格网
square map grid 地图格网
square matrix 方阵
squared Euclidean distance 欧式距离平方

squaring period 现蕾期
stability 稳定性
stable aggregate 稳定性团聚体
stable and high-yield field 高产稳产田
stable culture 稳产栽培
stable dung 厩肥
stable humus 稳定腐殖质
stable infiltration rate 稳定入渗率
stable manure 厩肥
stable organic nitrogen 稳定性有机氮
stably combined humus 稳结态腐殖质
stack 烟囱;堆;垛
stack altitude (= height of stacking) 堆垛高度
stacking 码垛
stacks resetting 翻垛
stage of diminishing returns 报酬递减阶段
stage of maturity 成熟阶段
staged harvest 分期采收
staged harvesting 分期采收
stagger cultivation 错期耕作
stagheaded (= dry-topped) 梢枯病;顶枯的(干顶病)
stagnic moisture regime 滞水土壤水分状况
stained 玷污的;污染的(烟叶)
stained group(s) 沾污烟叶组(分

级）

stalk 烟株;烟茎;拐头
stalk air-curing 整株晾制
stalk borer 普通蛀茎夜蛾
stalk circumference 茎粗;茎圆周
stalk curing 整株调制法
stalk cut-curing 带茎调制
stalk cutter 砍茎机
stalk cutting 砍茎;整株收割
stalk lodging 茎秆倒伏
stalk position 烟叶部位;叶位
stalk rot 茎腐
stalk-cut tobacco harvester 烟草砍茎收割机
stalk-cutting method 砍株法;砍茎法
stamen 雄蕊
stand density 种植密度
stand establishment percentage 成苗率
standard 标准
standard atmosphere 标准大气
standard atmosphere pressure 标准大气压
standard curve 标准曲线
standard deviation 标准偏差
standard deviation of a probability distribution 概率分布的标准差
standard deviation of a random variable 随机变量的标准差
standard error 标准误差

standard error of kurtosis 峰度系数标准误差
standard error of mean 均值标准误差
standard error of skewness 偏度系数的标准误差
standard for safe use of pesticides 农药安全使用标准
standard for safety application of pesticide 农药安全使用标准
standard hydrometer 标准比重计
standard method 标准方法
standard normal distribution 标准正态分布
standard pipette 标准吸管
standard pressure 标准压力
standard sample 标准样品
standard sample of flue-cured tobacco 烤烟实物标样
standard sample of tobacco leaf 烟叶标样
standard sieve 标准筛
standard solution 标准溶液
standard temperature 标准温度
standardization 标定;标准化
standardization management 标准化管理
standardized air curing barn 标准晾房
standardized variate 标准化变量
standards for irrigation water quality 农田灌溉水质标准

standards for safety application of pesticides 农药安全使用标准
standing power 抗倒伏能力
stand-up burley 直立型白肋烟
starch 淀粉;(淀粉)浆;(上)浆
starch accumulation 淀粉积累
starch grain 淀粉粒
starch synthesis 淀粉合成
starch-degrading enzymes 淀粉降解酶
starting temperature 起始温度
static burn 静态燃烧
static burning rate 静燃速率
static model 静态模型
statistic 统计的
statistic analysis system 统计分析系统
statistic model 统计模型
statistical analysis 统计分析
statistical coverage interval 统计覆盖区间
statistical distribution 统计分布
statistical hypothesis 统计假设
statistical test 统计检验
status of soil microelements 土壤微量元素状况
status of soil nitrogen 土壤氮素状况
status of soil phosphorus 土壤磷素状况
status of soil potassium 土壤钾素状况
steady state method 稳态流法(测土壤导水率)
steam distillation 蒸汽蒸馏
steam sterilization 蒸汽消毒;蒸汽灭菌
steam sterilizer 蒸汽灭菌器
steam treatment 蒸汽处理
steamed leaf 蒸片
stearic acid 硬脂酸;十八酸
steep slope 陡坡
steeping 浸种;浸湿
stem 茎;叶梗;烟梗
stem content in lamina 叶中带梗率
stem cutting 切梗丝
stem drying 烟梗干燥;干梗期;干筋期;干筋
stem drying stage 干筋期
stem extraction 去梗
stem in lamina 叶中含梗
stem percentage 含梗率
stem ratio 含梗率
stemmed 去梗的;抽梗的
stemmed leaf 去梗后叶片
stemmed tobacco 去梗烟叶
stem-rot 干腐;茎腐
step terrace 阶式梯田
stepwise 逐步的(分段的)
stepwise discriminant analysis 逐步判别分析
stepwise regression 逐步回归

stepwise regression procedure 逐步回归方法
stereo ecological agriculture 立体生态农业
sterile 贫瘠的
sterile culture 无菌培养(单一种纯净培养)
sterile culture experiment 无菌培养试验
sterile operation 无菌操作
sterilization 灭菌作用;消毒作用;杀菌;消毒;不育化
steroids 类固醇
sterol 甾醇;固醇
stick 烟竿;烟杆
stick barn 挂竿烤房
stick redrying 挂竿复烤
stiff soil 坚实土壤;粘结性土壤
stirrer 搅拌器
STMA (= State Tobacco Monopoly Administration) 国家烟草专卖局
stochastic function 随机函数
stochastic sampling 随机取样
stochastic variable 随机变量
stock fermentation 堆积发酵
stock manure 畜肥
stock rotation 翻垛
stock seed 原种
stockbreeding pollution 畜牧业污染
stoma 气孔
stomatal movement 气孔运动
stomatal regulation 气孔调节
stone 石块
stopper 塞子
stoppered volumetric flask 带塞容量瓶
storage capacity of soluble nutrients 可溶性养分库容量
store-pest 仓储害虫
storing rainwater in pit 窖蓄雨水
storm 暴风雨
storm runoff 暴雨径流
storm-runoff resources 暴雨径流资源
storm-water drain 雨水排水道
stout 肥厚(叶);黑暴烟
stove plant 温室植物
stoving 烘;烤
straight fertilizer 单质肥料
straight ploughing 顺坡耕(直耕)
stratification sampling 分层抽样
stratified sampling 分层抽样
stratographic analysis 色谱分析;色层分析
straw 秸秆
straw coverage 秸秆覆盖
straw incorporation 秸秆还田
straw manure 秸秆(厩)肥
straw manuring 秸秆还田
straw mulching 秸秆覆盖
straw resources 秸秆资源
streak 条纹病斑;条纹;色线;层;

加条纹
streak disease 条纹病(缺锰)
streak virus 条纹病毒
streaked necrosis 条纹状坏死
strength (烟叶)韧性;强度;劲头
stress 胁迫
stress avoidance 避逆性
stress dormancy 逆境休眠;胁迫休眠
stress ecology 逆境生态学
stress protein 逆境蛋白
stress resistance 抗逆性
stress resistance physiology 抗性生理
stress tolerance 耐逆性
stress-tolerant plant 逆境耐性植物
stringing 挂竿;穿绳;绑烟
stringing machine 绑烟机
stringy 叶脉显露;很窄
strip application 带状施肥
strip cropping 带状种植
strip intercropping 带状间作
strip sowing 带状播种
stripe protection 保护带
strips 去梗叶片;片烟
strip-till planting 条状耕播法
strong 强的(色度);浓(厚)的;(烟味)浓烈
strong acid 强酸
strong acidification 强酸化
strong base 强碱

strong-flavored type 浓香型
strong intensity color 色度强
strong sucker 强壮腋芽;优势腋芽
strong toxicity 强度毒性
strongly acid soil 强酸性土壤
strongly alkaline soil 强碱性土壤
strongly salinized 重盐渍化的
structural component 结构组分
structural polysaccharide 结构性多糖
structural soil 有结构土壤
structure water 结构水
structureless soil 无结构土壤
stubble 残株;留茬地;茬
stubble crop 茬地作物;留茬作物
stubble mulch 残茬覆盖
stubble mulch farming 留茬覆盖耕作
stubble mulch tillage 留茬覆盖耕作
stubble mulching 茬地覆盖
stubble ploughing 灭茬;粗耕灭茬
stubbling 灭茬
stunted growth 生长迟缓
subalimentation 营养不良(营养不足)
sub-arable layer 亚耕作层
Subgenus Rustica 黄花烟草亚属
Subgenus Tabacum 普通烟草亚属
subgroup 亚类
subirrigation 地下灌溉
submergence soil 淹水土壤

submicroscopic technique of soil 土壤超显微技术
subnutrition 营养不良(营养不足)
suborder 亚目
subsample 子样本
subsequent effect of fertilizer 肥料残效
subsoil 底土;心土
subsoil amelioration 底土改良
subsoil attachment 深耕器
subsoil crusher 心土破碎机
subsoil fertilization 深层施肥法
subsoil horizon 心土层
subsoil layer 心土层
subsoil plough 深耕犁
subsoiling 耕心土;耕底土;深耕
subspecies 亚种
substitution 替代作用
substrate mycelium 基内菌丝
subsurface irrigation 地下灌溉
subsurface layer 亚表层(心土层)
subsurface run-off 亚地表径流
subsurface soil 亚表土
subsurface stratum 亚表层(心土层)
sub-treatment 副处理
subwatering 地下给水;地下灌溉
subzero 零下;低温
subzero treatment 低温处理
succeeding crop 后茬作物;后作
succession of crops 复种(制);(农作物的)间作;轮作
successive cropping 连作
successive planting 连续种植
succinic acid 琥珀酸
sucker control 抑制腋芽;腋芽控制
sucker control agent 腋芽抑制剂
sucker killing agent 抑芽剂
sucker picking 抹杈
suckercide 抑芽剂
suckering 抹杈;打杈
suckering root 寄生根
sucrase 蔗糖酶
sucrose 蔗糖
suction bottle 吸滤瓶
suction flask 吸滤瓶
sufficiency and deficiency 丰缺
sufficient 充足
sugar 糖
sugar and nicotine ratio 糖/烟碱之比;糖/碱之比(率)
sugar content 糖分含量
sugar metabolism 糖类代谢
sugar synthesis 糖的合成
suitable moisture content 适宜含水量
sulfate 硫酸盐
sulfate of potash 硫酸钾
sulfide 硫化物
sulfur cycle 硫循环
sulfur dioxide 二氧化硫
sulfur oxidation 硫化作用

sulfur-containing amino acid 含硫氨基酸
sulfuric acid 硫酸
sulphate stress 硫酸盐危害
sulphur 硫
sulphur deficiency 缺硫
sulphur disturbance 硫素营养失调
sulphur fertilizer 硫肥
summary analysis 归纳分析
summer crop 夏收作物
summer drought 伏旱
summer dry region 夏旱地区
summer harvesting crop 夏收作物
summer tobacco 夏烟
sun burnt spots 日灼斑块
sun cured tobacco 晒烟
sun plants 阳生植物
sun shading network 遮阳网
sunburn 日灼;日烧
sun-cured 晒制
sun-cured tobacco 晒烟
sun-cured tobacco type cigarette 晒烟型卷烟
sun-cured type 晒烟型
sun-curing 晒制
sun-curing frame 晒烟架
sun-curing with pair of bamboo grates 折晒
sun-curing with string 索晒
sunlight greenhouse 日光温室
sunshine time 日照时间

sunshine-hour 日照小时(数)
super absorbent polymer 保水剂
superficial green 浮青
superficially gleyed soil 表层潜育土壤
superior 特级(烟叶;分级)
superior leaf(= high class leaf) 上等烟
supernutrition 过量营养
superoxide dismutase 超氧化物歧化酶
superphosphate 过磷酸钙
superphosphate fertilizer 过磷酸钙肥料
superphosphoric acid 过磷酸
supersaturated solution 过饱和溶液
supplementary fertilizer 补充肥料
supply constraint 供应限制因素
supply tank 供料罐
supplying capacity 供给能力
supplying capacity of plant essential nutrients 植物必需养分的供给能力
supplying capacity of soil phosphorus 土壤磷素供应能力
surface 表面
surface air temperature 地面温度;地面气温
surface application 表肥
surface area 表面积
surface broadcast 表面撒施

surface charge 表面电荷
surface configuration 地势
surface contamination 表面污染
surface crusting 结皮
surface cultivation 表土中耕;表土耕作
surface drainage 地面排水;明沟
surface drip irrigation 地表滴灌
surface flooding 地面漫灌
surface flow 地面水流;地面径流
surface horizon 表层(地表层)
surface humidity 表面湿度
surface irrigation 地面灌溉;地表灌溉
surface planting 浅栽
surface property 表面性质
surface runoff 地表径流;地表流失
surface runoff loss 地表水耗损量;地表径流损失
surface soil 表土
surface soil layer 表土层
surface source 面源
surface stratum 表层(地表层)
surface temperature 地表温度
surface tension 表面张力
surface water 地表水
surfactant 表面活性剂
surplus water 过剩水分
survival percentage of seedlings 出苗率;成苗率
susceptibility variety 感病品种

susceptible to lodging 易倒伏
suspended water 悬着水
suspension 悬浮(悬浮液;悬浊液)
suspension fertilizer 悬浮肥料
suspension state 悬浊态
suspension system 悬浮体系(悬浊体系)
sustainability 保持力;持续性
sustainable agriculture 持续性农业
sustainable development 可持续发展
sustained development 持续发展
sustained land use 持续的土地利用
sweat 出汗;发酵;回潮
sweeper harrow 浅耕松土耙
sweet 甜的;芳香的
sweetness 甜味
swelled stem 泅筋
swelling 膨胀
swelling water 膨胀水
swine excrement 猪粪
sylvogenic soil 森林土壤
symbiotic microorganism 共生微生物
symbiotic mutualism 共生互利
symbiotic N_2 fixation 共生固氮作用
symbiotic nitrogen fixation 共生固氮作用
symbiotic nitrogen fixation bacterium 共生固氮菌
symbiotic nitrogen fixer 共生固氮生

物

symbiotic nitrogen fixing bacterium 共生固氮菌

symbiotic nitrogen-fixing microorganism 共生固氮微生物

symbiotism 共生现象

symbiotrophism 共生营养

symbiotrophy 共生营养

symplast transport 共质体运输

symptom 症状

symptoms of deficiency （营养）缺乏症状

symptoms of poisoning 中毒症状

synecology 群落生态学；群体生态学

synergic effect 协同效应

synergism 协合作用；增效作用；协生性

synergist 增效剂

synergistic action 协合作用

synergistic effect 协同效应

synthesis 合成

synthesis ammonia 合成氨

synthesis of organic compounds 有机化合物的合成

synthetic amendment 人工合成改良剂

synthetic enduring capacity of resources and environment 资源与环境综合承载力

synthetic manure hotbed 人造厩肥温床

synthetic nitrogenous fertilizer 合成氮肥

synthetic organic pesticides 有机合成农药

synthetical benefit 综合效益

system analysis 系统分析

system of fertilization 施肥制度（系统）

system of safe use of pesticide 农药安全使用制度

systematic error 系统误差

systematic sampling 系统取样；系统选择；系统抽样

systematic variability 系统变异性

systemic disease 全株病；系统性病害

systemic fungicide 内吸性杀（真）菌剂

systemic herbicide 内吸（性）除草剂

systemic insecticide 内吸（性）杀虫剂

systemic invasion 系统性入侵

systemic pesticide 内吸性农药

systemic symptoms 全株性症状；系统性症状

T

T（=tips） 顶叶
T.D.P.（=thermal death point） 致死温度
tableland 台地
tan 淡红黄色（晾烟）
tank fermentation method 罐发酵法
tap root 直根；主根
tap root system 主（直）根系
taproot 直根；主根
tar 焦油
tar content 焦油量
tar delivery 焦油释放量
tar in smoke 烟气焦油；烟气中的焦油
tar/nicotine ratio 焦油/烟碱比
target population 目标总体
target spot（=Thanatephorus cucumeris Donk）（轮纹斑）靶斑病
target yield 目标产量
target yield fertilization 目标产量施肥
target yield oriented fertilization 目标产量施肥法
targeted analysis 目标分析
taste quality 吃味质量；吸味质量
TCMV（=tobacco chlorotic mottle virus）烟草褪绿斑驳病毒病
T-distribution T–分布

technical code 技术规范
technical data 技术数据；技术资料
technical design 技术设计
technical maturity 工艺成熟（度）
technical option of CO_2 emission mitigation CO_2 减排技术对策
technical overripe 工艺过熟
technical quality 工艺品质
technical ripe 工艺成熟
technical standard 技术标准
technique method 技术路线；技术方案
technique of fertilization 施肥技术
technological quality 工艺品质
technology of high-yield cultivation 高产栽培配套技术
telemetrical method 遥测法
TEM（=transmission electron microscope） 透射电子显微镜
temperate climate 温带气候
temperature 温度
temperature and humidity 温湿度
temperature and humidity recorder 温湿度记录仪
temperature compensation 温度补偿
temperature compensation point 温度补偿点
temperature control 温度控制

temperature control chamber 温度控制室
temperature difference 温(度)差
temperature difference of diurnal variation 昼夜温差
temperature distribution 温度分布
temperature for germination 发芽温度;萌发温度
temperature gradient 温度梯度
temperature summation 积温
tempest 暴风雨
temporal scale 时间尺度
temporal-spatial characteristics 时空特征
temporary dormancy 短暂休眠
temporary planting 假植
temporary standard 暂行标准
temporary transplantation 假植
temporary wilting 暂时萎蔫
tenacious consistency 韧性结持(度)
tensile strength 抗拉强度
tensiometer 张力计
tensiometer method 张力计法
tentative standard 试行标准
teratogenic 发生畸形的
terminal bud 顶芽
terminal growth 顶端生长
terminal inflorescence 顶端花序
terpene 萜(烯);萜(烃)
terpenoid 萜类化合物;类萜(烯)
terrace 梯田;阶地

terrace agriculture 梯田农业
terrace building 构筑梯田
terrace cropping 梯田种植
terrace cultivation 梯田耕作
terrace dry cropland 梯田旱耕地
terrace farming 梯田农业
terrace farmland 梯田农地
terrace field 梯田
terrace land 梯田
terrain 地势
terrestrial carbon sink 陆地碳汇
terrestrial ecology 陆地生态学
terrestrial ecosystem evaluation 陆地生态系统评价
test field 试验田
test of significance 显著性检验
test piece 检验片;试样
test plant 试验植物;试验装置
test plot 试验小区
test report 检测报告
test statistic 检验统计量
test tube 试管
testing 检测(检验)
testing method of water in pesticides 农药水分测定方法
tethelin 生长激素
tetrahydrofuran 四氢呋喃
TEV (= Tobacco etch virus) 烟草蚀纹病毒(病)
textural classification 质地分类
texture grade 质地等级(指土壤)

texture name　质地名称
texture separate　质地分级
t-factor　t-因子
thawing　融化
thawing-water irrigation　融水灌溉
the farmland nurture　耕地培育
thematic land use map　土地利用专题图
theodolites　经纬仪
theoretical available moisture　理论有效水量
theoretical organic carbon　理论有机碳量
theoretical production　理论产量
theoretical yield　理论产量
theory of humus nutrition　腐殖质营养学说
theory of mineral nutrition　矿质营养学说
theory of nutrient returns　养分归还学说
theory of organic nutrition　有机营养学说
theory of soil fertility　土壤肥力学说
thermal conductivity　热导率
thermal death point　热死点(致死温度)
thermal radiation　热辐射
thermal-process phosphate fertilizer　热法磷肥
thermal-radiating　热辐射

thermohygrostat　恒温恒湿箱
thermo-isopleth　等温线
thermolysis（=pyrolysis）　热解
thermometer　温度计
thermophiles　嗜热微生物
thermophilic microorganisms　嗜热微生物
thermophilic phase　高温期
thermoregulator　调温器
thermostat　恒温器
thermostatic　恒温的
thermostatic regulator　恒温调节器
thermotolerant bacteria　耐热菌
therophyte　一年生植物
thick　厚的;稠密的;浓的
thick planting　密植
thickened leave　叶片变厚
thicker leaf vein　粗叶脉;叶脉较粗
thickness of soil layer　土层厚度
Thielaviopsis basicola (Berk. & Br.) Ferraris　烟草根黑腐病菌
thin crop　身份欠佳的烟叶;马里兰烟叶第二组烟叶;亮色烟叶
thin film　薄膜
thin out　间苗
thin section of soil　土壤薄瘠层
thin soil　瘠土;薄土
thin to moderate　薄至适中
thin-medium　薄-中等
thinner leaf vein　叶脉较细
thinner planting　稀植

thinning 间苗;疏伐;稀化;稀释;冲淡
thinning seedling 间苗
thiophanate methyl 甲基托布津
thousand seed (grain) weight (种子)千粒重
three crops in two years 两年三熟
three elements requirement experiment 三要素适量试验
three essence of fertilizer 肥料三要素
three essential fertilizer ratio 三要素比例
three essential nutrients 养分三要素
three essentials of fertilizer 肥料三要素
three nutrients compound fertilizer 三元复合肥
three phase structure of soil 土壤三相结构
three to four leaves stage (烟苗)大十字期
three-plot system 三圃制
three-stage curing 三段式烘烤
three-way pipe 三通管
thremmatology 育种学;养育学
threonine 苏氨酸
threshing 打叶
threshing and redrying 打叶复烤
thresholds 阈值

thymol blue 百里酚蓝
tie cutting 解把
tie leaf buster 解把机
tier 绑烟工;扎把工;层;挂烟层次
tier frame 挂烟架
tier pole 挂烟杆
tier rails 挂烟杆导轨
tight combined humus 紧结态腐殖质
tight faced (烟叶)组织紧密
tight hanging 密挂
tight soil 紧实土壤
tight-grained 颗粒紧密的(分级)
till 耕种;耕耘
tillability 可耕性
tillable land 可耕地;适耕地
tillage 耕作
tillage dynamics 耕作动力学
tillage effectiveness 耕作效果
tillage erosion 耕作侵蚀
tillage implement 整地机具;耕作机具
tillage intensity 耕作强度
tillage layer 耕作层
tillage machinery 耕作机械
tillage measures of soil and water conservation 水土保持耕作措施
tillage operation 耕翻作业;整地作业
tillage system 耕作制
tilled soil 已耕地

tilling depth 耕深
tilling width 耕宽
tilth 耕作;耕层
time 时间
time domain reflectometry 时域反射法
time of application 施肥时期
time of seedling desired to plant 成苗期
timeliness of operation 适时管理
times of application 应(施)用次数
times of harvest 采收次数
times of pest generation 害虫发生代数
times of spraying 喷洒次数
times series 时间序列
timing of application 适时施用
timing transplanting 适时移植
tinsel 茸毛
tip 顶芽;蕾;苞;尖;端;滤嘴
tip angle 叶尖角
tip growth 顶端生长
tip pruning 打顶;摘心
tips 顶叶
tissue 组织;(烟叶)组织
tissue culture 组织培养
tissue test 组织测定
titrant 滴定标准液
titration curve 滴定曲线
titrimetric method 滴定法
titrimetry 滴定分析法

TLCV(= tobacco leaf curl virus) 烟草卷叶病毒;烟草曲叶病毒病
TMV(= tobacco mosaic virus) 烟草普通花叶病毒;烟草普通花叶病
TN(= total nitrogen) 总氮
TNV(= tobacco necrosis virus) 烟草坏死病毒病;烟草坏死病毒
toasting 烤制;烘焙
tobacco 烟草;烟叶;烟丝
tobacco additive 烟草添加剂
tobacco aging 烟草醇化
tobacco alkaloid 烟草生物碱
tobacco angular leaf spot (= black fire) 烟草角斑病
tobacco anthracnose 烟草炭疽病
tobacco aphid 烟蚜
tobacco aroma 烟草香气
tobacco ash 烟灰
tobacco bacterial black rot 烟草细菌性黑腐病
tobacco bacterial black stem disease 烟草细菌性黑茎病
tobacco bacterial leaf spot 烟草细菌性叶斑病
tobacco bacterial wilt 烟草青枯病
tobacco barn 烤烟房
tobacco black death 烟草低头黑病
tobacco black mildew 烟草黑霉病
tobacco black root rot 烟草根黑腐病
tobacco black shank 烟草黑胫病

tobacco black spot 烟草黑斑病
tobacco black spot stalk 烟草茎点病
tobacco blue mold 烟草霜霉病
tobacco broken leaf spot 烟草碎叶病
tobacco brown spot 烟草赤星病
tobacco budworm 烟草夜蛾;烟青虫
tobacco bundle 烟把
tobacco bundled regularly arranged 烟把排列整齐
tobacco capsid 烟草盲蝽象
tobacco caterpillar 烟草斜纹夜蛾
tobacco chemistry 烟草化学
tobacco clamping hanger 挂烟夹
tobacco classification 烟草分类;烟草分级
tobacco constituents 烟草组分
tobacco crop 烟草作物
tobacco curing 烟草(叶)调制
tobacco cutter 切丝机
tobacco cutting machine 切丝机
Tobacco Cyst Nematode 烟草胞囊线虫
tobacco damping off 烟草立枯病;烟草猝倒病
tobacco damping-off 烟草立枯病;烟草猝倒病
tobacco disease 烟草病害
tobacco duty 烟草税收
tobacco etch 烟草蚀纹病

tobacco extract 烟草提取物
tobacco fermentation 烟叶发酵
tobacco filling power 烟丝填充性(能力)
tobacco flavorants 烟草香味物;烟草香料
tobacco foil 烟草薄片
tobacco frenching 烟草剑叶病
tobacco frogeye 烟草蛙眼病
tobacco fusarium root rot 烟草镰刀菌根腐病
tobacco fusarium wilt 烟草枯萎病
tobacco grading 烟草分级
tobacco gray spot 烟草灰斑病
tobacco grower 烟草种植者;烟农
tobacco harvesting 烟叶收获;采收烟叶
tobacco hollow stalk 烟草空茎病
tobacco house 烟叶仓库
tobacco introductions 烟草引种
tobacco leaf 烟叶
tobacco leaf curl 烟草卷叶病
tobacco leaf grading factors 烟叶分级要素
tobacco leaf grouping factors 烟叶分组要素
tobacco leaf miner 烟草潜叶蛾
tobacco leaf moisture meter 烟叶测湿仪;烟叶含水率测定仪
tobacco leaf pigment 烟叶色素;烟叶色泽;烟叶颜色

tobacco leaf quality elements 烟叶品质因素
tobacco pest 烟草病虫害;烟草农业害虫
tobacco phyllosticta leaf spot 烟草斑点病
tobacco plant 烟株
tobacco powdery mildew 烟草白粉病
tobacco processing equipment 制丝设备;烟草加工设备
tobacco processing workshop 制丝车间;烟草加工车间
tobacco product 烟草制品
tobacco purchasing 烟叶收购
tobacco ragged leaf spot (= *Ascochyta* leaf spot) 烟草破烂叶斑病
tobacco root knot nematode 烟草根结线虫病
Tobacco Science College of Henan Agricultural University 河南农业大学烟草学院
tobacco sclerotinia rot 烟草菌核病
tobacco seedling 烟苗
tobacco shed 晾烟棚
tobacco sheet manufacturing factory 再造烟叶生产
tobacco shred 烟丝
tobacco smoke constituents 烟气组分
tobacco stem 烟梗
tobacco stem borer 烟蛀茎蛾;烟草麦蛾
tobacco stick 烟竿(杆)
tobacco strip 烟草叶片(不含梗)
tobacco taste 烟草吸味;烟草吃味
tobacco thrips 烟草蓟马
tobacco types 烟草类型
tobacco variety 烟草品种
tobacco vein mottle 叶脉病
tobacco vein-banding potato virus Y 烟草马铃薯Y病毒病
tobacco vein-distorting virus 烟草脉曲病毒病;烟草脉曲病毒
tobacco warehouse 烟叶仓库
tobacco warehouse fumigation 烟叶仓库熏蒸
tobacco weather fleck 烟草气候性病害
tobacco wild fire 烟草野火病
tobacco wireworm 烟草金针虫;烟草叩头虫
tobacco witches broom 烟草丛枝病
tobacco(brown)root rot nematode 烟草根腐线虫病;烟草根褐腐线虫病
TOC(= total organic carbon) 总有机碳
tolerance 耐受性;纯度允差(混等级)
tolerance dosage 允许(剂)量
tolerance species 忍耐物种
tolerance test 耐量试验

tolerance to fertilization 耐肥性
tolerance to heavy manuring 耐肥性
tolerance to low fertility 耐瘠性
top application 追肥；表施（指追肥）
top dressing 表施（追肥）；根外追肥
top dressing at different stages 分期追肥
top fertilization 追肥
top necrosis 顶部坏死
top off 打尖（去顶）
top removal 摘心；打尖
top soil 表土
top ventilation （烤房）顶部通风
topdressing 追肥；根外追肥
top-dressing at different stages 分期施肥
topographic drawing 地形图
topographic factor 地形因素
topographic feature 地形要素
topographic map 地形图
topographical distribution 地形分布
topographical factor 地形因素；地形因子
topographical identification 地形判读（地物识别）
topographical profile map 地形剖面图
topological matrix 拓扑矩阵
topometry 地形测量

topper 打顶器
topping 打顶；摘心
topping at button stage 现蕾期打顶
topping at initial flowering stage 初花期打顶
topping at peak flowering stage 盛花期打顶
topping at squaring stage 现蕾期打顶
topping height 打顶后株高
topping period 打顶期
topping stage 打顶期
topsin 托布津
topsoil 表层土；上层土
topsoil compaction 表土压实
torric moisture regime 干热水分状况
total alkalinity 总碱度
total alkaloids 总生物碱
total analysis 全量分析
total arable land resources 耕地总量
total area 总面积
total ash 总灰分
total carbon 全碳
total elemental analysis of soil 土壤全量元素分析
total leaf number 总叶数
total moisture 总湿度（总含水量）
total nitrogen 全氮；总氮
total nitrogen of soil 土壤全氮
total numbers of bacteria 细菌总数

total nutrient amount 养分总量
total organic carbon 总有机碳
total organic carbon analyzer 总有机碳分析仪
total output 总产量
total phosphorus 总磷
total pore space 总孔隙
total porosity 总孔隙度
total potassium 全钾
total potassium of soil 土壤全钾
total primary nutrient 总养分
total production 总产量
total residue 总残留量
total sugar 总糖(量)
total sulfur 全硫
total suspended particulate 总悬浮颗粒物
total uptake area （根系)总吸收面积
total use of the tobacco plant 烟草综合利用
total volatile acids 总挥发酸类
total volatile bases 总挥发碱
total water consumption 总耗水量
total yield 总产量
touch 手感
touching （烟气)呛喉
tough consistency 韧性结持(度)
toughness 韧度；柔韧性
toxic element 有毒元素
toxic substance 有毒物质

toxicity 毒性
TPM(= total particle matter) 总粒相物；总微粒物
trace analysis 痕量分析
trace components 痕量成分
trace constituent 痕量成分
trace contamination 痕量污染
trace element 痕量元素；微量元素
trace element fertilizer 微量元素肥料
trace fertilizer 微量元素肥料
trace gas 痕量气体
trace metal pollution 痕量金属污染
trace standard of agricultural chemicals 农药残留标准
traditional farming practice 传统农业技术(传统耕作措施)
traditional framing 传统农业
transamination 氨基交换作用
transfer 迁移
transfer pipet 移液管
transferase 转移酶
transferring culture 轮换栽培
transform values 数值转化
transformation of redox 氧化还原转化
transgenic crop 转基因作物
transgenic tobacco 转基因烟草
transient state method 瞬态流法(测土壤导水率)
transitional soil 过渡性土壤

translocation and storage of assimilates 同化物质的贮存和运输
translocation efficiency 运输效率
transparent film 透明薄膜
transpiration 蒸腾(作用)
transpiration coefficient 蒸腾系数
transpiration efficiency 蒸腾效率
transpiration intensity 蒸腾强度
transpiration rate 蒸腾速率
transpiration stream 蒸腾流
transpire 蒸发;发散
transplant 移植
transplantation 移栽
transplanter 移栽机;移栽者
transplanting 移栽;移栽的;移植的;假植的
transplanting density 移栽密度
transplanting equipment 移栽机具;移栽器
transplanting hole 移栽穴
transplanting injury 移栽损伤
transplanting period 移栽期
transplanting stage 移栽期
transport mechanism 迁移机制
transportation of pollutant 污染物迁移
transpose 转置矩阵
transversal slope 横坡
trap 捕集器,陷阱;诱捕器
trashy 干燥、缺油的烟叶
tray filler 装盘机;(育苗盘)基质装填机
tray method 盆栽法
tray-filling machine 装盘机
tray-filling unit 装盘装置;装盘机
tray-grown seedling 盘栽(烟)苗
treatment 处理
treatment combination 处理组合
treatment of solid wastes 固体废弃物处理
tree-crop intercropping 农林间作
trench culture 沟作
trench transplanting 沟栽;畦沟移栽
trencher 开沟器
trench-plough 深耕犁;深耕
trench-rooted 扎根深的
trend 倾向
trend of tobacco production 烟草生产发展趋势
tricalcium phosphate 磷酸三钙
tricarboxylic acid cycle TCA 循环;三羧酸循环
trichlorphon (= dipterex) 敌百虫
Trichoderma 木霉属
trichogramma 赤眼蜂;纹翅小蜂
trickle irrigation 滴灌;滴灌系统
trickling irrigation 滴灌;滴灌系统
trimanin <商>代森锰锌
trinary fertilizer 三元肥料
triple cropping 一年三熟制(三茬复种)
triple superphosphate 重过磷酸钙

trituration 研磨(作用)
triturator 研磨器
trophic interrelationship 营养相互关系
trophic signals 营养信号
trophyll 营养叶
tropical agriculture 热带农业
Trp(= tryptophan) 色氨酸
TRSV(= tobacco ring spotvirus) 烟草环斑病毒病;烟草环斑病毒
true leaf stage 真叶期
true value 真值
tryptophan 色氨酸
TSNA(= tobacco specific nitrosamine) 烟草特有亚硝胺
TSRC(= Tobacco Scientist Research Conference) 烟草科学家研究会议
TSV(= tobacco spot virus) 烟草斑点病毒
TSWV(= tomato spotted wilt virus) 烟草番茄斑萎病毒病;烟草番茄斑萎病毒
tuft(= cyme) 聚伞状花序
tumbler 平底大玻璃杯
turf-muck block 营养土块
turgescence 膨压
turgor pressure 膨压
Turkish tobacco(= aromatic tobacco; oriental tobacco) 土耳其烟(香料烟;东方型烟)

turkish tobacco 香料烟
turn over the stubble 翻茬
turn the soil 翻土
turn under 耕翻;翻入土内
turn up 耕翻上来;耕翻出土
turning rake 翻身耙;旋耕耙
turning stock 翻垛
turning the cultivated land into forests and grasslands 退耕还林还草
turnip moth 黄地老虎
turnover of organic nitrogen 有机氮转化
turnover period 周转期
turnover plow 翻转犁
turnover time 周转期
turpentine odour 松脂气
TVBMV(= tobacco vein-banding mosaic virus) 烟草脉带花叶病毒病;烟草脉带花叶病毒
TVMV(= tobacco vein-mottling virus) 烟草脉斑驳病毒(病)
twin-row cultivator 双行中耕机
two crops a year 一年二作
two crops annually 一年二作
two nutrients compound fertilizer 二元(复合)肥料
two true leaves stage 小十字期
two-sided confidence interval 双侧置信区间
two-sided test 双侧检验
two-tailed test 双尾检验

two-year-three-crop system 两年三熟制
2∶1 type mineral 2∶1型矿物
1∶1 type mineral 1∶1型矿物
type of ecosystem 生态系统类型
tyrosine 酪氨酸

U

udic moisture regime 湿润水分状况
ulmic acid 乌敏酸
ultimate analysis 元素分析;最后分析
ultimate wilting 永久凋萎;永久萎蔫
ultra far infrared 超远红外
ultra violet 紫外光
ultracentrifuge 超速离心机
ultrafiltering balloon 超滤瓶
ultrafiltration funnel 超滤漏斗
ultramicroscope 超显微镜
ultramicroscopic structure 超微结构
ultrasonic vibrator 超声波振动器
ultrastructure 超微结构
ultraviolet spectrophotometry 紫外分光光度法
ultraviolet-visible spectrophotometer 紫外-可见分光光度计
ultrophication 富营养化
umbelliform 伞形;伞型的
unactive carbon 稳定碳
unactive pore 非活性孔隙
unavailability 无效性
unavailable nutrient 无效养分
unavailable water 无效水
unbaling 解包
unbaling yield 开包完整率
unbiased estimation 无偏估计
unbiased estimator 无偏估计值
unblended cigarette 单料烟支;单料卷烟
unbound water 非结合水
uncombined oxide 非结合态氧化物
uncompletely-fermented compost 未腐熟堆肥
uncontaminated soil 未污染土;未感染土
unctuous soil 松软肥沃土壤
uncultivated 未耕的;未开垦的
under fertilizing 施肥量不足
under soil 底土;心土
under tip 次顶叶(分级)
underdevelopment 发育不全
underdrainage 地下排水
underground drainage 地下排水
underground irrigation 地下灌溉
underground percolation 地下渗漏

underground pollution 地下污染
underground runoff 地下径流
under-mulch-drip irrigation 膜下滴灌
undernourishment 营养不良(营养不足)
undernutrition 营养不良(营养不足)
under-row subsoiling 垄底深翻
undersized leaf 叶片稍小的
undersoil 底土
undersow 套种
undesirable taste (= offensive taste) 杂气;杂味
undisturbed soil 原状土;未扰动土
undisturbed structure 原状结构(未扰动结构)
undried 未干(烟叶);未烘的
uneffective precipitation 无效降水量
uneven ripening 成熟不整齐
uneven thickness 厚薄不一
unevenness curing 调制不均
unfermented 未发酵的
unfermented leaf 未发酵烟叶
unfermented tobacco 未发酵烟叶
unfree water 非自由水
unhumified organic matter 未腐解有机物质
unidentified form 未知态
unidirection transport 单向运输

unifactor 单因子
unifactor experiment 单因子试验
unifactor fertilizer experiment 单因素肥料试验
uniform distribution 均匀分布
uniform fertilizer application 均量施肥
uniform germinating 发芽整齐
uniform trial 空白试验
unimodal 单峰的
unit 单位
univariate 单变量的
univariate distribution 单变量分布
universal indicator 通用指示剂
unknown nitrogen 未知态氮
unpleasant 使人不愉快的;讨厌的
unpleasant odor 不愉快气息;难闻气味
unpleasant smell 不愉快的气味
unreclaimed field 未开垦地
unripe 未成熟;欠熟
unsaturated acid 不饱和酸
unsaturated fatty acid 不饱和脂肪酸
unsaturated soil water flow 土壤非饱和水流
unsaturation soil 不饱和土壤
unslaked lime 生石灰
unstable humus 不稳定性腐殖质
untied 不扎把(烟叶)
untied leaf 未扎把烟叶

237

untilled land 未耕地
untimely budding 异常现蕾
untimely defoliation 提前落叶
untimely mature 异常成熟
untying tobacco 解烟；下杆
unusually high temperature 异常高温
unusually low temperature 异常低温
up and down slope 顺坡
up-draft barn 气流上升式烤房
upland 旱地；高地；山地
upland field 旱地；高地田
upland soil 旱地土壤；高地土壤
upland water 地表水
upland weed 旱田杂草
upper bed 上层
upper hands （香料烟）上部叶
upper leaf 上部叶
upper leaf proportion 上部叶比率
upper limit slope for reverse cultivation 退耕上限坡度
upper plastic limit 塑性上限
upper soil layer 土壤表层
upper temperature limit 温度上限
uptake efficiency 吸收效率
upward irrigation 底水灌溉
urea 尿素；脲
urea nitrogen 尿素氮
urease 脲酶
urease inhibitor 脲酶抑制剂
urine 尿
usable tobacco transplants 可移栽烟苗
use of recycled nutrients 养分循环再利用
used plastic film 残膜
ustic moisture regime 半干润水分状况
utilization coefficient 利用系数
utilization efficiency 利用效率
utilization rate of fertilizer 肥料利用率
UV-vis spectrophotometer 紫外－可见分光光度计

V

vacant field 闲地；闲田
vacuum condition 真空回潮
vacuum desiccator 真空干燥器
vacuum distillation 真空蒸馏
vacuum dryer 真空干燥器
vacuum drying oven 真空干燥箱
vacuum filter 真空滤器
vacuum pump 真空泵
valerianic acid 戊酸
valid period 有效期限

validation 确认;批准;生效
valine 缬氨酸
valley 谷地
valley land 谷底田
vanadium 矾
Vapam 威百亩
vaporization 汽化;蒸发作用
vaporous water 气态水
variability 变异性
variable 变量
variable charge 可变电荷
variable charge surface 可变电荷表面
variable cost 可变成本
variable rate fertilization 变量施肥
variable rate fertilizer application 变量施肥
variable-rate application technology 变量施肥技术
variable-rate fertilizer recommendation 变量施肥推荐
variance 方差
variance analysis 方差分析
variance and standard deviation 方差与标准差
variance of a probability distribution 概率分布的方差
variance of a random variable 随机变量的方差
variate 变量
variation 变种

variation coefficient of a random variable 随机变量的变异系数
variation of pathogenicity 致病性分化
variegated 杂色(烟叶);青块
variegated color 杂色斑
variegated leaf(= dingy) 杂色叶
variegated scorch 烤红
variegated strain 杂色品系
variegated 杂色:代号 K
variegation 有斑;花片
varietal characteristic 品种特征
varietal complexity 品种混杂
variety 品种
variety certification 品种鉴定;品种审定
variety degeneration 品种退化
variety improvement 品种改良
variety purification 品种提纯
variety rejuvenation 品种复壮
variogram 方差图;变异函数
VCR(= value/cost ratio) 产投比
vector data 矢量数据
vector insect 媒介昆虫
vector map 矢量地图
vegetable fiber 植物纤维
vegetable layer 植被层
vegetable mould 腐殖土
vegetal cover 植被(植物覆盖层)
vegetation carbon 植被碳
vegetation resources 植被资源

vegetational cover 植被覆盖
vegetative cycle 生长周期；营养周期
vegetative growth 营养生长
vegetative organ 营养器官
vegetative point 生长点
vegetative stage 营养生长期
vein 支脉；叶脉
vein chlorosis 叶脉褪绿
vein clearing 脉明；明脉症
vein mosaic （叶）脉花叶
velvety 绒毛状的
venation 脉络；叶脉
ventilation 通风
verification 验证
vermicompost 蚯蚓粪
vermiculite 蛭石
vertical 垂直的；直立的；竖的；顶端的
vertical profile 垂直剖面
verticillate 轮生的
verticillium wilt 黄萎病
vesicular arbuscular mycorrhizae 泡囊丛枝菌根
vetch 苕子
vibrating subsoiler 振动式碎底土机；振动式翻底土机
vibratory tillage 振动耕作
Vicia villosa Roth 毛叶苕子
Vicia villosa Roth var 光叶紫花苕子

vicious cycle 恶性循环
vicious cycle of agroecosystem 农业生态恶性循环
vigo(u)r of germination 发芽势
vigo(u)r of plant 植株长势
vigo(u)r seed 健壮种子
vigorous growth 生长发育旺盛；旺盛生长
vigorous plant 壮株
vigorous seedling 壮苗
vinyl plastic seeding tray 乙烯基塑料育苗盘
virescence 绿化；绿色；变绿
virescent 带绿色的；淡绿色的
virgin soil 原始土壤；未开垦的土壤
Virginia aroma type 烤烟型香气
Virginia tobacco 弗吉尼亚型烟；烤烟
virginia type cigarette 烤烟型卷烟
virus 病毒
virus diseases of tobacco 烟草病毒病害
virus-free seed 无病毒种子
viscosity coefficient 黏度系数
viscosity factor 黏度系数
viscous consistency 黏滞结持(度)
visible and ultraviolet spectrophotometry 可见紫外分光光度法
visible light absorption 可见光吸收值

visual symptom 可见症状
vitamin 维生素
void volume 孔隙容积
voiding 排泄物
volatile acid 挥发酸
volatile alkali 挥发碱
volatile base 挥发碱
volatile component 挥发性成分
volatile organic acid 挥发性有机酸
volatile organic pollutant 挥发性有机污染物
volcanic ash 火山灰
volcanic ash soil 火山灰土
volcanic soil 火山土

volume heat capacity 容积热容量
volume of aroma 香气量
volume of output 产量
volume of production 产量
volume of root system 根系体积
volume water percentage 水分容积百分率
volume weight 容重
volumetric cylinder 量筒
volumetric flask 容量瓶
volumetric heat capacity 容积热容量
volumetric water content 体积含水量

W

walking tractor 手扶拖拉机
walnut 暗褐色;栗色(叶);胡桃;胡桃香
warm house 温室
warm manure 热性肥料
warmth odor 浓香气
warp 淤积物
warp soil 淤积土
wash erosion 冲蚀
washing salinity by irrigation 灌水洗盐
waste 残伤;残损;废料(分级);废叶;损耗;损失;消耗

waste disposal 废物处理
waste land 荒地
waste ratio of bale 烟包内的损耗率;包内废物率
waste recovery 废物回收
waste tolerance 残损允许度;废料允许极限
waste treatment 废物处理
waste water 污水
wastewater irrigation 污水灌溉
water affinity 润湿性;亲水性
water agency 水力
water and fertilizer coupling 水肥耦

合

water and nutrient transport 水和养分的运输
water and soil erosion 水土流失
water and soil loss 水土流失
water balance 水平衡
water body pollution 水体污染
water budget 水分平衡
water capacity （土壤）持水量；保水量
water cellar 水窖
water circulation 水分循环
water compensation 水分补偿
water conservancy for farmland 农田水利
water conservation 水分保持（保水）
water conservation agriculture 节水农业
water consumption 耗水量
water-consumption of crop 作物耗水量
water contamination 水污染
water content 含水率；水分含量
water content at wilting point 萎蔫点含水量
water critical period 水分临界期
water culture 水培（溶液培养）
water culture experiment 水培试验
water culture method 水培法
water cycle 水分循环
water damaged leaf 水浸渍坏叶

water deficiency 水分亏缺
water deficit 水分亏缺
water demand 需水量
water duty 用水率；灌溉率
water environmental quality assessment 水环境质量评价
water equilibrium 水分平衡
water erosion 水蚀
water erosion process 水蚀过程
water examination 水质检验
water extraction depth 吸水深度
water extraction rate （根系）提水率
water flow 水流
water for analytical laboratory 实验用水
water furrow 排灌毛沟；排灌犁沟
water holding capacity 持水量；持水能力
water holding pore 持水孔隙
water holding power 持水力
water humus 水溶性腐殖质
water insoluble nitrogen compounds 水不溶性含氮化合物
water layer 水层
water log 积水；水涝
water loss 水分损失（水分流失）
water loss and soil erosion 水土流失
water metabolism 水分代谢
water of infiltration 入渗水（湿润水）
water of saturation 饱和水

water of supersaturation 过饱和水
water penetrability 透水性
water percolating capacity 渗水力（水分渗透能力）
water percolation 水分渗漏
water permeability 透水性
water pollution 水污染
water preserving capability 保水性
water productivity 水分生产率
water quality 水质
water quality assessment 水质评价
water quality control 水质控制
water quality evaluation 水质评价
water quality index 水质指标
water quality monitoring 水质监测
water quality standard 水质标准
water quality standard for irrigation 灌溉水质标准
water reception and sediment reduction 蓄水减沙
water regime 水分状况
water release capability of soil 土壤释水性
water repellency 斥水性；疏水性
water requirement 需水量
water resource allocation 水资源配置
water resource information system 水资源信息系统
water resource management 水资源管理

water resources 水资源
water resources shortage 水资源短缺
water retaining capacity 保水能力；持水量
water retention 水分保持
water retention ability 持水力
water retention capability of soil 土壤持水性
water retention property 持水性
water ripening 多雨季节的假熟
water root 水生根
water sample 水样
water sample collection and preservation 水样的采集与保存
water sampling 水样采集
water saturation content 饱和含水量
water saving agriculture 节水农业
water saving irrigation 节水灌溉
water scarcity 缺水
water shortage 水分亏缺
water shortage type 缺水类型
water soluble acid 水溶性酸
water soluble boron 水溶态硼
water soluble calcium 溶液态钙
water soluble copper 水溶态铜
water soluble magnesium 溶液态镁
water soluble manganese 水溶态锰
water soluble molybdenum 水溶态钼
water soluble nitrogen 水溶性氮
water soluble nutrient 水溶性养分

water soluble organic carbon 水溶性有机碳

water soluble phosphate fertilizer 水溶性磷肥

water soluble potassium 水溶性钾

water soluble zinc 水溶态锌

water stability 水稳性

water stable aggregate 水稳性团聚体

water stained 水渍

water storage pit 蓄水坑

water storage pit irrigation 蓄水坑灌

water stress 水分胁迫

water supply 水分供应;供水

water supply-demand equilibrium 水分供需平衡

water supplying capability 供水性

water surplus 水分盈余

water surplus and shortage 水分盈亏量

water temperature 水温

water transpiration 水分蒸腾

water transport pathway 水分运输途径

water uptake and release （根系）吸水和失水

water use coefficient 水分利用系数

water use efficiency 水分利用效率

water utilization 水分利用

water-bath 水浴

water-collecting irrigation 集水灌溉

water-drain 排水沟

water-drop penetration time(= WDPT) 水滴入渗时间

water-holding agent 保水剂

water-holding capacity 持水量(容水量;保水量)

water-holding pore 持水孔隙

water-holding power 持水力

watering 浇灌

watering period 灌溉(周)期

waterlogged compost 沤肥

water-logged farmland 水涝池;水浸地

waterlogged field 水涝地

waterlogged lowland 涝洼地

waterlogged tolerance 耐渍性

waterlogging 水涝;渍水

water-logging resistance 抗涝性;抗淹性

waterloggogenic horizon 潴育层

water-saving agricultural model 节水型农业模式

water-saving agricultural technique 节水农业技术

water-saving agriculture 节水农业

water-saving engineering 节水工程

water-saving irrigation 节水灌溉

water-soil conserving agriculture 水土保持农业

water-soil environment quality 水土环境质量

water-soil-plant relations 水－土－

植物关系
water-soluble pesticide 水溶性农药
water-soluble phosphate fertilizer 水溶性磷肥
water-soluble phosphatic 水溶性磷肥
water-soluble phosphorus 水溶性磷
water-soluble potassium 水溶性钾
water-soluble salt 水溶性盐类
water-stable aggregate 水稳性团聚体
wave spectrum 波谱
wavy （烟叶）稍皱;稍耙松
wax coat 蜡质层
weak （烟叶）淡（色）;弱（的）;淡（的）
weak acid 弱酸
weak acidification 弱酸化
weak base 弱碱
weak flavo(u)r 微弱香味
weak growth 生长发育弱
weak in aroma (= weak flavor) 香气弱
weak seedlings 弱苗
wealthy （香味）丰满
wealthy flavor 香味浓郁
weather fleck 气候斑点;气候斑点病
weatherable mineral 可风化矿物
weathering 风化作用
weathering and eluviation 风化淋溶作用
weathering erosion 风化侵蚀

weathering feature 风化特征
weathering intensity 风化强度
weathering process 风化过程
weathering product 风化产物
weathering rate 风化速率
weathering residue 风化残积物
WebGIS 网络地理信息系统
weed 杂草
weed control 防治杂草;除草
weed encroachment 杂草为害
weed eradication 除草
weed infestation 杂草丛生
weed killer 除草剂
weedicide 除草剂
weeding 除草
weeding machine 除草机
weed-killer spray 除草剂喷洒
weighing 称重
weighing bottle 称量瓶
weighing dish 称量皿
weighing glass 量杯
weight 权（权重）
weight potential 重力势
weight unit area 单位面积重量
weighted arithmetic average 加权算术平均
weighted average 加权平均
weighted mean 加权平均值
weighting arithmetic mean 加权算数平均值
weighting function method 加权函数

· 245 ·

法
weighting methods 权重方法
well-aerated soil 通气良好土壤
well-balanced 谐调
well-drained 排水良好
well-drained land 排水良好的土地
well-structured soil 结构良好土壤
wet and dry bulb hygrometer 干湿球湿度计
wet bulb 湿球(温度计)
wet bulb temperature 湿球温度
wet combustion method 湿烧法
wet deposition 湿沉降
wet digestion 湿法消化(解)
wet method 湿法
wet rib 泗筋
wet year 丰水年
wet-bulb depression 干湿球温差
wet-bulb temperature 湿球温度
wetland 湿地
wet-process phosphate fertilizer 湿法磷肥
wettability 可湿性
wetting and drying cycle 干湿交替
wetting and drying effect 干湿效应
wheat field 麦田
wheat straw 麦秆
wheat-rice double cropping 麦稻两熟
white back 叶背发青
white pollution 白色污染
white stripe 白色条斑

white tobacco 发白烟叶
white veins 白脉病
whittle 削减;削弱
whole leaf 全叶;整叶
whole plant 全植株
whole-layer placement of fertilization 全层施肥
whole-plant slicing 整株砍收
wide-narrow row 宽窄行(种植)
wide-row planting 宽行栽植
wild growth 疯长
wildfire disease 野火病
wilt 萎蔫;调萎;枯萎
wilt disease 萎蔫病
wilting 萎蔫
wilting coefficient 凋萎系数;萎蔫系数
wilting moisture 萎蔫湿度
wilting percent 萎蔫百分率
wilting percentage 萎蔫含水率
wilting phenomena 萎蔫现象
wilting point 凋萎点;凋蔫点;萎蔫含水量
wilting symptom 凋枯症状
wilting water content 萎蔫含水量
wind abrasion 风蚀(作用)
wind agency 风力
wind burn 风害;风灾
wind damage 风伤(俗称风磨);风致伤(分级);风害
wind damaged leaf 风磨烟叶

wind deposits 风积物
wind drift 风积
wind erosion 风蚀(作用)
wind laid deposit 风积物
wind resistance 抗风性
windblown soil 风积土
wind-deposit soil 风积土
windrow 条垛式
windstorm 暴风
winnowing 风选
winter crop 越冬作物;冬作
winter dormancy 冬季休眠
winter fallow 冬闲地
winter hardiness 耐冬性;抗寒性;耐寒性
winter injury 冻害;冬害
winter ploughing 冬耕
winter resistance 抗寒性
wintering 越冬
winterkill 冻死;冻坏
winterness plant 冬播作物;冬性作物

wireworm 金针虫
witches broom of tobacco 烟草丛枝病
withered 枯萎
withering 变黄(烟叶);凋萎
withering of leaves 叶萎蔫
within row 行内
within-row spacing 株距
woody aroma 木香
workability of soil 土壤适耕性
working curve 工作曲线
World Conference on Smoking and Health 世界吸烟与健康会议
worm 虫;蠕虫
worm cast 蚯蚓粪
worm counts 虫体计数
wound healing 创伤修复,愈合;伤口愈合
WP(=wettable powder) 可湿性粉剂
wrapper leaf 外包叶(雪茄烟)
wrinkled 起皱
wrinkly (烟叶)皱缩

X

X(lugs) 下部叶组;下二棚;脚叶(白肋烟)
xanthin 花(叶)黄素;植物黄素
xanthochroism 黄化现象
xanthophyll 叶黄素;胡萝卜醇
xanthozem 黄壤

xenobiotic pollutants 外源污染物
xenobiotic substance 外源性物质
xeric 干燥的;干旱的
xeric moisture regime 夏旱水分状况
xeromorph 旱生植物

xeromorphic soil 旱成土
xerophilous plant 喜旱植物；适旱植物
xerophytes 旱生植物
xerophytic plant （＝xerophyte） 喜旱植物

xerosis 干燥
X-ray diffraction X射线衍射
X-ray diffraction pattern X射线衍射图谱
xylem transport 木质部运输
xylophyta 木本植物

Y

yard manure 厩肥；圈肥
year old seed 陈年种子
year types of precipitation 降雨年型
yearly plant 一年生植物
yeast 酵母菌
yellow earth 黄壤
yellow leaf group 黄烟组
yellow leaf with green stem 青筋黄片
yellow podzolic soil 灰化黄壤
yellow soil 黄壤；黄土
yellow-brown earth 黄棕壤
yellow-brown soil 黄棕壤
yellow-cinnamon soil 黄褐土
yellowing 落黄；变黄；黄化
yellowing degree 变黄程度
yellowing of leaves 叶发黄
yellowing stage 变黄期

Yellowish red soils 黄红壤
yield 产量；收益
yield component 产量组成
yield curve 产量曲线
yield estimation by remote sensing 遥感估产
yield goal 目标产量
yield index 产量指标
yield level 产量水平
yield output 产量
yield per hectare 每公顷产量
yield per mu 亩产量
yield per plant 单株产量
yield potential 生产潜力
yield prediction 产量预测
yield response to fertilization 肥料产量效应
young soil 幼年土壤

Z

zero burette 自满滴定管
zero point charge 电荷零点
zero point of charge 电荷零点
zero point of titration 滴定零点
zero slope 平坡
zero tillage 免耕(法);免耕种植
zero tillage system 免耕法
zero-tillage seed placement 免耕播种
zheltozem 黄壤
zinc 锌
zinc chelates 螯合态锌
zinc deficiency 缺锌
zinc disturbance 锌素营养失调
zinc fertilizer 锌肥
zinc pollution 锌污染
zinc sulfate 硫酸锌
zinc sulfate heptahydrate 七水硫酸锌
zinc sulphate 硫酸锌
zinc sulphate monohydrate 一水硫酸锌
Zn-deficient 缺锌症
zonal soil 地带性土壤

汉文索引

A

ATP 合酶 18
阿拉伯树胶 94
阿魏酸 79
矮秆的 125
矮化病 64
矮化的 64
矮生株型;低矮植株;
　矮株型烟草 125
矮态;矮生性;矮化现
　象 64
安全的土壤健康质量
　213
安全剂量 117,120,
　192
氨 12
氨毒害 12
氨化(作用) 12
氨化能力(氨化量)
　12
氨化强度 12

氨化细菌 12
氨挥发 12
氨挥发损失 12
氨基丙酸 10
氨基氮 12
氨基氮;胺氮 12
氨基交换作用 233
氨基酸 12
氨基酸分析仪 12
氨基酸态氮 12
氨基酸自动分析仪
　18
氨基酸组成;氨基酸成
　分 12
氨基糖 12
氨基糖氮 12
氨解(作用) 13
氨水 12,122
氨态氮 12
氨态氮肥 12

氨氧化细菌 12
铵 12
铵的黏土矿物固定作
　用 13,81
铵的吸附和解吸 213
铵固定 13
铵态氮 13
铵态氮肥 13
铵盐同化作用 17
按烟叶颜色分组 93
胺 12
胺化作用 12
暗(光泽);香气沉闷;
　无光的;单调的 64
暗代谢 50
暗淡色 75
暗褐色;栗色(叶);胡
　桃;胡桃香 241
暗黑地老虎 73
暗呼吸作用 50

暗绿色 50
暗色 50
(烟叶)暗色 64
(烟叶)暗色;淡色 159
暗色土(火山灰土) 14
暗棕壤 50,51

凹斑;陷斑 167
凹玻片 99
凹载玻片 99
螯合(作用) 32
螯合肥料 32
螯合环 32
螯合剂 32

螯合剂诱导的植物修复 32
螯合态锌 249
螯合态养分 32
螯合铁 109
螯合铜 32
螯合物 32

B

(轮纹斑)靶斑病 225
巴氏消毒法 160
拔根机 190
把烟;扎把烟叶 95
白氨酸;亮氨酸 120
白边地老虎;白边切夜蛾 73
白炽灯 104
白粉病 175
白化(现象);白化病 10
白化;褪绿(病);缺绿;萎黄病 34
白化烟草组织 10
白化叶绿素缺乏症 10
白浆化作用 10
白浆土 10,20
白肋烟 27

白肋烟、烤烟最底部脚叶;脚叶;飘叶(分级,指淡晾烟) 83
白肋烟型 27
白脉病 246
白色条斑 246
白色污染 246
白天光照 51
白云母 141
白云石 61
白云石质;石灰石 61
百分温度;摄氏温度 31
百里酚蓝 228
败育的 2
败育花粉 2
败育胚珠 2
败育种子 2
斑驳腐烂 139

斑点 126
斑点;花斑 139
斑点状坏死 215
斑萎病 215
(蜱类的)斑须蜱象 61
半定量的 198
半方差 198
半方差函数建模 198
半方差函数值 198
半方差图 198
半分解土壤有机物质 97
半分解有机层 108
半分解有机质层 132
半腐解腐殖质 64
半腐殖质(半腐熟腐殖质;酸性腐殖质) 137

·251·

半干旱 198
半干旱(地)区 198
半干旱地 198
半干旱地区生态系统 198
半干旱气候 198
半干润水分状况 238
半胱氨酸 49
半挥发(性)的 198
半挥发性成分;半挥发性物质 198
半挥发性有机化合物 198
半晾半烤(烟叶) 198
半微量 198
半微量法 198
半纤维素 97,198
半纤维素降解 97
半香料烟 198
半香料烟(草) 198
半叶测定 94
半叶法 94
半叶卷雪茄 198
伴生杂草 3
拌药的种子 62
拌种 197
绑烟工;扎把工;层;挂烟层次 228
绑烟机 220
绑叶;编叶 118
(烟)包;打包 20

包膜 37
包膜肥料;包衣肥料 37
包膜控释 37
包衣种子 37,161
包装肥料 159
苞;托叶 26
孢子丝 215
雹;冰雹 94
雹害 94,106
雹害;雹灾 94
雹伤(害) 94
雹伤的(烟叶) 94
薄层;薄板;薄片 114
薄地 172
薄荷香;晾香 136
薄荷香的;晾香的 136
薄膜 227
薄片状的;层状的 114
薄土;浅层土 199
薄至适中 227
薄-中等 227
饱和持水量 195
饱和导水率 194
饱和点 195
饱和度 52
饱和含水量 195,243
饱和亏缺 195
饱和率 161

饱和曲线 195
饱和湿度 194
饱和水 242
饱和酸 194
饱和土壤 194
饱和脂肪酸 194
饱满香气 86
饱满种子 86
保本价格 26
(菌种)保藏 41,127,176
保持(固持;吸持) 188
保持力;持续性 223
保存;饲养;维护 112
保肥性 150
保护措施 178
保护带 220
保护地气体毒害 87
保护地盐害 193
保护地栽培 48,178
保护行 26,94
保护性开发 178
保护性农业 41
保苗 41
保湿力 1
保湿性能;保水能力 138
保水剂 222,244
保水能力;持水量 243

保水性 243
保土耕作(保护性耕作) 41
保土作物 212
保温材料 96
(烟气)抱团 26
报酬递减 58
报酬递减规律;报酬递减法则 116
报酬递减阶段 216
鲍尧科斯比重计(甲种比重计) 26
暴风 247
暴风雨 97,219,226
暴风雨灾害 50
暴晒作用 107
暴雨 37,107,182
暴雨径流 219
暴雨径流资源 219
北方根结线虫 132
北纬 148
备耕 127
背景含量 20
背景浓度 20
背景值 20
钡 21
被测定物 54
被动吸收 160
被遏制状态;静止;休眠 181
被侵蚀土地 72

被侵蚀土壤 72
本底浓度 20
本底水平 20
本地品种 105
本地品种(土种) 105
本地种;土种 142
苯 22
苯丙氨酸 163
苯丙蒽 22
苯丙酸 101
苯并[α]芘 22
苯并芘 22
3,4-苯并芘 22
苯甘氨酸 163
苯甲醇 22
苯甲醛 22
苯甲酸 22
苯醌;醌 181
苯乙醇 163
苯乙酸 163
泵 179
鼻烟 203
比表面 214
比表面积 214
比较试验 39
比较优势 39
比孔容 214
比例抽样 178
比热 214
比色法 38
比色计 34

比重 214
比重测定 92
比重计;比重瓶 179
比重计;重力仪;重力计 91
(液体)比重计 101
比重计法 101
比重瓶 92
吡啶;氮(杂)苯 179
吡咯 103,179
必需氨基酸 72
必需矿物质元素 72
(生命)必需元素 72
必需脂肪酸 72
闭管法 195
闭蓄态颗粒有机物 151
壁式铧犁耕作 138
避病机制 59
避病性 59
避免日晒 112
避免受潮 112
避逆性 220
边行作物 128
边际报酬递减 58
边际产量 129
边际成本 128
边际分析 128
边际贡献;边际利润;边际所得 43
边际贡献率 128

· 253 ·

边际生产率递减规律 116
边际效应 128
边际效应;边际收益 129
边际效用递减 58
边缘分布 128
边缘腐烂 129
边缘坏死 128
边缘失绿 128
苄醇 22
变褐率 52
变黄(烟叶);凋萎 247
变黄程度 248
变黄期 161,248
变黄天数 51
变量 239
变量施肥 239
变量施肥技术 239
变量施肥推荐 239
变温发芽箱 12
变异函数 239
变异系数 28,37
变异性 239
变窄 142
变种 239
标本袋 193
标定;标准化 217
(同位素)标记的 113
标记化合物 113

标记同位素 113
标记原子 113
标明量 51
标签 113
标签内容 113
标志土层(关键土层) 112
标准 217
标准比重计 217
标准大气 217
标准大气压 217
标准方法 217
标准化变量 217
标准化管理 217
标准晾房 217
标准偏差 217
标准曲线 28,217
标准溶液 217
标准筛 217
标准试验 43
标准温度 217
标准温度计 148
标准物质;标准样品;参考物质(标准物质) 185
标准误差 217
标准吸管 217
标准压力 217
标准样品 217
标准样品;标准试样 129

标准正态分布 217
表层(地表层) 223
表层潜育土壤 222
表层土;上层土 232
表肥 222
表观同化率 15
表观阳离子交换量 15
表面 222
表面电荷 223
表面活性剂 223
表面积 222
表面撒施 222
表面湿度 223
表面污染 223
表面性质 223
表面张力 223
表皮(复) 49
表皮;护膜;角质层 49
表皮蜡质 49
表施(追肥);根外追肥 232
表土 223,232
表土层 223
表土压实 232
表土中耕;表土耕作 223
表现型 163
瘪籽 2
冰雹灾害 94

冰爆 102
冰点;凝固点 84
冰箱 102
丙氨酸 10
丙二酸 127,178
丙酮 3
丙酮酸 112,179
丙酮酸激酶 179
饼肥 21,28,196
饼肥种类 112
饼粕堆肥 28
饼图;圆形图 166
病斑 59
病虫害 50
病虫害监测 162
病虫害综合防治 107
病虫驱避性 162
病毒 240
病害防治 59
病害分级 59
病害共生复合体 59
病害监测 59
病害强度 59
病害循环 59
病害严重度 199
病害预测 59
病害诊断 59
病害治理 59
病害种类 112
病情指数 59
病势 59

病态;发病率 139
病原;病原体;病菌 160
病原菌 59
病原微生物 160
病原物 160
病原物抑制性土壤 160
病原细菌 160
病原性细菌 160
病株 59
波尔多液 25
波谱 245
波谱分析 215
玻璃棒 90
玻璃电极 90
玻璃电极电位计法 90
玻璃器皿;玻璃仪器 90
剥蚀 53
播幅 63
播前处理 176
播前灌水 110
播前施用 176
播前整地;整地 176
播种 197,213
播种板 196
播种法;播种技术 197
播种方法 197

播种机 168,197,214
播种机;播种者 197
播种机种子箱 63
播种季节 214
播种精度 3
播种量 181
播种量;种子量 197
播种密度 53,197
播种面积 214
播种期 169,197,214
播种期试验 51
播种深度 53,197,214
播种深度;种子覆盖深度 197
播种时间;播种期 197
播种时期;播种期;播种时间 197
播种速率;发芽率 197
播种质量 197
泊松分布 170
补播;重播;追播 186
补充处理 7
补充肥料 222
补苗 80
补栽;补植 186
补植;补栽;间作 108
捕集器;陷阱;诱捕器 234
不饱和酸 237

· 255 ·

不饱和土壤　237
不饱和脂肪酸　237
不成熟的种子　175
不打顶　147
不带滤嘴的卷烟　167
不定根　6
不定芽　6
不发芽的;无发育的　147
不覆盖　147
不规则　110
不规则的失绿　110
不规则分布的坏死斑点　110
不均匀燃烧　97
不可逆的　147
不可逆机械压实　110
不可逆吸附(作用)　110

不利的生长条件　6
不利的土壤条件　6
不利因素　6
不列级烟叶;级外烟　146
不毛之地　21
不耐肥的　78
不能发芽的种子　147
不能生物降解的　146
不溶(解)的　147
不溶化;不溶解　107
不溶解的　105
不适的气味;异味　83
不透水土壤;渗透性不良土壤　103
不完全肥料　104
不完全花　104
不完全胚珠　2

不完全区组设计　104
不完整种子;不饱满籽粒　103
不稳定性腐殖质　237
不相似　60
不休闲耕作　162
不愉快的气味　237
不愉快气息　59,237
不育性;不育;不肥沃　105
不扎把(烟叶)　237
不整地播种　147
不足　107
布朗运动　27
布氏漏斗　27
(烟叶)部位(分级);位置　173
部位特征　31

C

CO_2　减排技术对策　225
CO_2　同化作用　37
采前喷药　175
采收操作　96
采收成熟度　95
采收次数　229
采收辅助机(设备)　96
采收机　176
采收机;收割机　95
采收期　95,166
采收前的　175
采收前期　175
采收适宜期　76
采收叶　95

采样点　194
采样过程　194
采样技术　194
采样间隔　194
采样率　194
采样频率　85
采样瓶　194
采样器　193,194

采样设备 194
采样时间 194
采样系统 194
采样周期 194
采叶;逐叶采摘 176
采摘 170
采摘;初期;优质的;原始的;最初 176
采摘成熟度 166
采种田;种子繁殖场;种子繁殖田;种子田 178
菜籽饼 38,183
菜籽饼发酵 183
参数 160
残茬覆盖 220
残差 186,187
残积土 142,187
残积物 186
残留氮 116
残留肥料 30
残留量 186
残留率 161,187
残留期 186
残留酸度 186
残留态 186
残留性农药 187
残膜 187,238
残伤;残损;废料(分级);废叶;损耗;损失;消耗 241

残损允许度;废料允许极限 241
残烟 185
残烟;次烟;疵烟 52
残株;留茬地;茬 220
仓储害虫 219
仓式调制 39
操作准确性;操作精确性 3
草本的;草质的;药草香的;干草气的 97
草地 91
草甘磷 90
草克死(除草剂) 34
草木灰 91,167
草乃敌(除莠剂) 70
草酸 158
草酸钠 204
草田轮作 91
草酰乙酸 158
侧根 18,104,16
侧灌 116
侧流烟气 200
侧漏 116
侧脉 116
侧生的;侧面的;横向的 116
侧生根 19
侧施 200
侧施肥料 200
侧芽(腋芽) 116

侧枝 116
测定范围 56
测定含水量;湿度测定 138
测高计(高差表) 30
测量仪器 131
测渗仪 125
测土施肥;测土配方施肥 78
测土推荐施肥 211
(烟叶)层 116
层次分化 (程)度 52
层次分析法 14,98
层厚度 117
层级系统理论 98
层间表面 108
层间电荷 108
层间固定 108
层间钾 108
层流 114
插图;图解;例证说明;实例;解说 103
茬地覆盖 220
茬地作物 220
茬口 176
茬口安排 46
查苗 79
差别沉淀法 57
差异显著水平 120
差异显著性 200

257

掺合肥料　25
掺假(烟叶)　143
产甲烷菌　132
产量　177,241,248
产量;产品;排出物　157
产量;收益　248
产量低　125
产量定额　157
产量曲线　248
产量水平　248
产量预测　248
产量指标　157,248
产量组成　248
产流　192
产流量　192
产品价格　176
产品质量　180
产酸细菌　4
产投比　157,239
产值　157
长度　113,120
长花期　124
长距离运输　123
长宽比　57
长满杂草的　105
长期储存;长期储藏　124
长期定位试验　124
长期定位研究　123
长期肥料试验　124

长期试验　124
长期试验;永久性试验　162
长期天气预报　124
长期田间试验　124
长期土壤肥力试验　124
长期预测;长期预报　124
长日照　123
长日照型　123
长日照植物　123
长日照作物　123
长势监测　94
长效的;(香气)持久　124
长效肥料　43,123
常规播种(移栽);常规栽培　185
常规分析;例行分析　191
常规耕作　148,185
常规灌溉　148
常规烘烤　43
常规化学肥料　43
常规检测　43
常规施肥　185
常规施药(肥)　43
常量;大量　126
常量矿物质　126
常年收成　148

常湿润土壤水分状况　162
常数　42
常温　154
常温;室温　190
常用农业技术(传统栽培技术)　43
超富集　102
超积累植物　102,132
超积累植物(超富集体)　102
超净工作台　36
超滤漏斗　236
超滤瓶　236
超声波振动器　236
超嗜热菌　102
超速离心机　236
超微结构　236
超细微结构;超精细结构　102
超显微镜　236
超氧化物歧化酶　222
超硬实　75
超远红外　236
潮褐土　15,130
潮红(专指烤烟调制变黄期过长所产生的红褐色斑)　215
潮红烟叶　215
潮湿　50
潮湿水分状况　15

潮土 31,83
沉淀 53
沉淀物 175
沉淀物;沉积 196
沉积(作用);沉降 196
沉积黏土 196
沉积土壤 196
沉降参数 196
沉降反应 175
沉降分析 196
沉降平衡 196
沉降容积 196
沉降时间 196
沉降速度 196
沉降筒 196
沉降系数 196
沉闷 64
沉溢 158
陈年种子 248
陈烟(叶) 152
称量皿 245
称量瓶 245
称重 245
成本效益分析 44
成本要素 69
成虫 103
成虫(复, imagines, imagos) 103
成虫期;成株期 6
成苗率 168,198,217

成苗期 229
成苗数(潜力);植株潜力 168
成熟 129
成熟;成熟度 130
(烟叶)成熟;成熟度 189
成熟;充分成熟;工艺成熟 189
成熟;熟化;陈化 189
成熟不整齐 237
成熟催化剂;促熟剂 189
成熟的;工艺成熟;尚熟 129
成熟度 130
成熟度;松软性 132
成熟度不足;未成熟的(烟叶) 103,104
成熟度指数 130
成熟过程 189
成熟阶段 216
成熟期 51,129,130,161
成熟期;成熟阶段 129
成熟前落叶 176
成熟蒴果 189
成熟特征 31
成熟土;熟土 129
成熟组织 129

成熟组织抗性 129
成土过程(土壤发生过程) 161
成土环境 212
成土母质 157,212
成土作用;成土过程 207
成瘾剂 5
成株抗性 6,129
承兑;接受;认付;验收 3
乘坐式采烟机 189
橙色的 151
吃味粗糙;刺激味 95
吃味质量;吸味质量 225
持肥力 150
持火力 27
持久香气 64
持久性有机污染物 162
持水当量 138
持水孔隙 242,244
持水力 242,243,244
(土壤)持水量;保水量 242
持水量(容水量;保水量) 138,244
持水量;持水能力 242
持水能力 138
持水曲线 138

· 259 ·

持水性 138,243
持续的土地利用 223
持续发展 162,223
持续改进 42
持续性干旱 178
持续性农业 223
尺寸分布;(颗粒)大小分布;粒径分布 202
尺度 195
尺度数据 195
斥水性;疏水性 243
赤霉素 89
赤眼蜂;纹翅小蜂 234
冲积的 11
冲积壤土 82
冲积土 11,83
冲积土壤 82
冲积物 11
冲积物;冲积层;冲积土 11
冲蚀 195,241
充足 6,221
(香气)充实 86
(香气)充足 189
虫;蠕虫 247
虫传的 70
虫害 50,107,162
虫害发生季 88
虫害防治 43

虫口密度 172
虫媒 71
虫体计数 247
抽梗;除梗 119
抽水;排水沟 179
(对烟支限定长度)抽吸次数 178
(卷烟)抽吸长度 178
抽吸(烟)特性 203
抽吸间隔;抽吸间隔时间 178
抽吸口数 178
抽吸频率 178
抽吸曲线 178
抽吸容量;每口抽吸烟量 178
抽烟时不把烟气咽下 203
抽样 194
抽样;取样 193
抽样比例 194
抽样单位 193
抽样方案 194
抽样分布 194
抽样区间 194
抽样误差 194
稠环芳烃 171
臭气;臭味;杂气;异常气息 152
臭味;杂气;杂味 152
臭氧 159

臭氧害 50
出汗;发酵;回潮 223
(烟叶)出炕 59
出苗 70,197
出苗;成苗 197
出苗率;成苗率 223
出苗前猝倒 175
出苗情况不好 172
出苗始期 21
出丝率 183
出土子叶 72
出叶率 183
出叶日期 118
出叶速率 118
初步调查 175
初花;第一朵花 81
初花期打顶 232
初烤烟叶;原烟 48
初龄幼虫 106
(烟叶调制)初期凋萎 53
初生根;主根 176
初生叶 176
初生叶;初生茎 5
初始入渗率 106
初霜 64
初霜;早霜;霜降 81
初萎 104
除草 36,245
除草机 245
除草剂 97,245

除草剂降解 52
除草剂喷洒 245
除草剂污染 97
除虫菊酯 179
<商>除芽通 3
锄地 99
储存防霉 140
储存库 186
处理 234
处理组合 234
畜肥 219
畜粪 31
畜粪尿 62,122
畜牧业污染 219
畜禽粪污染 171
畜舍污水 21
触杀剂 42
(接)触(使)杀作用 42
穿透 161
穿透阻力 161,187
传播 60
传粉 171
传粉媒介 171
传染病 105
传统农业 233
传统农业技术(传统耕作措施) 233
传统调制;常规烘烤;一般烘烤;习惯烘烤 43

创伤修复,愈合;伤口愈合 247
吹;吹风;送风;鼓风 25
垂直的;直立的;竖的;顶端的 240
垂直剖面 240
春播作物 216
春地 80
春耕 216
春耕;春季栽培 216
春化作用;春化处理 111
春烟 216
纯度;纯洁 179
纯度;纯净 179
纯度检验 179
纯化 179
(烟气)纯净 179
纯块金效应模型 179
纯收入 143
纯水;去矿质水;软化水 53
纯系品种 179
纯系选择;品系选择 122
纯种 179
醇和的 132
醇和型;柔和性;温(缓)和;(柔)软适度 135

醇和型烤烟 135
醇和型烤烟品种 135
(香气)醇厚 132
醇化 132
醇化过程 8,132
醇化时间;老化时间 7,8
醇化条件 8
醇化烟草;陈化烟草 7
醇化烟叶;陈化烟叶 7
醇酸 10
瓷坩埚 172
瓷蒸发皿 172
磁搅拌器 126
磁性搅拌棒 126
雌蕊 167
雌性 77
雌雄同花(的) 139
雌雄同株(的);雌雄同体 139
次顶叶(分级) 236
次氯酸钙 121
次生代谢产物 196
次生根 116,196
次生碱度 196
次生碱化(作用) 196
次生矿物 196
次生黏化层 196
次生潜育作用 196

261

次生土壤 97,196
次生污染影响;二次污染效应 196
次生盐渍化 196
次要营养 196
次要症状 136
刺激;辛辣(烟味) 179
刺激味;尖刺;辛辣味 179
刺激性 110
刺激性气味;辛辣味 179
刺状产卵器 5
从茎部生根的 182
丛枝菌根 16
丛顶(症) 139
丛枝 16
(烟气)粗糙;硬度 95
粗糙(指烟叶、烟气) 95
粗糙气息 95
粗糙性;刺激性 95
粗的;粗糙的 37
粗放农业 74
粗放型 74
粗腐殖质 64,183
粗腐殖质(地面枯叶层) 37,191
粗腐殖质;酸性有机质 139
粗腐殖质层 64,139
粗骨土 202
粗灰分 46
粗孔筛 37
粗粒质土壤 37
粗沙 16,93
粗沙砾 37
粗沙粒 37
粗沙土 37
粗沙质 16
粗沙质壤土 37
粗沙质土 37
粗筛 37
粗碎片(粗碎块) 37
粗提出物 46
粗纤维素 46
粗叶脉;叶脉较粗 227
粗支烟 126
促进发芽 178
促进萌发 96
促进效应 3
促进作用 178
猝倒病 50
醋酸 3
醋酸短纤维 3
醋酸纤维 3
醋酸纤维丝束工艺 3
醋酸纤维丝束型滤嘴 3
醋酸盐 3
醋酸乙酯;乙酸乙酯 73
醋纤滤棒 3
醋纤滤棒成型机 3
醋纤滤嘴 3
醋纤丝束 3
簇叶症(缺锌或缺硼) 191
催化降解 30
催化酶 112
催化作用 30
催熟剂 189
催芽 3,175
催芽播种 216
催芽种子 176
脆(性)的;易碎的 26
脆性 84
脆性破坏试验 26
萃取;浸提;提取 74
萃取反应 2
萃取器 75
错期耕作 216

D

DNA 酶足迹法 61
DNA 大小决定基因 61
DNA 大小筛分 61
DNA 分型 61
DNA 合成仪 61
DNA 环(环状 DNA) 61
DNA 克隆(化) 61
DNA 连环 61
DNA 连接仪 61
DNA 裂解;DNA 切割 61
DNA 螺距 61
DNA 酶 61
DNA 切口形成 61
DNA 体外扩增 61
DNA 序列测定;DNA 测序 61
DNA 序列测定仪;DNA 测序仪 61
DNA 序列分析 61
DNA 印迹(法) 61
DNA 转折;弯曲 61
(卷烟)搭口宽度 115
(卷烟)搭口内折 115
(卷烟)搭口外折 115
(卷烟)搭口印码 115
(卷烟)搭口粘封 115
打包机 20
打包机;包装机 159

打包麻袋布 27
打包烟叶密度 118
打顶;摘心 229,232
打顶高度 97
打顶后株高 232
打顶期 232
打顶器 232
打顶早的矮株烟草 125
打分技术 195
打尖(去顶)51,232
打捆;扎把 27
打捆机 113
打破休眠;破除休眠 62
打叶 228
打叶复烤 228
大比例尺调查 115
大比例尺土壤图 115
大豆饼 214
大规模生产 115
大旱 92
大号卷烟(84~85 mm 长) 112
大环境 126
大量元素(常量元素) 126,127
大量元素;主要营养

176
大量元素肥料 126
大量元素养分;大量营养物 126
大苗 115
大农业 115
大批量分析 129
大气层 17
大气沉降 10
大气氮沉降 17
大气干旱 17
大气候 126
大气圈 17
大气水 18,132
大气污染 6,9,10,17,18
大气污染物 17,18
大气 - 植物界面 10
大气质量 18
大气质量控制标准 10
大气组成 17
大群繁育 129
(烟苗)大十字期 228
大田 114
大田病害 79
大田生育期 51,94
大田生长发育 79

263

大田生长期 79
大田栽培 79
大团聚体 126
大小;尺寸;量;数值 126
大小分级 202
大型叶 115
大雪茄 115
大循环 92
大样本 115
代表性样品 186
代森锰 60
〈商〉代森锰 113
〈商〉代森锰锌 64, 127,234
〈商〉代森锌 113
代谢;新陈代谢 132
代谢产物 132
代谢过程 132
代谢平衡 132
代谢失调 132
代谢途径 132
带病种子 59
带播 197
带电表面(荷电表面) 31
带茎调制 217
带绿色的;淡绿色的 240
带塞容量瓶 219
带有压土轮的条播机 176
带状播种 220
带状耕种 80
带状间作 220
带状施肥 78,220
带状种植 220
待测成分 14
袋装种植 170
单变量的 237
单变量分布 237
(香气)单薄 139
单侧检验 153
单侧置信区间 153
单峰的 237
单环芳烃 139
单季试验 201
单交杂种 108
单粒种子;种子个体 105
单料烟支;单料卷烟 236
单偏光 167
单人采叶机;单人操作烟叶打顶机 153
单色光 139
单糖 139
单尾检验 153
单位 237
单位面积产量 161
单位面积利润 177
单位面积年产量 14

单位面积重量 245
单向运输 237
单循环病害;单一性病害 201
单一肥料 201
单一作物农业 152
单因素肥料试验 237
单因子 237
单因子试验 201,237
单元素肥料 201
单质肥料 201,219
单株产量 248
单株选择;个体选择 105
单作 139,152,201, 213
单作多熟型 212
单作农业(单一经营) 139
单作一熟型 213
单作制 152
胆甾醇;胆固醇 34
(香气)淡薄 120
淡橙色;淡橘黄 121
淡红黄色(晾烟) 225
淡黄(色,分级) 121
淡青色(分级) 120
淡色覆盖物 121
淡色烟(烤烟) 121
(烟叶)淡(色);弱(的);淡(的) 245

淡色烟叶(烤烟) 26
蛋氨酸 132
蛋白胨 161
蛋白酶 178
蛋白质 178
蛋白质氮 178
蛋白质合成 178
氮(素)过量 145
氮(素)过量症 145
氮(素)含量 145
氮(素)平衡 145
氮(素)同化作用 145
氮(素)吸收 145
氮(素)循环 145
氮沉降 145
氮代谢作用 145
氮的径流损失 192
氮的淋洗损失 145
氮的生物固持作用 23
氮对环境的影响 103
氮肥 144,146
氮肥优化管理 153
氮固定 145
氮固定;固氮作用 145
氮过量 73
氮回收 146
氮钾比 146
氮钾复混肥料 146
氮径流损失 146

氮矿化势 145
氮磷复混肥料 148
氮磷钾比例 178
氮磷钾肥 19,146,148
氮磷钾复混肥料 148
氮磷施肥制 164
氮硫比 146
氮素肥料;(有机)氮肥 145
氮素过量;氮盈余 146
氮素活度指标 145
氮素矿化固持过程 145
氮素矿物化 145
氮素流动 137
氮素生产力 146
氮素吸收 149
氮素营养(吸氮作用) 57
氮素营养失调 145
氮同位素比值分析 145
氮氧化合物 145
氮氧化物大气污染 145
氮用量;含氮率 148
氮源 146
氮周转 146
当地肥料 123
当地技术 105

当地品种 142
刀齿耙 113
刀式覆土器 113
刀式回转耙 113
刀式耙 113
导电性(传导性) 41
导电性;电导率 69
导热率 96
倒春寒 38,116
倒伏 123
稻草 189
稻茬麦 93
稻麦轮作 189
稻田 159,189
稻田生态系统 66
等高带状间作 43
等高耕作 42,43
等高灌溉 21,42
等高农地 43
等高梯田 43
等高线 43
等高线图 43
等高植生带 111
等高种植 43
等级;繁茂的;腐臭(烟叶);过于肥沃的;列;排 183
(烟叶)等级合格率 3
等级图 98
等位(基因)多样化 11

等位(基因)互补 11
等位基因连锁反应(分析) 11
等位基因特异性限制内切酶分析 11
等位染色体(的) 11
等温的;等温线(的) 111
等温线 111,227
等物候线 111
等物候线;等始花线 111
低残留 125
低残留农药 120
低产;歉收 172
低产田 125
低产土壤 125
低成分肥料 124
低成分肥料;低浓度肥料 125
低垂 63
低纯度复合肥 124
低档烟 125
低的;贫乏的;营养不足的;下部的 124
低地;洼地 124,125
低地;洼地;盆地 125
低地砖红壤 124
低毒农药 125
低害卷烟 120
低架厩肥撒施机 125

低焦油 125
低焦油卷烟 125
低能耗农业 125
低平地 124
低燃烧性 124
低山丘陵区 124
低生物碱品种 124
低纬度 124
低温处理 221
低温恒温器 47
低温菌;嗜冷细菌 178
低温冷害 33,38
低温室;冷藏室 44
低温诱导休眠(再度休眠) 125
低温灾害 50
低限水量(指植物生长所必需的) 2
低烟碱 124
低烟碱品种 124
低养分水平 124
低植物碱烤烟 114,124
滴滴涕 51
滴定标准液 229
滴定法 229
滴定分析法 229
滴定管 27
滴定零点 249
滴定曲线 229

滴定终点 70
滴灌 63
滴灌;滴灌系统 234
滴灌带 63
滴灌系统 63
敌百虫 58,234
<商>敌克松 77
底(部)脚叶 93
底层 26
(烟叶)底烘 176
底脚叶 26
底脚叶;沙土叶 194
底水灌溉 238
底土 26,237
底土;心土 221,236
底土改良 221
地表滴灌 223
地表径流 93,158
地表径流;地表流失 223
地表水 223,238
地表水耗损量;地表径流损失 223
地表温度 223
地带性规律 116
地带性土壤 249
地方品种 142
地方气候;局部气候 123
地方种;地方种族 123
地老虎 49

地理空间数据库 89
地理信息 89
地理信息科学 89
地理信息系统 89
地力保持 176
地力衰退 51
地貌类型图 115
地貌图 89
地面分辨率 93,151
地面覆盖;覆盖 140
地面灌溉;地表灌溉 223
地面漫灌 223
地面排水;明沟 223
地面排水;土地排水 114
地面气温;地温 93
地面施用(药) 93
地面水流;地面径流 223
地面温度;地面气温 222
地面信息 93
地面蒸发 73
地面植被 93
地面植被层 80,93
地膜覆盖 44
地膜覆盖栽培 80
地球化学循环 89
地球化学元素 89
地上部分 6

地上部分的生长 6
地上部与地下部比率 191
地上害虫 2
地势 223,226
地统计学 89
地统计制图 89
地图格网 216
地图数据库 30
地图数字化数据库 30
地温 65
地温计 65,89
地温梯度 89
地下给水;地下灌溉 221
地下灌溉 220,221,236
地下害虫 208
地下径流 237
地下排水 236
地下渗漏 236
地下水 93,165
地下水埋深 93
地下水侵蚀 93,192
地下水位 93
地下水污染 93
地下水硝酸盐污染 144
地下水质量标准 180

地下水资源 93
地下污染 237
地形测量 232
地形分布 232
地形判读(地物识别) 232
地形剖面图 232
地形图 232
地形图;地势图 186
地形要素 232
地形因素;地形因子 232
地中海气候 131
典范分析 29
典范相关分析 29
典型分布 29
典型相关 29
点播 57,99,214,215
点播;穴播 197
点估计 170
点灌 215
点克立格法 179
点燃 121
点源 170
点源污染 170
电导 69
电导池 41
电导池常数 41
电导法 69
电导分析法 41
电导率;导电性 69

· 267 ·

电导仪 41
电感耦合等离子发射光谱仪(ICP – AES) 105
电感耦合等离子体 105
电感耦合等离子体质谱分析法 105
电荷 31,69
电荷零点 170,249
电荷平衡的维持 188
电烘箱 69
电化学分析 69
电化学位 69
电极 69
电离 109
电炉 69
电热板 69
电热苗床 69
电位(电势) 69
电泳 69
电子传递链 69
电子扫描显微镜 69
电子扫描显微镜;扫描电镜 69,72
电子探针 69
电子探针诊断法 69
电子显微技术 69
电子显微镜 69
玷污的;污染的(烟叶) 216

淀粉 13
淀粉;(淀粉)浆;(上)浆 218
淀粉分解 13
淀粉合成 218
淀粉积累 218
淀粉降解酶 218
淀粉粒 218
淀粉酶 13
淀粉酶;淀粉酶制剂 57
淀粉水解酶 13
淀积层 103
淀积土壤 103
淀积作用 103
凋枯症状 246
凋落物(枯枝落叶) 122
凋萎点;凋蔫点;萎蔫含水量 246
凋萎系数;萎蔫系数 246
顶部坏死 232
顶部生长点生长停滞 188
(烤房)顶部通风 232
顶点;顶生组织;翅尖;顶端;叶尖 15
顶端分生组织 15
顶端花序 226
顶端坏死 4

顶端生长 15,226,229
顶端优势 15
顶喷式喷灌机 158
顶梢枯死 15
顶生的 46
顶生叶 4
顶芽 226
顶芽;蕾;苞;尖;端;滤嘴 229
顶芽;叶原 165
顶叶 92,225,229
订单农业 154
订单生产;分批生产 154
定氮仪 19
定根 148
定量测定 181
定量方法 181
定量滤纸 181
定苗 72,199
定苗距离;定苗间距 199
定色 38
定色期 38,161
定位氮素管理 202
定位农业(定点农业) 201
定位施肥 201
定位施肥管理 201
定位养分管理 80,202

定向变异 59
定向改良 154
定向培育 59
定向选择;定向育种 59
定向杂交;合理交配 111
定性方法 180
定性分析 180
定性滤纸 180
定植苗;栽植材料 169
定植穴 169
东方型香(烟叶);东方型香料 156
东方型烟(香料烟,土耳其烟) 156
冬播作物;冬性作物 247
冬耕 169,247
冬季的 97
冬季休眠 247
冬眠状态;冬眠滞育 98
冬闲地 247
动力耕耘机 175
动力镇压器 139
动态模型 64
动态平衡 64,99,112,137,200
动态特征 64

动物粪便 14
动物生态学 14
动物性废弃物有机肥 100
动物性杂肥;畜肥 14
动植物残体 167
冻干保藏法;冷冻干燥 125
冻害 50,84,85
冻害;冬害 247
冻害;霜害 85
冻害;霜害;霜冻害 85
冻结土壤 86
冻涝害 81
冻融 84
冻融过程 84
冻融交替 11
冻融交替(作用) 12
冻融侵蚀 84
冻融土壤 85
冻融作用 84
冻伤(霜害) 85
冻伤(烟叶) 85
冻伤;霜害 85
冻死;冻坏 247
冻死率 38
冻土;冻地 85
冻土效应 68
冻雨 85
陡坡 218

斗烟丝;烟斗用的烟丝 167
豆饼 21
豆科 119,178
豆科绿肥 119
豆科绿肥作物 119
豆科植物根瘤菌 119
豆科植物根瘤菌接种 119
豆科作物 119
豆类作物 178
豆香 21
毒性 233
独立样本的检验 104
度;程度;等级 52
短距离运输 200
短期轮作 200
短期天气展望 87
短日照 200
短日照植物 200
短日照作物 200
(香气)短少 200
短效农药 200
短暂休眠 226
断键水 26
锻苗 198
锻苗;蹲苗 95
堆(烟)房 159
堆垫 48
堆垛方法 167
堆垛高度 216

269

堆肥 40,64
堆肥;厩肥 21
堆肥;人造粪肥 16
堆肥发热 128
堆肥熟化 49
堆积(作用) 3,53
堆积变黄 27,166
堆积发酵 27,219
堆积烤房;密集烤房 27
堆积密度;体积密度,单位体积重量 27
堆积调制 39
堆制;堆沤;堆肥法 40
(土壤)对动物健康的促进 178
(土壤)对动物健康的影响 68
(土壤)对人类健康的促进 178
(土壤)对人类健康的影响 68
(土壤)对植物健康的促进 178
(土壤)对植物健康的影响 68
对甲基苯甲醛 178
对抗生物 15
对立假设 12
对硫磷 160

对数残差 123
对数模型 123
对数线性回归 123
对数正态分布 123
对照(物);控制;调节;操纵;管理 43
对照;核对 35
对照分析 32
对照品种 32,43
对照区 32,43
对照试验 32,43
对照样本 32
对照样品 43
对照植株 32
钝化(作用) 160
钝化剂 160
多变量 141
多成分肥料 140
多次抽样 141
多次耕翻 186
多次收获 141
多点试验 141
多段式 141
多分枝式 169
多酚类化合物 172
多功能性肥料 140
多光谱扫描系统 141
多行中耕机 141
多铧犁 87,128,140,141
多环芳烃 171

多级抽样 141
多菌灵 29,194
多抗(病)性 141
多抗霉素 172
多孔的 172
多孔团聚体 173
多孔性;孔隙度;透气度;孔性 173
多孔性;疏松性 172
多磷酸盐 172
多氯联苯 171
多氯联苯污染物 171
多年生牧草 161
多年生杂草 161
多年生植物 162
多羟(基)的 172
多区轮作 140
多熟;复种;多茬;种植 140
多糖 99,172
多糖类;聚糖 172
多萜(烯) 172
多维尺度分析 140
多维分布 141
多系杂交 171
多系杂交法 171
多项分布 140
多项式 172
多效霉素 175
多样性 61
多样性指数 61

· 270 ·

多叶的 119,169,170
多因素肥料试验 140
多因子试验 140,141
多雨的;洪水的;雨成的;雨期 170
多雨地区 170
多雨季节的假熟 243
多雨气候 182

多元(复合)肥料 141
多元分析(多变量分析) 141
多元回归 141
多元回归分析 141
多元酸 172
多元线性回归 141
多云天气 37

多种营养元素肥料 140
多重比较 140
多重相关系数 140
多重营养缺陷型 171
多作多熟型 140
多作一熟型 137
惰性腐殖质 105

E

Eh 值 69
EMP 途径 70
恶性循环 240
蒽酮 14,15
儿茶酚 30
二次抽样 62
二次发酵;再次发酵;后发酵 7,185
二次侵(感)染 196
二个独立样本检验 104
二个相关样本检验 185
二甲基吡咯 58
二甲基呋喃 58
2,5-二甲基呋喃 58
二氯二苯三氯乙烷 51
二氯甲烷 57

二氯乙烷 57
二嗪农 57
二氢大马酮 50
二氢猕猴桃内酯 58
二氢香豆素 58
二氢紫罗兰酮 58
二水磷酸二钙 57
2,4-二硝基苯酚 58
2,4-二硝基苯衍生物 61
二糖;双糖 59
二维分布 25
二维高斯分布 25
二维正态分布 25
二项分布 22
二项式检验 22
二氧化氮 145
二氧化硫 221

二氧化碳 29
二氧化碳当量 37
二氧化碳当量浓度 37
二氧化碳当量排放 37
二氧化碳肥料 37
二氧化碳固定(作用) 29
二氧化碳计 14
二氧化碳排放 37
二氧化碳施肥 29,37
二氧化碳同化作用 29
二氧化碳循环 37
二元(复合)肥料 235
二元变量 22
二元肥料 22

271

F

F-分布 76
发白烟叶 246
发病率 104
发病烟田 59
发光持续时间 90
发酵 77
发酵反应 77
发酵罐 77,111
发酵过程 77
发酵时间 77
发酵室 77
发霉烟叶 141
发射光谱仪 70
发生畸形的 226
(病虫害)发生预测 152
发香基团 152
发芽 216
发芽;萌发;增殖 178
发芽迟缓 53
发芽率 89
发芽能力 1,89
发芽前处理 175
发芽势 89,240
发芽试验 89
发芽室 190
发芽速度 89
发芽天数 51

发芽温度 89,226
发芽整齐 237
发芽周期 162
发育不良;畸形 127
发育不良;细胞减生的 102
发育不全 102,236
发育不全的 2
发育激素 19
发育期 56
发育受阻 103
发育条件 56
发育停滞 16
法定品种鉴定试验 152
翻茬 235
翻堆 27,166
翻垛 216,219,235
翻耕 170
翻耕土地 184
翻埋绿肥 169
翻入;耕翻 169
翻身耙;旋耕耙 235
翻土 191,235
翻转犁 140,190,235
翻转犁;转向犁 188
矾 239
繁茂;肥沃 183

繁殖苗圃 141
繁殖芽;花芽 186
反射率 185
反射因子 185
反硝化势 53
反硝化损失 53
反硝化细菌(脱氮细菌) 53
反硝化抑制剂 53
反硝化作用(脱氮作用) 53
返苗期;还苗期 198
泛酸 160
方差 239
方差分量 40
方差分析 14,239
方差图 239
方差与标准差 239
方格计数法 45
方格网 216
方形图 38
方阵 216
芳香;香味;香气;香料 84
芳香油类;精油 73
防潮的;防水的;抗湿的 138
防霉技术 139

防渗 198
防治效果 68
防治杂草;除草 245
仿制样品 44
放射性碳同位素 182
放射性同位素 182
放松;疏松（土壤）
   124
放线菌 5
非必需氨基酸 147
非必需元素 147
非参数检验 147
非持久性农药 147
非豆科绿肥 147
非放射性碳 147
非腐殖化物质 147
非腐殖物质 147
非复合的有机物质
   146
非共生固氮生物 17
非共生固氮作用 17,
   147
非共生微生物 17
非挥发性成分 148
非挥发性的 147
非挥发性类萜 148
非挥发性酸 148
非挥发性有机酸 148
非活性孔隙 236
非活性养分 103
非降解性污染物 147

非交换性钾 147
非交换性离子 147
非交换性阳离子 147
非交换性养分 147
非结合水 236
非结合态氧化物 236
非竞争性抑制作用
   146
非毛管孔隙 146
非毛管孔隙度 146
非酶促棕色化反应
   147
非酶催化分解 147
非黏结性土壤 146
非农用地 146
非侵染性病害 147
非生物环境 1
非生物降解 1,146
非生物性病害 1
非生物性挥发 146
非生物因素 1
非生物组分 1
非水解性氮 147
非水解性胡敏酸 147
(烟支中)非填充空间;
   孔隙体积 172
(养分)缺乏症状 52
非线性回归 147
非盐渍化的 147
非有机营养的 14
非有效性钾 146

非正常苗 1
非正常芽 1
非专化的;非专性的
   147
非专性吸附 147
非自由取样 147
非自由水 237
肥害 79
肥厚 159
(烟叶)肥厚;稍厚 81
肥厚(叶)；黑暴烟
   219
肥力 77
(土壤)肥力分类 77
肥力保持 127
肥力等级 77
肥力级差 91
肥力监测 77
肥力减退 58
肥力鉴定 77
肥力评价 77
肥力要素 77
肥力指标 77
肥料 77
肥料;粪便;施肥;厩肥
   128
肥料标识 78
肥料残效 186,221
肥料产量效应 248
肥料成分 78
肥料成分和配合式 78

273

肥料单位 79
肥料的产量效应 68
肥料的有效利用 68
肥料对环境的污染 78
肥料反应曲线 46
肥料分析 77
肥料分析式 77
肥料管理 78
肥料过量伤害 50
肥料合理使用准则 192
肥料价值 128
肥料经济学 67,78
肥料净重 143
肥料利用率 184,238
肥料流失 78
肥料配方 78
肥料配合比例 78
肥料品位 78
肥料缺乏 78
肥料溶解度 213
肥料三要素 228
肥料伤害 78
肥料烧伤 78
肥料试验 78,79
肥料试验设计 78
肥料输入 107
肥料水分测定 54
肥料田间试验 128
肥料投入 78

肥料污染 79
肥料效率 78
肥料效应 46,78,128
肥料效应;施肥反应 188
肥料效应函数 46,78
肥料盐害 78,79
肥料养分 78
肥料样品的采集与制备 38
肥料因子试验 75
肥料用量;施肥量 78
肥料有效性 78
肥料元素 78
肥料增效剂 78
肥料种类 79
肥料主要成分 127
肥料主要元素 127
肥水灌溉 79,128
肥田的农作物 44
肥土(沃土) 76,77,88
肥土层 76
肥沃的 73,77
肥沃的可耕地 177
肥沃黏土 189
肥沃土壤 77,177,189
肥效 68
肥效;肥料效应 128
肥效模型 78
肥效试验 128

肥源 78
废弃物肥料 203
废物变能源 185
废物处理 241
废物回收 241
废物再利用 184
废物资源化技术 187
分辨率 187
分布函数 60
分部位烟叶 117
分层抽样 219
分层施肥 116,199
分次施用 215
分根(分株) 190
分根法 215
分根技术 215
分根培养 215
分根培养试验 215
分根试验 215
分根装置 215
分光光度法 215
分级 91
分级沉淀 84
分级和整理 91
分级练习 91
分级烟叶样品 91
分级因素 91
分级指导 91
分解 51
分解(代谢)产物 30
分解产物 51

分解程度 51
分解代谢 30,112
分解过程 51
分解者 51
分解蒸馏 54
分类 30
分类;分级 36
分类法 30
分类轴;尺度轴 30
分离 111,198,199
分离器 199
分离系数 199
分裂;劈(烟茎);劈(叶);丝束分裂 215
分馏;分馏作用 84
分泌物(渗出物) 196
分配(定)律 116
分批发酵 21
分期采收 216
分期施肥 232
分期追肥 232
分散度 52
分散剂 60
分散调查法 61
分散系数 37
分生孢子 41,215
分生组织 88
分析报告 14
分析纯试剂 14
分析数据 14

分析天平 14
分析仪器 14
分析质量控制 14
分液漏斗 199
分钟 136
分株 167
分子大 138
分子量 138
分子农业 138
分子微生物生态学 138
分子形状 138
分组 21,93
分组代号 93
分组取样 37
芬芳;芳香的;美味的 53
酚酞 163
焚烧 104
焚烧炉 104
粉(沙)粒 201
粉(沙)质沙土 201
粉块(粉状) 178
粉沙(土) 130
粉沙颗粒 201
粉沙黏土 201
粉沙黏壤土 201
粉沙壤土 201
粉沙土;粉粒土 201
粉碎(作用);研磨(作用) 178

粉碎;磨成粉状;喷雾;松土 178
粉碎;磨细 39
粉碎机 39,45,60
粉碎机;喷雾器 179
粉碎机;砂轮机;研磨机 93
粉状肥料 175
粪 68,77,154
粪;粪肥 64
粪;粪肥;腐泥土;腐殖土 140
粪便污染 77
粪池;粪坑 128
粪堆 128,140
粪肥 77
粪坑;积粪窖 128
粪粒团聚体 44
粪水,厩液 122
丰产田 99
丰富;充足(指香气);多(油分)丰满;身份充实 189
丰满;成熟度好;完熟度;熟透的;松软的;肥沃的(指土壤);柔和 132
(香气)丰满 2
(香味)丰满 245
(叶态)丰满;身份充实 86

丰缺　221
丰收　97
丰收;多实;多产　86
丰收;丰产　115
丰收;高产　96
丰水　2
丰水年　182,246
风成土　72
风分　170
风分室;风分箱　170
风干　10
风干土　10
风干样品　10
风干重　10
风干状态　10
风害　50
风害;风灾　246
风化残积物　245
风化产物　245
风化过程　245
风化淋溶作用　245
风化强度　245
风化侵蚀　245
风化速率　245
风化特征　245
风化作用　245
风积　247
风积的(风成的)　72
风积过程　6
风积土　6,247
风积物　6,247

风力　246
风力堆积　6
风磨烟叶　246
风伤(俗称风磨);风致伤(分级);风害　246
风蚀(作用)　6,52,72,246,247
风选　10,170,247
风灾　59
封闭水　151
疯长　73,246
峰顶　45
峰度　160
峰度(陡度)　113
峰度系数标准误差　217
峰值　160
峰值;峰顶;顶点;波峰;螺纹牙顶;牙顶;齿顶　45
呋喃　87
呋喃甲醇;糠醇　87
2-呋喃甲醛　87
6-呋喃甲基腺嘌呤　112
孵化箱　104
弗吉尼亚型烟;烤烟　240
伏旱　222
氟　82

氟化物　82
氟污染　171
<商>福尔马林;甲醛水溶液　83
浮青　222
辐射　181
辐射率(辐射度)　181
辐射热　181
辐射育种　181
辐照度　110
辅酶　37
腐败;腐烂　179
腐败发酵　179
腐败气味　179
腐解过程　51
腐烂;衰变;衰减　51
腐烂味　179
腐霉属　179
腐生细菌　132
腐生线虫　194
腐生营养　132
腐蚀;侵蚀　72
腐熟促进剂　40
腐熟堆肥　86,129
腐熟腐殖质;细腐殖质;混土腐殖质　140
腐熟厩肥　51,191
腐心症(缺硼)　96
腐殖　100
腐殖化　100

· 276 ·

腐殖化(作用) 100
腐殖化程度 52
腐殖化过程 100
腐殖化系数 100
腐殖化有机物质 100
腐殖煤;褐煤;腐质煤 100
腐殖酸;胡敏酸 100
腐殖酸铵 13
腐殖酸复合肥 100
腐殖酸钾 174
腐殖酸磷肥 164
腐殖酸盐 100
腐殖土 140,239
腐殖土;腐殖质 100
腐殖物质 100
腐殖质 100
腐殖质层 100
腐殖质层(淋溶层;A层) 1
腐殖质储量 100
腐殖质的木质素-蛋白质学说 121
(暗色)腐殖质淀积层 213
腐殖质分级 100
腐殖质富集层 100
腐殖质含量 100
腐殖质积累作用 100
腐殖质类似物 100
腐殖质黏粒复合体 100
腐殖质品质 100
腐殖质土壤 140
腐殖质土壤;腐殖土 100
腐殖质形成过程 100
腐殖质营养学说 227
腐殖质组成 100
腐殖质组型 100
负电荷 143
负激发效应 143
负相关 143
附生微生物 72
复耕 196
复耕;再耕 196
复合的有机物质 40
复合肥料 38,40
复合化肥;复合肥料 40
复合假设 40
复合试验;复因子试验 40
复合土壤结构体 40
复合团聚体 7,40
复合污染 38,40,111
复合胁迫 40
复混肥料养分分析 149
复极差检测法 141
复烤 184
复烤厂 118
复烤烟叶 184
复因子 40
复杂性 40
复种(制) 221
复种轮作 141
复种面积 141
复种模式 141
复种指数 46,140
复种制 141
副处理 221
副芽;次生芽 196
傅里叶变换 84
傅里叶变换红外光谱(法) 84
傅里叶变换 109
傅里叶值 84
傅里叶变换红外光谱仪(计) 84
富啡酸(黄腐酸) 86
富含有机质土壤 155
富积层 70
富集 70
富集系数 70
富钾植物 174
富铝化(作用) 11
富马酸;反式丁烯二酸;延胡索酸 86
富碳有机废物 30
富营养化 73,158,236
覆层;表层 158
覆盖;埋藏 103

覆盖;遮掩;覆盖物;覆盖面 158
覆盖薄膜 140
覆盖层 158
覆盖层;覆盖物 140

覆盖稻草 44
覆盖机;覆盖层 140
覆盖物 140
覆盖效应 44

覆盖栽培 44,140
覆盖作物 44
覆盖作物;保护作物 149

G

(土壤)改良;改善 132
改良措施 12,103
改良地膜覆盖 103
改良剂(调节剂) 138
改良品种 103
改良效应 12
改良种系;良种 103
改性磷肥 138
改造中低产田 104
钙 28
钙肥 28
钙积土 28
钙镁比 28
钙镁磷肥 28,87
钙素失调症 28
钙调素 28
钙质土 28,121
盖玻片 44
概查 184
概率 177
概率抽样 177

概率分布 177
概率分布的标准差 217
概率分布的方差 239
概率分析 177
概率密度函数 177
概率图 177
概率误差 177
甘氨酸 90
甘油磷脂 164
甘油酸 90
甘油吸附法(测比表面) 90
坩埚 46
坩埚钳 46
杆腐病;梗腐病 170
感病品种 223
感病性 59
感官分析 199
感官检验;感官评定 156
感官品质 199

感官评定 199
感官评价 156
感官特征 156
感光分解作用 164
感光期;光照阶段;光阶段;(植物发育)光期 164
感觉 199
感应;响应;反应 188
干草气息;药草香 97
干草香 96
干沉降 63
干法 63
干腐;茎腐 218
干果香型 62
干旱 63
干旱;干燥性;干旱性 16
干旱地区生态系统 16
干旱壤质沙土 63
干旱水分状况 16

干旱土壤 63
干旱灾害 63
干旱周期 63
干季 63
(调制过程中的)干筋 112
干筋期 161,218
(烟气)干净 36
干馏 179
干馏的 179
干牛粪 62
干禽粪 62
干球温度 63
干扰 108
干扰组分 108
干热水分状况 232
干烧法 63
干湿交替 246
干湿交替(作用) 12
干湿球湿度计 246
干湿球温差 246
干湿球温度差 57
干湿球温度计 63
干湿效应 246
干土效应 68
干物质 62,63
干物质产量 63
干物质分配率 63
干物质积累 63
干羊粪 62
干叶重 48

干燥 16,54,63,64,200,248
干燥、缺油的烟叶 234
干燥程度 64
干燥的 16
干燥的;干旱的 247
干燥法(测水分) 54
干燥感 64
干燥器 54
干燥室 112
干燥箱 63
干燥指数 104
干燥装置;干燥设备 63
干重 63,138
岗地 140
高 98
高;海拔;顶点 97
高半胱氨酸 99
高产 98
高产高效 99
高产粮区 99
高产农业 177
高产施肥技术 78
高产田 98,99
高产稳产 98
高产稳产田 94,216
高产优质 99
高产栽培 98
高产栽培技术 98

高产栽培配套技术 43,225
高成分肥料 98
高程地图 102
高纯试剂 74
高低图 98
高度计(高程计) 12
高度显著性 99
高度相关 98
高发病率 98
高肥力 98
高分辨率凝胶成像系统 98
高分解有机层 99
高腐有机土壤物质 194
高秆品种 124
高光谱传感器 102
高光谱数据 102
高光谱遥感 102
高级脂肪酸 98
高架喷灌系统 158
高岭石 112
高岭土 112
高垄;高畦 98
高氯烟叶 193
高锰酸钾 174
<商>高灭磷 3
高浓度肥料 98
高浓度复合化肥 98
高浓度培养基 98

· 279 ·

高山地区 98
高山气候 11
高山土壤 98
高丝氨酸 99
高斯分布 88
高速离心机 99
高速喷雾器 9,10
高温 132
高温发酵 98,100
高温发酵法 99
高温分解 179
高温干燥;烤焦 195
高温灭菌室 160
高温期 227
高温气候 132
高温伤害 98
高温胁迫 98
高温蒸馏 179
高温蒸馏的 179
高效持续农业 98
高效集约化利用 98
高效农药 99
高效生态农业 98
高效液相层析;高效液相色谱(法) 98
高效液相色谱－质谱仪 98
高压灭菌法 18
高压灭菌锅;高压灭菌器 18
高压蒸汽灭菌器 98

高营养水平 98
高原气候 169
割草机 140
隔行栽培 202
隔离沟 111
隔离行;保护行 111
隔离田 111
隔离栽培;隔离培养;离体培养 110
隔热材料;保温材料;热绝缘体 96
镉(Cd) 28
镉污染 28
镉污染肥料 28
个体发育;个体发生 153
个体发育;个体发展 105
个体生态学 18,103
个体植株;单株 105
个体株型 16
各营养元素间的相互作用 108
各组质地特征 31
铬 34
铬酸钾 173
铬污染 34
根 190
根表面积 190
根层上部灌溉 110
根产物 190

根的吸收力 190
根端 190
根腐 190
根腐病 190,191
根腐病;基腐病 83
根冠 29,167,190
根冠比 191
根冠层 46
根冠腐烂 46
根黑腐病 25
根际 188
根际 pH 163
根际淀积 188
根际动态 188
根际技术 189
根际酸化(作用) 188
根际调控 188
根际微生态系统 188
根际微生物 133,188,189
根际微生物区系 188
根际微生物区系;根围区系 188
根际细菌 188
根际效应 188
根际修复 188
根际营养;根圈营养 188
根尖 190
根结 191
根结(病) 113

根结病;根瘤病 190
根结线虫 191
根结线虫病 190
根结线虫属 132
根茎;初生主根 190
根茎;地下茎 188
根茎病害 190
根颈 190
根径 190
根连续洗涤法 199
根瘤 146,190
根瘤固氮法 145
根瘤菌 146,188,190
根瘤菌肥料 188
根瘤联合共生(根际联合作用) 188
根瘤形成(作用) 146
根毛 190
根毛层 167
根毛密度 190
根毛区 190
根密度 190
根面;根表 188
根内微生物 133
根培养 190
根区 190
根圈灌溉 159
根群 190
根容重 190
根深 53
根生长受抑制 106

根室法 189
根体积 190
根土比 191
(微生物)根土比 181
根土比率 189
根－土界面 108
根土界面 191
根外施肥 74,75,157
根外施肥;根外营养 74
根外追肥;叶面施肥 83
根外追肥;叶面施肥;叶面喷洒 83
根围 190
根系 182,190
根系;根丛 143
根系发育 190
根系分布 190
根系分泌物 190
根系干重 190
根系构型 190
根系呼吸 190
根系活力 190
根系交织 129
根系伸展 190
根系生物量 24
根系生物学 181,190
根系生长 190
根系生长空间 190
根系提水作用 101

根系体积 241
根系吸收面积 190
根系形态学特征 139
根系阳离子交换量 30
根系长度 190
根鲜重 190
根压 190
根域 190
根长 190
根长密度 190
根状;根形 182
庚二酸 167
耕层;耕作层 169
耕地 169
耕地;可耕地 169
耕地;犁;投资 169
耕地;犁地 169
耕地;农田 46,47,76,169
耕地保护 47,76
耕地等级 91
耕地非农化 146
耕地合理利用 183
耕地流失 47
耕地面积 8,16,47
耕地培育 227
耕地生产力 47
耕地使用过度 158
耕地熟化过程 79
耕地数量 47,181

· 281 ·

耕地用养结合原则 177
耕地资源 47,76
耕地总量 158,232
耕翻;翻入土内 235
耕翻前撒施 26
耕翻上来;耕翻出土 235
耕翻作业;整地作业 228
耕宽 229
耕埋厩肥 28
耕期 169
耕深 47,53,170,229
耕深调节装置 53
耕心土;耕底土;深耕 221
耕性(可耕度、熟化度) 47
耕性良好的土地 114
耕种;耕耘 228
耕种过度;滥垦;过度开垦 157
耕作 76,228
耕作;耕层 229
耕作;犁地 170
耕作层 16,47,48,169,170,228
耕作淀积层 8
耕作动力学 228
耕作防治 48

耕作机械 228
耕作技术 76
耕作强度 228
耕作侵蚀 228
耕作试验 48
耕作土 48
耕作土壤;熟土 47
耕作效果 228
耕作制 16,76,228
耕作制度 46
梗丝 49,189
梗中带叶率 114
工厂化育苗 75
工程侵蚀 70
工业污染 105
工业用腐殖酸盐 39
工艺成熟 225
工艺成熟(度) 225
工艺过熟 225
工艺品质 225
工艺质量 225
工作曲线 247
公共因子 39
公害 178
功能性避病 113
功能养分 86
汞 132
汞污染 132
共存 37
共固定化作用 37
共生(现象) 39

共生固氮菌 223,224
共生固氮生物 223
共生固氮微生物 224
共生固氮作用 223
共生关系 67
共生互利 223
共生微生物 223
共生现象 224
共生营养 224
共质体运输 224
供给不足 107
供给能力 222
供料罐 222
供水量 13
供水性 244
供需缺口 87
供应不足 107
供应限制因素 222
沟;畦;垄沟;槽;开沟 87
沟播 87,214
沟播地中耕机 122
沟播机 122
沟灌 60,87
沟金针虫 169
沟垄耕作 87
沟施;条施 87
沟栽;畦沟移栽 234
沟植 87
沟作 234
枸不溶性磷 35

构溶性磷　35
构溶性磷肥　35
构筑梯田　226
购进验查　179,184
估计　72
估计量的偏差　22
谷氨酸　90
谷氨酸合酶　90
谷氨酸盐　90
谷氨酰胺　90
谷氨酰胺合成酶　90
谷氨酰胺酶　90
谷底田　239
谷地　239
谷胱甘肽　90
谷胱甘肽过氧化物酶　90
骨粉　25
鼓风干燥箱　62
鼓风弥雾机　137
固醇　219
固氮共生体　146
固氮共生现象　145
固氮过程　177
固氮活力　145
固氮菌　19
固氮菌肥料　19
固氮菌剂　19
固氮酶　146
固氮能力　146
固氮生物　57

固氮速率　145
固氮体系　57
固氮微生物　145
固氮细菌　19,57,146
固氮藻类　146
固氮者　145
固氮植物　146
固氮作用　19
固定　81,103
固定(的)大气氮素　81
固定;(烤烟过程)定色;卷烟纸、水松纸的上色处理　81
固定成本　81
固定态铵　81
固定态钾　81
固定性　103
固态发酵　212
固碳　29
固体部分　213
固体残渣　213
固体肥料　213
固体废弃物处理　213,234
固体废物　213
固相　213
固相容积　213
固有有机质　142
瓜氨酸　35
挂(烟)密度　95

挂竿;穿绳;绑烟　220
挂竿复烤　219
挂竿烤房　219
挂灰(叶)　195
挂灰;糊片;蒸片　21
挂灰烟叶　195
挂烟层数　95
挂烟杆　228
挂烟杆导轨　228
挂烟杆数　148
挂烟工　95
挂烟工具　118
挂烟夹　230
挂烟架　228
挂烟间隔　60
挂烟重量　95
挂叶　95
怪味;杂气　152
关联分析　17
关联系数　17
关系式数据库管理系统　185
观测值　151
官能团　86
管架式调制棚　167
管式加热器　167
管式排水　167
冠;根茎;副(花)冠　46
惯例　49
灌溉　110

· 283 ·

灌溉(用)水 110
灌溉(周)期 244
灌溉标准 110
灌溉地;水浇地 110
灌溉定额 110
灌溉方式 110
灌溉工程;灌溉计划 110
灌溉过度 158
灌溉间隔(距) 110
灌溉农田 110
灌溉农业 100,110
灌溉侵蚀 110
灌溉渠 110
灌溉设备 110
灌溉施肥;肥水灌溉;加肥灌溉 77
灌溉时间 110
灌溉水田 110
灌溉水质标准 243
灌溉网 110
灌溉系统 110
灌溉效率 110
灌溉需水量;灌水定额 110
灌溉需水指数 104
灌溉需水总量 64
灌溉原则 110
灌溉制度 110
灌溉作物 110
灌水次数 149

灌水工具 110
灌水技术 110
灌水量 110
灌水频率 110
灌水洗盐 241
罐发酵法 225
光(周)期;光照期;日长;日照长度 164
光饱和点 121
光补偿点 120
光促进发芽的种子 121
光催化降解 164
光电分光光度计 164
光度计 125,164
光发芽;光萌发 121
光反应 121
光害 120
光合的;光合作用的 165
光合功能 165
光合活性 165
光合磷酸化作用 164
光合强度 165
光合生产率 165
光合细菌 165
光合效率 165
光合有效辐射 165
光合自养生物 165
光合作用 164
光合作用的光化系统 164

光合作用过程 165
光合作用率;光合速率 165
光合作用强度 104,107
光合作用系统 165
光合作用抑制因子 164
光呼吸作用 164
光滑(叶):代号 S 202
光滑(叶组);光滑叶 202
光滑烟叶 202
光化学降解 164
光降解膜 164
光解;光(分)解作用 164
光解作用 164
光利用效率 68
光敏色素 164
(植物)光敏素;(植物)光敏色素 166
光能利用率 120
光能自养生物 165
光谱 153
光谱反射 215
光谱范围 215
光谱分析法 215
光谱辐射率;光谱辐射强度 215

284

光谱曲线 215
光谱数据 215
光谱数据库 215
光谱特性 215
光谱图 215
光谱吸收 214
光谱响应 215
光强度 120
光同化 164
光温度指数 121
光温生产潜力 120
光线 121
光胁迫 121
光学性质 153
光叶紫花苕子 240
光抑制 164
光抑制发芽种子;暗发芽种子 120
光诱导;光感应 164
光诱导周期 164
光泽 26
光泽(雪茄烟叶) 125
光泽暗;(叶色)很暗 58
光照处理 121
光照气候 120
光照强度 107,125
光照栽培 103
光周期反应 164
光资源 121
胱氨酸 49

广谱性除草剂 26
广谱性杀虫剂 26
广谱性杀菌剂 26
归还 188
归纳分析 222
归一化处理 148
归一化法 148
归一化植被指数 148
硅 201
硅肥 201
硅胶 200
硅铝层 192,200
硅酸盐分解细菌 201
硅酸盐细菌(钾细菌) 200
硅酸盐细菌肥料 201
辊子;卷筒;卷(雪茄)外包皮工;滚压器 189
滚刀式切丝机 190
国产烟叶 62
国际标准化组织(ISO) 108
国际单位制(SI) 108
国家标准 142
国家烟草栽培生理生化研究基地 142
国家烟草质量监督检验中心 33,47
国家烟草专卖局 219
果胶 161

果胶物质 161
果聚糖 86
(水)果酸 86
果糖 86
果香 86
过饱和溶液 222
过饱和水 243
过大烟叶;憨烟 157
过度放牧 73
过度湿润(过湿) 73
过度收获 158
过渡土壤;中间土壤 108
过渡性土壤 108,233
过量 73
过量施氮 73
过量营养 222
过磷酸 162,222
过磷酸钙 28,121,222
过磷酸钙(普钙) 148,201
过磷酸钙肥料 222
过密的 158
过黏土壤 157
过热;加热过度 157
过剩水分 223
过剩与不足 73
过熟 158
过熟枯萎的烟叶气味 157
过氧化氢 101

· 285 ·

过氧化氢酶 13,17,112
过氧化氢酶(触酶) 30
过氧化物酶 162
过早开花的;过早熟的 175
过早萌发;提前萌发 175

H

海拔 2,69
海绵组织 215
海泡石 195
害虫爆发 162
害虫发生代数 229
含氨基酸类肥料 79
含铵肥料 13
含氮化合物 145
含腐殖酸类肥料 100
含梗率 218
含梗率的测定 55
含钾矿物 174
含硫氨基酸 222
含氯化肥 34
含沙量 196
含水率 161
含水率;含水量;水分 138
含水率;水分含量 242
含水特征 31
含碳物质 30
含盐的 192
含盐度 193
含氧官能团 159
函数值 86
寒潮 38
寒害;冷害;冻害 38
寒流 38
旱斑;干斑 63
旱成土 248
旱地 63
旱地;高地;山地 238
旱地;高地田 238
旱地耕作制 76
旱地金针虫 64
旱地农业 64
旱地土壤;高地土壤 238
旱耕地,旱作农田 63
旱情监测 63
旱生植物 247
旱田杂草 238
旱灾 63
旱作农田 63
旱作农业 63
行;排 191
行边条施 20
行播;条播 214
行间撒施 26
行间中耕 108
行间中耕机 191
行间作物 191
行距 60,191,214
行内 247
行施(肥料) 191
行为信号 21
行政区划图 6
行株距 60
航空测量(航空调查) 10
航空摄影调查 7
航空遥感 6
航天遥感 214
好光性种子 121
好气反硝化作用(好气脱硝作用) 7
好气固氮细菌 7
好气培养 7
好气性处理 7
好气性发酵 7
好气性分解 7

好气性细菌　6
好气性种子　7
好氧降解　7
好氧条件,好气条件　7
好氧微生物　7
好氧微生物(好氧菌)　6
耗地作物　45
耗竭法　53
耗竭土壤　74
耗水量　242
禾本科绿肥　91
禾本科作物;谷类作物;谷香　31
(烟)盒　159
合成　224
合成氨　224
合成代谢;同化作用　13
合成氮肥　224
合格品种　3
合格质量水平　3
合格种子;认可种子　31
合理灌溉　6
合理开发利用　183
合理轮作　68,178,183
合理密植　183
合理施肥　6,178,183
合同制农业　43

(烟气)和顺;(烟气)柔和;光滑化;平整　203
(烟味)和顺;醇和　135
河床冲积物;河滩荒地　189
河谷　189
河流侵蚀　82,189
河南农业大学烟草学院　231
河滩地　81
核磁共振　148
核磁共振波谱法　148
核苷酸　148
核酸　148
褐斑叶;蒸片　119
褐煤　100
褐片(烟叶)　50
褐色;棕色;褐色的;棕色的　27
褐色斑点　27,34
褐色坏死部分　27
褐土　35,62
褐糟叶　50
黑暗发芽器　50
黑暴烟　219
黑钙土　33
黑胫病　25
黑胫病;黑脚病　25
黑垆土　97

黑麦草　192
黑麦草;多花黑麦草　123
黑炭　25
黑土　25,163
黑云母　25
黑糟叶　50
痕量成分　233
痕量分析　233
痕量金属污染　233
痕量气体　233
痕量污染　233
痕量元素;微量元素　233
很容易生物降解的　99
恒温的　227
恒温干燥箱　111
恒温恒湿箱　41,227
恒温烘箱　42
恒温烘箱法　42
恒温器　227
恒温器;恒温箱　29
恒温调节器　227
恒温箱;培养箱;孵化箱;细菌培养器　104
恒温浴　41
横坡　234
横坡耕作　5
横向的;杂交;杂种

46
横坐标 2
烘;烤 219
烘干 157
烘干法 157
烘干土 157
烘干样 157
烘干重 157
烘烤 82
(烟叶)烘烤性能 48
烘烤;焙烧;干燥 20
烘烤技术 82
烘烤速度 49
烘烤组件 49
烘丝机 49
烘箱 157
烘箱;干燥炉;干燥箱 64
烘箱;加热炉 96
烘箱;烤箱 10
红褐色;(烟叶颜色)赤黄 126
红花烟草 144
红黏土 184
红壤 113,184
红壤;红土 184
红外光谱 106
红外烘烤;红外焙烤 106
红外吸收光谱 106
红外线 106

红外线(的) 103
红锈(缺铜,烟叶斑点病) 184
红棕色 184,185
(烟叶)红棕色 184
洪积土壤 58,158
洪积物 58,178
洪水 81
后茬作物;后作 221
后处理 7
后代;结果;幼苗;后裔 152
后发酵 7,185
后期 14
后期感病 116
后期抗病 116
后熟 7,173
后熟种子 7
后天抗病性 4
后味 7
后味;余味 187
后效 7,186
后作 7
后作;后茬作物 83
厚薄不一 237
厚的;稠密的;浓的 227
厚叶(分级) 96
厚叶的 159
呼吸代谢(作用) 188
呼吸根;气孔 170

呼吸强度 187
呼吸商 187
呼吸系数;呼吸商 37
呼吸作用 187
胡萝卜素;叶红素 30
胡敏–富啡酸 100
胡敏素(HM) 100
胡敏酸–富啡酸比(HA/FA) 100
胡敏酸组分 100
胡桃醌;胡桃酮 111
琥珀酸 221
互补离子 39
互补离子(补偿离子) 39
互补作用 39
互惠共生(现象)39
互惠共生;共栖;共生性 141
互惠共生;共生 141
互生叶序 11
互作 108
花(序)芽 81
花(叶)黄素;植物黄素 247
花,开花 25
花斑叶 139
花萼 29
花而不实 81,82
花粉 170
花粉败育 170

288

花粉采集器　171
花粉培养法;花粉培养　171
花粉气　171
花粉刷　170
花岗岩　91
花梗　161
花冠　44
花冠长度　120
花蕾初期　82
花蕾脱落　82
花期天数　51
花色　82
花生饼　160
花生酸　16
花香　81
花序;开花期　105
花芽;花蕾　82
花芽分化　81,82
花叶　81
花枝　82
花轴　81,82
花轴长度　82
铧式犁　140
滑坡　115,202
化肥　32
化肥污染　32
化肥污染;肥料污染　78
化感作用　11
化能自养生物　33

化学沉淀　32
化学除草剂　32
化学纯　32
化学氮肥含氮量测定　54
化学毒物　33
化学反硝化作用;化学脱氮　32
化学防除杂草　33
化学防御　32
化学分解　32
化学风化　33
化学固定(作用)　32
化学剂打顶　33
化学钾肥含钾量测定　54
化学碱性肥料　33
化学降解　32
化学浸提技术　32
化学磷肥含磷量测定　54
化学灭菌　32
化学农药　32
化学侵蚀　32
化学杀虫剂　32
化学生长调节剂　32
化学试剂罐;药剂桶　32
化学酸性肥料　33
化学提取　32
化学吸附　32,33

化学信号　32
化学修复　32
化学抑芽　32
化学引诱物;化学吸引物　33
化学元素　32
化学制品污染　32
化学中性肥料　33
坏死　143
坏死斑(枯斑)　143
坏死花叶病　143
坏死陷斑　143
还原糖　184
还原性三羧酸循环　185
环斑病毒　189
环割　189
环沟施肥　35
环节动物　14
环境　71
环境背景值　71
环境承受力　71
环境恶化　71
环境净化机制　71
环境可持续性　71
环境空气质量　12
环境空气质量标准　12
环境评价　71
环境容量　71
环境生态学　18

环境生物技术 71
环境湿度 71
环境适应性 81
环境衰退;环境破坏 71
环境挑战 71
环境条件 12
环境退化 71
环境温度 71
环境问题 71
环境污染 71,171
环境污染评价 71
环境污染损失 71
环境污染物 71
环境污染综合防治 107
环境效益 71
环境修复 71
环境烟气 73
环境因子 71
环境友好 71
环境友好型可持续发展技术 71
环境有毒化学物质 71
环境有效性 71
环境指示物 71
环境质量 71
环施 35
环形施肥 35
环氧乙烷 72

缓冲(缓冲剂) 27
缓冲(作用) 27
缓冲能力 27
缓冲容量(缓冲力) 27
缓冲体系 27
缓岗地 189
缓坡 89
缓坡梯田 91
缓释 202
缓释氮肥 202
缓释方法 202
缓释肥(长效肥) 202
缓释肥料;缓效肥料 202
缓释技术 202
缓释型农药 202
缓释有机氮肥 203
缓效氮肥 202
缓效钾 202
缓效农肥 202
缓效性养分 202
换气;通风 6
荒地 241
荒地土壤(未垦地土壤) 85
荒漠化 54
荒漠化过程 54
黄带浮青;(烟叶)带青色;微带青 92
黄地老虎 73,235

黄多青少 139
黄瓜花叶病 37,47
黄瓜花叶病毒 47
黄褐土 248
黄红壤 248
黄花烟 192
黄花烟草 142,144,192
黄花烟草亚属 220
黄化现象 7,247
黄绿相间条纹 11
黄壤 247,248,249
黄壤;黄土 248
黄色烟草;浅色烟草;淡色烟草
黄土 123
黄土高原 123
黄土性土;黄绵土 123
黄萎病 240
黄烟组 248
黄棕壤 248
灰 17
灰斑病 92
灰分(含)量 17
灰分比;含灰率 17
灰分测定 17
灰分分析 17
灰分元素 17
灰化 17
灰化层 170

灰化黄壤　248
灰化土　170
灰化土;灰壤　170
灰化砖红壤　170,184
灰化棕壤　170
(烟叶)灰黄(色,分级)　159
灰漠钙土　200
灰漠土　93
灰壤的　170
灰色　38
灰色关联分析　92
灰叶　92
灰棕漠土　93
挥发碱　241
挥发酸　241
挥发性成分　241
挥发性有机酸　241
挥发性有机污染物　241
回潮　138
回潮;加湿　138
回潮;加湿;烟叶湿润;排列次序;次序关系;调整　154
(烟叶)回潮;品级;品味(指烟叶含水率);秩序;级;次;次序;等级　154
回归;退化　185
回归方程　185

回归分析　14,185
回归函数　185
回归模型　137
回归设计　185
回归系数　37,185
回归系数的显著性　200
回归线的置信区间　41
回归正交设计　185
回交(轮回)亲本　184
回收率　184
茴香苷　83
毁林　52
毁灭性病害　54
毁灭性灾害　54
混耕犁　116
混合　137
混合的;混杂;混级(烟叶)　137
混合堆肥　37
混合肥料　137
混合类肥;复合类肥　40
混合侵染　137
混合侵蚀　137
混合取样　40
混合施肥　137
混合土样　40
混合型卷烟　25
混合育种;群体育种;

大群繁育　129
混合指示剂　137
混级　91
混色　137
混杂种子　3
混组(烟叶)　137
混作　137
(香气)浑厚　86
活动温室　137
活力单位　5
活体培养　48
活性成分　5
活性腐殖质　5
活性孔隙　5
活性磷　113
活性酸度　5
活性炭　5
活性炭纤维　5
活性物质　5
活性养分　5,137
活性乙酸途径　5
活性有机磷　113
活性有机质　5
活性组分　5
(根系)活跃吸收面积　5
活跃性微生物区系　5
火成岩　103
(烤房)火管;烟道　82
火管烘烤(的);烘烤(的)　82

291

火管烘烤;火管烤制;
烘烤 82
火管烤房;炕房;烤房
82
火山灰 241
火山灰土 241

火山灰土(暗色土)
14
火山土 241
火焰发射分光光度计
81

火焰分光光度计 81
火焰光度法 81
火焰光度计 81
火焰光度检测器 81
获得抗病性 4

J

机会成本 153
机械采收 131
机械防御 131
机械耕作 131
机械化采收(机械化收获) 131
机械化堆肥 131
机械化灌溉 131
机械化集约农业 131
机械化农业 131
机械化生产 131
机械化栽培;机械化耕作 131
机械接种;摩擦接种 131
机械式烤房 131
机械损伤(分级) 131
机械损伤(分级) 131
机械通风干燥 131
机械移栽 126
机械阻力(机械阻抗)

131
(土壤)机械组成 131
(土壤)机械组成分析 131
机械组织 131
机械作业 131
鸡粪 33
积肥 128
积累率 183
积累植物 3
积年流行病害 172
积水;水涝 242
积温 3,48,226
姬蜂科;地老虎寄生姬蜂 102
基本电荷 69
基本发生层(基本土层) 129,177
基本农田 21,176
基本农田保护 176
基本微结构 69

(植物生长的三种)基本温度(即最低,最高,最适) 30
基本组分 69
基肥 21,93
基肥;底肥 21
基内菌丝 221
基施 21
基因表达 88
基因操作 88
基因档案 89
基因定位 88
基因多样性 88
基因复制;基因重复 88
基因工程学 88
基因互作 108
基因加性作用 5
基因库;基因源 88
基因库材料 88
基因免疫接种 89

·292·

基因替代　88

基因污染　89

基因芯片杂交系统　88

基因型　89

基因修饰细胞;基因改造细胞　89

基因修饰烟草;基因改造烟草　89

基因修饰植物;基因改造植物　90

基因重组　88

基因转换　88

基质　93

基质势(衬质势)　129

畸形的　137

畸形卷烟　137

畸形叶　127

畸长;疯长　102

激动素　112

激发效应　176

激酶;致活酶　112

激素　99

激素平衡　100

激素诱捕剂;激素诱捕器　163

激肽;细胞分裂素　112

级外烟　146

级外烟组(N)　147

极差;变程(范围)　183

极差法　183

极端干旱　75

极端嗜热菌　75

极端嗜盐菌　75

极缺　73

极性有机分子吸附法(测定比表面)　170

极值　75

急性危害　5

集水灌溉　244

集水农业　76,182

集雨;雨水集蓄　182

集雨补灌　182

集约化农区　107

集约农业　107

集约型　107

集约栽培　107

集约栽培;精耕细作　107

集中施肥　41,123

集中条施　15

瘠薄土壤　101

瘠地　105

瘠土;薄土　227

几丁质酶　34

几何级数增长　89

己醛　97

己酸　97

己酸己酯　97

己糖　97

己糖二磷酸途径　97

计算机施肥系统　40

计算科学　40

记分表　195

技术标准　225

技术规范　225

技术规范;说明书　214

技术规格;质量指标　180

技术结构　42

技术路线;技术方案　225

技术设计　225

技术数据;技术资料　225

季节性变化　195

季节性冻土　195

季节性干旱　161

(病害的)季节循环　195

剂量;(用)药量　62

剂量效应　62

剂型　62

寄生根　221

寄生现象　160

寄生植物　6

寄主植物　100

寄主专门化;寄主特异化　100

寄主专一性　100

· 293 ·

加标样回收率 184
加工工艺 177
加工性能 177
加过料的烟叶 30
加活性炭复合滤嘴 31
加权函数法 245
加权平均 245
加权平均值 245
加权算术平均 245
加权算数平均值 245
加热管 96
加热器控制 96
加热器套管 96
加热设备 96
加湿器(机);润湿器 100
加速发芽试验 2
加速风化(作用) 3
加香过量 157
加香率 81
加性基因 5
加性效应 5
加性遗传变量;加性遗传方差 5
加压灌溉系统 176
加长卷烟 124
家畜 61,122
家畜粪 122
家庭式农业 75
夹烟架;烟架 181

甲醇 132
甲基橙 133
甲基红 133
甲基化作用 133
甲基托布津 31,70,135,228
2-甲氧基-3-甲基吡嗪 133
6-甲基-3,5-庚二烯-2-酮 133
甲醛 83
甲酸 83,133
甲酸酯;甲酸甲酯 83
甲烷;沼气 132
钾 173
钾的固定;钾固定 173
钾的临界浓度 45
钾的释放 174
钾肥 173
钾肥;草碱;钾碱;碳酸钾 173
钾钙活比性 112
钾供应强度 174
钾固定能力 173
钾缓冲容量 173
钾离子通道 173
钾平衡 173
钾素营养失调 173
钾素状况 173
钾位 174

钾吸收转运系统 174
钾循环 173
钾盐 112,174
钾盐镁矾(肥料) 112
假设 102
假设检验 102
假熟 176
假熟;欠成熟;未成熟的;早熟的 175
假植 226
嫁接 91,106
尖锐气息;辛辣气息 200
坚果香 151
坚实土壤 219
坚实叶 95
间断 26
间隔;间距 108
间隔;间距;株行距 214
间混作;间作 108
间接肥料 105
间接排序 105
间接作用 105
间距分离 60
间苗 59,227
间苗;疏伐;稀化;稀释;冲淡 228
间种;套作 108
间作 39,191
(农作物的)间作;轮作

221
间作;套种 108
间作模式 108
间作套种 108
间作物 108
间作作物 108
肩负型喷雾器 95
兼性反硝化(作用) 75
兼性腐生 41
兼性寄生 41
兼性嗜热菌 75
兼性需氧菌 75
兼性厌氧菌 7,75
拣叶台 166
检测(检验) 226
检测报告 226
检测能力 175
检测限 54
检验片;试样 226
检验统计量 226
减量化;减量 184
减排 137
减排能力 137
减少污染物的技术 171
剪切 200
剪切强度 200
剪切应力 200
简单假设 201
简单随机抽样 201

简单相关 201
简易比重计法 201
碱 21
碱;强碱 10
碱地 11
碱度;碱性 21
碱化 11
碱化(作用) 213
碱化土壤 11
碱化作用 11
碱解氮 11
碱金属 11
碱熔融 11
碱土 213
碱土金属 11
碱性(碱度) 11,21
碱性的 11
碱性肥料;基肥 21
碱性钾;氢氧化钾 31
碱性磷酸(脂)酶 11
碱性溶液 21
碱性石灰质土 11
碱性土 11,21
碱性氧化物(成碱氧化物) 21
健壮种子 240
渐变 91
鉴定 102
箭舌豌豆 39,216
降低污染的措施 171
降解产物 52

降解机制 52
降解碳 52
降霜 85
降水;降水量;降雨量;沉淀(作用) 175
降水充沛 13
降水强度 175
降水使用效率 182
降水效率 175
降水蒸发比 175
降水指数 175
降水资源 175,182
降烟碱;去甲基烟碱 148
降雨;降雨量 182
降雨持续时间 182
降雨径流(地面径流) 182
降雨量 13,182
降雨年型 248
降雨频率;降雨强度 182
降雨期 82
降雨侵蚀 182
降雨侵蚀程度 182
降雨入渗 182
交换价值 39
交换扩散 73
交换容量 73
交换态铁 74
交换态铜 74

交换吸附　73,74
交换现象　73
交换性铵　74
交换性钙　74
交换性钾　74
交换性金属阳离子　74
交换性离子　74
交换性镁　74
交换性锰　74
交换性钠　74
交换性氢　74
交换性酸度　73
交换性盐基　74
交换性阳离子　74
交换性阳离子百分率　74
交换性阴离子　74
浇灌　244
浇水量　216
胶(霉)毒素　90
胶体　38
胶体的　38
胶体溶胶　38
胶体物质　38
胶体形态　38
胶体性质　38
焦糊味　195
焦磷酸酶　179
焦磷酸钠　204
焦磷酸盐　179

焦甜香　28
焦香　28
焦油　225
焦油/烟碱比　225
焦油量　225
焦油释放量　225
焦灼病(缺锰)　85
焦灼症(缺钾)　28
嚼烟　33
嚼烟;烟饼　170
嚼烟饼　33
角斑病;(叶)角斑　14
脚叶　83,176
脚叶(分级)　194
脚叶组　159
搅拌器　8,219
窖灌农业　31
窖蓄雨水　219
酵母菌　248
阶式梯田　218
接触面积　42
接触吸收　42
接触性除草剂　42
接受性;合格率;可接受性;可承兑性　3
接受域　3
接头;黏合　111
(盘纸)接头检测器　111
接种　106
(人工)接种;芽接

106
接种剂　106
接种量　106
秸秆　219
秸秆(厩)肥　219
秸秆堆肥　40
秸秆覆盖　219
秸秆还田　219
秸秆资源　219
节间;节距　108
节间缩短　200
节间长度;节距　108
节略的　2
节能技术　70
节水措施　131
节水工程　244
节水灌溉　243,244
节水农业　242,243,244
节水农业技术　244
节水型农业模式　244
节约地使用资源　41,67
节约能源的　70
节约资源型技术　187
拮抗生物　15
拮抗效应　14
拮抗性土壤　14
拮抗作用　14
结冰温度　86
结持性(稠性)　41

结构差的土壤 172
结构良好土壤 246
结构水 220
结构松散 124
结构性多糖 220
结构组分 220
结果;结实 86
结果率 86
结合水 101
结合态胡敏酸 38
结合态养分 25,38
结晶水 47
结块(结构)土壤 125
结块(结块性) 28
结皮 223
结实;结果;子实体 86
结实;坐果;着果 86
结实率 197
结团烟丝 125
截获 108
截留效应 68
解把 95,228
解把;松把 27
解把机 228
解包 236
解磷微生物 163
解磷细菌 163
解磷作用 163
解吸;脱附;解吸作用 54

解吸过程 54
解烟;下杆 238
戒烟 2
界 112
金属、有机复合污染 38
金属-胡敏酸络合物 132
金属气 132
金属污染物 132
金针虫 247
紧固不良 172
紧结态腐殖质 228
紧实 95
紧实层 39
紧实土壤 228
进风口 10
近红外 143
近红外反射光谱仪 143
近红外分光光谱仪 (NIRS) 143
近红外线 143
近似计算 15
近紫外 143
劲头 103
浸泡(润湿作用) 126
浸润;浸湿 62
浸润灌溉 110
浸水(洪水;泛滥) 109
浸提 58,74

浸提器 75
浸提液 117
浸析(的);淋溶(的);沥滤;浸提;淋洗 117
浸种 58,197,198
浸种;浸湿 218
茎;叶梗;烟梗 218
茎粗;茎圆周 217
茎腐 217
茎秆倒伏 217
茎皮堆肥 128
茎破裂 45
茎伸长缓慢 125
茎叶处理 83
茎叶角度 118
京都议定书 113
经常施肥 41
经度 124
经过鉴定的品种 15
经过特殊处理的深色烟叶 25
经济产量 67
经济成本;节约的成本 67
经济肥力 67
经济价值 39
经济施肥 67
经济效益 67
经济最佳施肥量 67
经济作物 67

297

经济作物;工业原料作物 105
经纬仪 227
经验常数 70
经验分布 70
经验回归方程 70
经验回归系数 70
经验模型 70
经营成本 153
晶格结构 116
精氨酸 16
精耕细作 107,175
精确播种 175
精确定位管理 201
精选种子;优良种子 69
精制;净化;提纯 179
精准农业(精确农业) 175
精准施肥 175
径流 82,192
径流含沙量 201
径流汇集 192
径流量 192
径流率 192
径流模数 192
径流强度;流速 82
径流侵蚀 192
径流侵蚀力 192
径流深度 53
径流系数 192

径流小区;径流场 192
径向运输 181
净成本;净价 143
净固定作用 143
净灌溉需水量 143
净光合作用速率 183
净光能合成;净光合作用 143
净耗水量 143
净化能力 179
净化效率 36,179
净化因素;去污系数 51
净化作用 36,179
净矿化作用 143
净利润 143
净生产力 143
净生产量 143,184
净生产率 143
净同化率 143
竞争离子 39
竞争平衡 39
竞争性抑制 39
竞争优势 39
静燃速率 218
静态模型 218
静态燃烧 218
静止的;休眠的 188
久耕地;熟地 123,152

酒精发酵 10
酒石酸钾钠 174
酒香 26
厩肥 75,113,213,216
厩肥;圈肥 248
厩肥层 76
厩肥发酵 96
厩肥撒施机 128
厩肥条施 128
厩肥液汁 128
厩液施肥机 122
局部坏死 123
局部基台 160
局部枯斑 123
局部侵染 123
局部施用 123
局部通风;局部排气 123
局部熏蒸剂 215
局部症状 123
桔黄色褪绿 154
橘红 154
橘黄色;橙黄色;橙色 154
橘色烟叶 126
咀嚼式口器(昆虫) 127
矩阵 129
矩阵方程 129
矩阵方法 129
矩阵分析 129

· 298 ·

矩阵计算　129
矩阵特征值　129
矩阵相关　129
巨豆三烯酮　132
巨脉病;巨脉　22
剧毒农药　5
飓风　101
聚胺(PA)　171
聚丙烯　172
聚丙烯薄膜　172
聚丙烯酰胺　171
聚合(作用)　172
聚合酶链式反应　160
聚集(作用);群聚的;
　族聚;团聚作用　8
聚类　37
聚类分析　98
聚类分析法　37
聚类统计　37
聚伞花序　49
聚伞状花序　235
聚乙烯(PE)　172
聚乙烯薄膜　172
卷曲叶;扭叶　118
卷烟　35
卷烟车间;制造部
　128
卷烟工艺　35

卷烟规格　35
卷烟机　35
卷烟检测器;卷烟检测
　装置　35
卷烟静燃仪　35
卷烟物理性能　35
卷烟烟气　35
卷烟纸;盘纸　35
卷烟主流烟气　127
卷叶　49,118
卷叶病　118
卷叶病毒病　49
决策树　51
决策支持系统　51
决定系数　54
决定因素　54
绝对差　2
绝对产量　2
绝对持水量　2
绝对干旱　2
绝对干重　2
绝对关联　39
绝对含水量　2
绝对频数　2
绝对容积　2
绝对湿度　2
绝对真空　2
绝对最低温度　2

绝对最高温度　2
绝热;保温;热绝缘
　96
均方误差　130
均量施肥　237
均匀度　73
均匀分布　237
均值标准误差　217
均值图　19
均质烟叶调制　99
菌根　20,141
菌根根际(菌根圈)
　141
菌根共生　141
菌根群丛　141
菌根形成　141
菌根营养　141
菌根营养的　141
菌根真菌　141
菌落　38
菌落计数器　38
菌落形成单位　38
菌丝　102
菌丝际　102
菌丝体　141
菌丝状的　141
菌种保藏　48

K

喀斯特;岩溶;水蚀石灰岩地区　112
喀斯特地貌　112
喀斯特地区　112
卡尔文循环　29
卡方分布　34
卡方交叉　34
卡庆斯基土粒分级制　111
卡庆斯基土壤质地分类制　112
卡庆斯基吸管　112
开包完整率　236
开放传粉;自然传粉　153
开沟　60
开沟机　87
开沟犁　94
开沟器　234
开沟器;松土器　153
开沟条施　15
开花　25,68,82
开花阶段;花期　82
开花期　81,82
开花期(指单花)　14
开花期长的　123
开花日期　82
开花时期　82

开花诱导　82
开花早期　64
开垦荒地;土壤改良;驯养　184
开垄　80
开穴　58
开穴机　57
凯氏定氮法　112
凯氏烧瓶;长颈烧瓶　112
勘测(查勘;考察)　74,192
砍茎;整株收割　217
砍茎机　217
砍株法;砍茎法　217
砍株收割　96
糠醛　87
2-糠醛　87
糠醛渣　87
抗病品种　59,187
抗病性;抗病力　59
抗病育种　59
抗虫性　162,187
抗虫性;昆虫抗药性　107
抗虫育种　26
抗倒伏能力　218
抗倒伏性　123

抗冻性;耐冻性　85
抗风性　247
抗寒性　33,247
抗寒植物　38
抗旱保水剂　63
抗旱节水　63
抗旱品系　63
抗旱性;耐旱性　63
抗坏血酸氧化酶　17
抗碱性　11
抗拉强度　226
抗涝性　81
抗涝性;抗淹性　244
抗逆性　187,220
抗逆育种　26
抗破碎性(分级)　200
抗破碎性;耐碎性　200
抗生素　15
抗酸的　4
抗酸性;耐酸性　4
抗碎强度　47
抗碎性;耐破度　26
抗性　187
抗性;耐性　95
抗性品系　187
抗性生理　95,220
抗盐性　193

抗药性 32,63,163,187
抗蒸腾剂 15
烤房 82
烤房;晾房;堆房;谷仓 21
烤房;烟炕 112
烤房斑点病 21
烤房漏气 10
烤房水珠凝结 56
烤房维修 21
烤红 239
烤焦味 195
烤焦叶;烫伤烟叶;高温损害叶;烤红 195
烤烟 82
烤烟房 229
烤烟国家标准 142
烤烟基本烘烤技术规程 192
烤烟实物标样 217
烤烟型 82
烤烟型卷烟 70,240
烤烟型香气 240
烤制;烘焙 229
苛性钾;氢氧化钾 31
颗粒大小分布 160
颗粒大小组成 160
颗粒肥料 91
颗粒紧密的(分级) 228
颗粒状物 160
可变成本 239
可变电荷 239
可变电荷表面 239
可彻底降解的 40
可持续发展 223
可风化矿物 245
可耕层 16
可耕地 16
可耕地;适耕地 228
可耕地面积 16
可耕地区 47
可耕土壤 16
可耕性 228
可见光吸收值 240
可见症状 241
可见紫外分光光度法 240
可降解的 52
可降解地膜 52
可降解地膜覆盖 52
可降解污染物 52
可交换态 74
可浸提腐殖质 74
可开发性;可利用性 74
可靠区间 41
可靠性 186
可靠性分析 186
可控制费用 43
可矿化氮 136
可矿化碳 136
可矿化养分 136
可逆的 188
可燃的;易燃的 39
可溶性氮 213
可溶性腐殖质 213
可溶性灰分 213
可溶性碳水化合物 213
可溶性糖 213
可溶性盐(溶性盐) 213
可溶性养分 213
可溶性养分库容量 219
可溶性有机氮 213
可溶性有机碳 213
可溶性有机物;可溶性有机质 60
可生物降解的 23
可湿性 246
可湿性粉剂 247
可塑性 169
可吸入颗粒物 106
可氧化养分 159
可移栽烟苗 238
可用微生物降解的污染物 134
可蒸馏的 60
可制堆肥垃圾 40

可置换离子 186
克雷伯氏循环 113
克立格法(空间局部插值法) 113
克立格法制图 113
克氏瓶 113
客土法 210
肯塔基深色明火烤烟 192
空白 25
空白试验 25,237
空的;(烟味)平淡 70
空间比较 214
空间变异性 214
空间尺度 214
空间的 214
空间分析 214
空间连续性 214
空间特征 214
空间熏蒸剂 214
空间自相关指数 214
空茎病 99
空气传染 10
空气分离 10
空气净化 9,10
空气侵染 10
空气湿度 10
空气污染 10
空气污染控制 10
空气循环密集烤房 9
空气质量 10

空中喷洒 6
空籽;不饱满籽粒 99
孔;毛孔;气孔 172
孔度计;孔隙度测定仪;透气度测定仪 172
孔径 99,172
孔径;孔径大小 172
孔隙度测定仪 173
孔隙分布 172
孔隙空间 172
孔隙容积 241
孔隙水 172
控释 43
控释肥料(长效肥) 43
控释机理 43
控释农药 43
口针(昆虫) 140
叩头虫 36,203
枯焦气 195
枯水期 125
枯水期;旱季 63
枯死 50
枯萎 162,200,247
枯萎状 195
枯心 51
枯叶 51
枯枝落叶层 51,56,93,118,122
苦(味)的 25

苦味 25
库(汇) 201
块金 148
块金方差 148
块金效应 148
块状播种 169
快速分析方法 183
快速聚类 181
快速土壤分析 181
宽行栽培 214
宽行栽植 246
宽窄行(种植) 246
矿化氮量 13
矿化过程 136
矿化势 136
矿化速率 136
矿化速率常数 136
矿化碳 136
矿化特征 32
矿化系数 136
矿化作用 136
矿物;矿物的;无机的 135
矿物风化 136
矿物钾 135
矿物结合态有机物 135
矿物态铁 135
矿物营养生理失调 166
矿物源农药 163

矿物质从叶片中的溢流 157
矿物质过量 135
矿物质流失 135
矿物质缺乏 135
矿物质缺乏症状 135
矿物组成;矿物成分 135
矿质层 135
矿质代谢 135
矿质肥料;无机肥料 135

矿质颗粒 135
矿质土壤 136
矿质营养 135
矿质营养学说 227
矿质元素 135
亏缺元素 52
昆虫 107
昆虫病原线虫体 71
昆虫传播 70
昆虫抗药性 107
昆虫天敌 107
昆虫种群质量 172

扩散 58
扩散(作用) 58
扩散过程 58
扩散控制过程 58
扩散路径 58
扩散速率 183
扩散条件 58
扩散系数 58
扩散运输 58
蛞蝓;蛞蝓型幼虫 203
阔叶型烟草 26

L

垃圾堆肥 87,185
拉丁方试验设计 116
蜡质层 245
辣椒 33
辣口的 195
赖氨酸 125
赖百当 113
赖百当类化合物 113
蓝细菌 49
蓝藻 25
烂根 191
烂苗病 21
烂泥田 140,203
劳动产出比率 113
劳动生产率 113

劳动效率;人工效率 113
劳务用工 113
老化;陈化;醇化 7
老化;陈化;年龄;醇化 7
老化的种子;陈种子 7
老龄土 152
老龄幼虫 152
老式烤房;传统烤房 43
老种子;陈种子 152
涝害 81
涝洼地 244

酪氨酸 236
乐果 58
蕾期 27
累积分布函数 48
累积和图 48
累积频数 48
累加效应 5
类固醇 219
类胡萝卜醇 166
类胡萝卜素;类叶红素 30
类似脱氧核糖核酸的核糖核酸 61
类萜(烯) 226
类脂 122

· 303 ·

类脂;脂类 122
类脂;脂类;脂质 122
冷点;冻点 38
冷冻处理 33
冷冻干燥 84
冷冻温度 33
冷害 44,50
冷浸提;冷消解 38
冷凝管 41
冷杉醇 1
冷杉酸 1
冷性肥料 38
离差 56,60
离群值(异常值) 157
离散 59
离散方差 60
离散分布 59
离散高程矩阵 59
离散随机变量 59
离体繁殖 104
离体培养 48
离体培养法 104
离体叶片 54
离心管 31
离心过滤器 31
离心机 31
离子半径 109
离子泵 109
离子大小 109
离子辐射 110
离子价 109

离子间关系 108
离子交换 109
离子交换电泳(法) 109
离子交换色谱(法) 109
离子交换树脂 109
离子交换吸附 109
离子交换柱 109
离子拮抗作用 109
离子扩散 109
离子浓度 109
离子排斥 109
离子平衡 109
离子强度 109
离子亲和力 109
离子取代 109
离子色谱法 109
离子态 109
离子态养分 109
离子通道 109
离子协同作用 109
离子选择性 109
离子选择性电极法 109
离子主动吸收 5
犁底层 170
犁底层;犁磐 170
犁幅宽度 170
犁耕区 170
李比希型限制因子 120

李比希最低定律 120
理论产量 227
理论有机碳量 227
理论有效水量 227
理想根构型 153
理想构型 102
理想气候 102
理想栽培系统 102
理想质地 102
力稳性团聚体 131
立地条件 201
立地质量评价 73
立方体,(育苗的)营养钵 47
立体农业 141
立体生态农业 219
利润增量 177
利用系数 238
利用效率 238
砾土(沙砾土) 91
砾质 91
砾质粗沙土 91
砾质黏土 91
砾质壤土 91
砾质沙土 91
砾质土 91
砾质细沙土 91
粒度分级法 160
粒度分析 91,202
粒径分布 202

粒径分级 160
粒重 91
粒状农药 91
连接酶 120
连年耕作制(常年耕作制) 162
连通性 41
连续(土)层 42
连续的 42
连续的植物提取 42
连续发酵 42
连续分布 42
连续分析 42
连续灌溉 42
连续化学提取 199
连续监测 42
连续开花的;永久的;永恒的;不间断的;多年生植物 162
连续流动法 42
连续流动分析 42
连续流动分析仪 42
连续流动培养 42
连续轮作制 42
连续漫灌(持续淹没) 42
连续培养法 42
连续随机变量 42
连续吸收 42
连续性分析 42
连续性中雨 42

连续雨;绵雨 42
连续种植 221
连雨天气 123
连作 42,48,152,186,199,221
连作危害 106
连作障碍 42,186
连作制 153
联苯胺 22
联合共生固氮 57
联合固氮菌 17
联合固氮作用 17
联合国粮农组织 83
联合收割机 38
联合修复 38
联合育苗 44
镰刀菌属;镰孢菌属 87
链格孢菌(烟草赤星病菌) 11
良田 77
良性循环 22
良种 22
良种(子) 103
粮食安全 91
粮田 91
量杯 245
量瓶 91
量筒 91,131,241
两年三熟 228
两年三熟制 236

两熟连作 42
两相侵蚀 62
两性花(的) 139
两性氧化物 13,108
亮度;光泽;鲜明度 26
亮色烟叶(指白肋烟) 26
晾房 10
晾烟 10
晾烟棚 231
晾烟品种 10
晾烟型 10
晾制 10
晾制的 10
撂荒 200
列当属 157
劣等的;低等的;低级的;低质量的 125
劣地 20
劣质种子 172
裂解酶;裂合酶 125
裂区 215
裂区设计 215
裂区试验 215
邻苯二酚 30
邻苯二甲酸 165
邻苯二甲酸二丁酯 57
邻苯二甲酸二甲酯 58
林农间作 83

·305·

林烟草　142
临界点　45
临界含水率　45
临界剪应力　45
临界浓度　45
临界区间　45
临界湿度（临界含水量）　45
临界水平　45
临界温度　45
临界域　45
临界值　45
临界值诊断法　57
淋溶层　69,117
淋溶褐土　117
淋溶损失　124
淋溶土　10,70,125
淋溶作用;残积作用　70
淋失　117
淋失输出　157
磷的活化　137
磷肥　79,146,163,164
磷肥残效　186
磷固定　163,164
磷化氢熏蒸　164
磷灰石　15
磷钾复混肥料　167
磷钾施肥制　164
磷解吸　163
磷矿粉　93,175

磷矿石　189
磷素循环　164
磷酸　164
（正）磷酸　157
磷酸铵　13
磷酸铵镁　126
磷酸二钙　57
磷酸二氢铵　13,28
磷酸二氢钾　112
磷酸二氢钾（磷酸一钾）　139,173
磷酸二氢钠　203
磷酸化作用　164
磷酸钾　174
磷酸酶　163
磷酸镁　126
磷酸葡萄糖酸途径　164
磷酸氢二铵　57
磷酸氢二钾　174
磷酸氢二钾（磷酸二钾）　58
磷酸三钙　234
磷酸戊糖途径　161
磷酸吸持　163
磷酸烯醇式丙酮酸　164
磷酸盐　163
（正）磷酸盐　157
磷酸盐肥料　163
磷酸盐缓冲液;磷酸缓冲剂　163

磷酸盐吸附　163
磷酸一铵;磷酸二氢铵　139
磷酸一钙　139
磷位　163
磷细菌　164
磷细菌肥料　163
磷脂　164
磷脂酰胆碱(卵磷脂)　164
磷脂酰乙醇胺　164
磷脂脂肪酸分析　164
磷质石灰土　164
(虫)龄;龄期　107
灵敏的;准确的;优美的;细致的;细腻的;(烟叶)组织细致　53
零假设　148
零下;低温　221
留茬　186
留茬覆盖耕作　220
留茬作物　220
留权　112
流动注射分析　82
流动注射分析仪　82
流行病;病害流行　72
流量　82
流速　82
硫　222
硫醇　132

306

硫肥　222
硫化氢　101
硫化物　221
硫化作用　221
硫素营养失调　222
硫酸　222
硫酸铵　13
硫酸钙　28
硫酸钾　174,221
硫酸钾镁　174
硫酸镁　126
硫酸锰　128
硫酸氢钾　4
硫酸氢钠　4
硫酸铁　77,110
硫酸铜　44
硫酸锌　249
硫酸亚铁　77
硫酸盐　221
硫酸盐危害　222
硫循环　221
馏出液　60
馏分　60
六六六　22
六氯化苯　22
垄;田埂　189
垄播　189
垄底深翻　237
垄高;畦高　97
垄耕;培土　189
垄沟　189

垄沟灌溉　44,189
垄沟渗漏　44
垄间作业机　203
垄上栽培;垄作栽培　48
垄栽;垄作　189
垄作　169,189
垄作(垄畦耕作)　189
垄作;墩植　140
垄作播种机　189
垄作免耕　189
蝼蛄　138
蝼蛄科　94
墣土　124
漏斗　87
漏栽率　136
陆地生态系统评价　226
陆地生态学　226
陆地碳汇　226
陆圈　89
露地播种　157
露地苗床　157
露地苗床　93
露地栽培　48,79,153,157
露地植物　153
露点　56
露点传感器　56
露点记录仪　56
露点湿度计　56

露天晒干　157
掠夺式开采　54
掠夺性开发　175
掠夺性农业　74
轮耕地(轮作田)　191
轮灌方式　191
轮灌制度　191
轮换施药　191
轮换栽培　233
轮回杂交　191
轮牧草地　191
轮生的　240
轮休地　114
轮作　43,45,46
轮作;旋转　191
轮作方式　160
轮作规划　191
轮作顺序　46,191
轮作栽培　191
轮作栽培;轮作种植　191
轮作制　191
轮作周期　161,191
轮作作物　46
罗盘　39
逻辑斯谛增长模型　123
螺旋根;气生根　215
螺旋菌　215
螺旋形上升叶序　215
裸露土壤　21

307

洛桑试验站 191
络合滴定法 40
络合反应 40
络合剂 40
络合态铁 32,40
络合物 40
络合作用 40
落花 82
落黄;变黄;黄化 248
落叶 200
落叶病 118,119
铝 12
铝毒害 12
铝硅酸盐 12
绿肥 92
绿肥休闲地 92
绿肥作物 92
绿化;绿色;变绿 240

绿色;新鲜 93
绿色的;新鲜的;未成熟的;青(叶);青色;青黄(分级术语) 92
绿色经济 92
绿色农业 92
绿色食品 92
绿原酸 34
氯 34
氯(代)甲烷 34
氯仿 34
氯仿熏蒸法 34
氯害 34
氯化铵 13
氯化钙 28
氯化钾 141,173

氯化苦 34,57
氯化镁 126
氯化钠;食盐 203
氯化氢 34
氯化物 34
氯素失调症 34
氯酸盐;氯化 34
2-氯乙基磷酸(CEPA) 34
氯乙烯 34
滤出液 161
滤光光度计 2
滤嘴卷烟 80
略图 2
略硬实 202
(烟叶)劣级;(烟叶)次级 172

M

马弗炉 140
马来酸 127
马来酰肼 127
马里兰烟 129
马铃薯 212
马铃薯Y(型)病毒属;马铃薯Y病毒属 174
马铃薯块茎蛾 174

马铃薯块茎蛾 90
码垛 216
蚂蚁 14
埋生的;包埋的 70
麦茬稻 189
麦稻两熟 246
麦秆 246
麦田 246
(叶)脉花叶 240

脉间坏死 109
脉间黄化 109
脉间失绿 109
脉络;叶脉 240
脉明;明脉症 240
(叶脉)脉色 79
漫灌;洪水;溢流 81
漫灌;淹灌 81
漫灌法 109

毛管持水量 29
毛管孔隙 29
毛管水 29
毛管吸附水 29
毛管现象 29
毛细(管)作用(毛细现象) 29
毛细管 29,112
毛细管的 29
毛细管电泳 29
毛细管气相色谱法 29
毛叶苔子 94,240
媒介昆虫 239
煤烟病 50
酶 72
酶促反应 71
酶促降解 71
酶反应 72
酶活性 72
酶活性失调 60
酶联免疫吸附实验;酶联免疫吸收分析,酶结合抗体法 69
酶选择性 72
酶学 72
酶学诊断 72
酶专一性 72
霉;霉菌 138
霉;霉菌;松软土地 140

霉变 138
霉的;霉变 140
霉味 140
霉味烟叶 86
霉烟 135,140
霉烟处理 140
每个烤房适用种烟面积 44
每公顷产量 157,248
每支烟的烟碱 144
美国病理学家及细菌学家协会 1
美国有机及合成肥料协会 1
美好香气 32
美花烟草 142
美拉德反应;棕色化反应 126
美式混合型卷烟 12
镁(Mg) 126
镁肥 126
镁磷混合肥 87
镁缺乏 126
镁石灰 126
镁素营养失调 126
门;类群 165
萌发;突出体 70
萌发后;出土后 173
萌发后处理 173
萌发力 89
萌发前处理 175

萌发前的种子 175
萌发温度 226
萌芽率 183
蒙脱矿 139
蒙脱石(微晶高岭石) 139
锰(Mn) 127
锰毒 137
锰毒害 128
锰肥 127
锰缺乏症 127
锰营养失调 127
(稠)密的;浓厚的 53
密度 53
密度函数 53
密封堆积发酵 10
密封剂;密封胶;黏合剂 111
密封圈 123
密挂 228
密集烘烤 27
密集烤房 27
密栽;密植 36
密植 36,39,227
密植株行距 37
幂函数模型 175
免耕 148
免耕(法) 147
免耕(法);免耕种植 249
免耕播种 249

·309·

免耕法 148,249
免耕覆盖 146
免耕农业 148
免耕栽培 148
面积图 16
面向对象数据库系统 151
面源 223
面源污染 16,147
苗床 149
苗床;畦 129
苗床;下种床;播种床 196
苗床;植床 167
苗床病害 197
苗床病害防治 59
苗床播种机 149
苗床覆盖 21
苗床覆盖物 196
苗床感染 149
苗床管理 127
苗床害虫防治 163
苗床结构 42
苗床期 196,198
苗床施肥 197
苗床土层 117
苗肥 78
苗高 97
苗根带土 198
苗期病害 59
苗期病害;幼苗病害 197
苗期生长势 93
描述性统计 54
秒 195
灭茬 36,54,220
灭茬;粗耕灭茬 220
灭菌作用;杀菌;不育化 219
敏感植物 199
明沟 153
明沟排水 153
明火烘烤;明火烤制 80
明火烘烤的;明火烘烤 80
明火烤烟 51,80
明火调制 80
模糊理论 87
模糊数学方法 87
模拟降雨 182,201
模式分析 160
模式基因 160
模型试验 137
模型验证 137
膜翅目 102
膜下滴灌 63,237
摩擦(力) 85
摩擦系数 85
摩擦阻力 85
抹杈 221
抹杈;打杈 221
末级 115
莫合烟 127
母本 77
母岩 139
母质 139,160
母质层(C层) 28,160
母质层(土壤) 99
亩产量 248
木本植物 248
木霉属 234
木霉素 90
木炭;炭;活性炭 31
木香 247
木质部运输 248
木质化作用 121
木质气 121
木质素 121
目标产量 225,248
目标产量施肥 225
目标产量施肥法 132,225
目标分析 225
目标总体 225
苜蓿,紫花苜蓿 10
牧畜废水 123
钼 137
钼(Mo) 139
钼肥 139
钼辅因子 139
钼黄素蛋白 139

钼酸铵 13　　钼酸钠 204　　钼铁蛋白 137

N

钠 203
钠盐 204
耐病性 59
耐藏性;保存性 112
耐除草剂 97
耐储藏的 124
耐冬性;抗寒性;耐寒性 247
耐冻性;抗冻性 84
耐肥品种 79
耐肥性 232
耐光性 121
耐寒性 33,38,85
耐寒性;抗寒性 38
耐寒性;耐低温性 125
耐旱性 63
耐瘠性 232
耐碱微生物 11
耐碱性 11
耐冷的 178
耐量试验 231
耐逆性 220
耐热菌 227
耐热性 96
耐受性;纯度允差(混等级) 231
耐霜性;抗霜性 85
耐酸微生物 4
耐酸细菌 4
耐酸性 4
耐性 70
耐压微生物 21
耐盐的 193
耐盐水性;耐海水性 193
耐盐性 95,193
耐盐植物 193
耐氧的 7,159
耐氧细菌 7
耐药性 63
耐药性;农药允许量 163
耐阴的 199
耐阴植物 199
耐渍性 244
萘甲酸 142
萘乙酸 142
南方根结线虫 132
难分解养分 187
难降解的 148
难降解污染物 184
难降解有机污染物 185
难溶性磷肥 58,95,125
难溶性养分 57,95,214
难溶性氧化物 107
难闻气味 237
内包皮叶组 22
内标法 108
内表面 108
内表面积 108
内翻耕 88
内寄生;内寄生物;内寄生虫 70
内生菌根 70
内外生菌根 67
内吸(性)除草剂 224
内吸(性)杀虫剂 224
内吸性农药 224
内吸性杀(真)菌剂 224
内循环 108
内酯酶 114
能量元素 70
能流 70

能源利用 70
能源消耗 70
能源作物 70
尼龙袋 151
尼龙袋法 151
尼龙网 151
泥灰质土壤;泥灰土 129
泥炭;泥煤 160
泥炭肥料 160
泥炭土 161
泥炭形成(作用) 160
泥炭形成(作用);泥炭化过程 159
泥炭形成过程 159
泥炭资源 161
泥土气味 65
泥杂肥 203
拟合优度;吻合度 90
拟合优度检验 91
逆境 71
逆境蛋白 220
逆境耐性植物 220
逆境生态学 220
逆境休眠;胁迫休眠 220
逆矩阵 109
逆相关(负相关) 109
年产量 14
年降水量 14
年较差 13,14

年平均 14
年平均降雨量 14,130
年平均气温 14
年平均温度 130
年总生物量;年营业额;年周转额 14
黏度系数 240
黏附(现象) 6
黏附水 6
黏化层 36
黏结性土壤 219
黏粒 36,44
黏粒;黏土 36
黏粒矿物 36
黏壤土 36
黏土 36
黏土含量 36
黏土矿物 36
黏土质 36
黏土组 36
黏质的(含黏土的) 36
黏质底土 16,36
黏质粉沙壤土 36
黏质沙土 36
黏质细沙土 36
黏滞结持(度) 240
黏重土壤 96,97
黏重质地 97
酿热物 77

鸟氨酸 157
尿 238
尿素 238
尿素(脲) 29
尿素氮 238
脲 238
脲酶 238
脲酶抑制剂 238
镍 144
镍坩埚 144
镍污染 144
柠檬黄色;柠檬 119
柠檬色;淡黄色(分级) 113
柠檬酸 35
柠檬酸铵溶性磷 13
柠檬酸钾 173
柠檬酸循环 35
柠檬叶 120
凝集;烧结 7
凝胶 88
凝胶电泳(法) 88
牛粪 31,44,45
牛肉浸膏 21
农产品产量;农业产出 8
农产品成本 44
农场等级和卷烟加工分级 76
农村能源 192
农村生活污水 192

农地土壤保持 75
农地资源 8
农杆菌 9
农耕地资源信息系统 8
农活 76
农家肥 75,76
农家品种;地方品种 115
农家调制;农家烘烤 64
农具 75
农林复合系统 11
农林间作 234
农牧交错地 9,46,76
农区环境空气质量监测技术规范 177
农事操作体系 75
农田氮素循环 145
农田覆盖 76
农田给水工程 76
农田灌溉 76
农田灌溉水质标准 217
农田耗水量 80
农田排水 68,75
农田排水沟 8
农田评价 8
农田生态系统 76,79
农田生态系统土壤养分循环 49

农田生态系统中的氮素循环 145
农田生态系统中肥料氮的去向 76
农田水分平衡 80
农田水利 242
农田土壤环境质量监测技术规范 177
农田土壤水分特性 8
农田污染 75
农田小气候 8,79,134
农田需水量 76
农田杂草 198
农田质量评价 76
农闲季节 202
农闲期 76
农学效率 9
农药 8,9
农药pH值的测定方法 54
农药安全使用标准 217,218
农药安全使用规定 185
农药安全使用间隔期 109
农药安全使用说明书 59
农药安全使用制度 224

农药残留;残留性农药 186
农药残留标准 233
农药残留的危害 162
农药残留量 162
农药残留量分析 162
农药残留物 162
农药残留限量;农药允许残留量 163
农药残效 186
农药持久性 162
农药代谢 162
农药的沉降 75
农药的滥用 105
农药的微生物降解 133
农药毒性 163
农药毒性等级 36
农药肥料 163
(含)农药肥料;农药化肥复合剂 162
农药分解代谢 60
农药合理使用准则 94
农药剂型 162
农药降解 162
农药伤害 32
农药水分测定方法 226
农药污染 42,162
农药药害 162

313

农药中毒 170
农业残渣 8
农业产量预报系统 9
农业产业化 105
农业的规范试样 129
农业的物质循环 49
农业地图 8
农业地形气候学 9
农业地质图 9
农业废弃物 8
农业废弃物处理 8
农业废水 9
农业废物还田 188
农业化学 8,9
农业化学图 9
农业化学研究方法 9
农业环境污染评价信息系统 8
农业加工废物 8
农业节水 9
农业结构调整 9
农业经济 75
农业科技园 9
农业轮作制 11
农业面源污染 8
农业气候图 8,9
农业气象灾害 9
农业区划 61
农业生态恶性循环 240
农业生态良性循环 90
农业生态模式 9
农业生态区 9
农业生态系统 8,9
农业生态系统生态学 9
农业生态系统中养分循环 149
农业生态学 8,9
农业生物学 9
农业试验 8
农业塑料薄膜污染 9
农业投资 8
农业土壤 9
农业土壤发生 9
农业土壤改良 8,9
农业土壤退化 9
农业微生物学 8
农业温室气体 8
农业污染 8
农业污染物 8
农业现代化 138
农业效率 75
农业信息科学 8
农业信息系统 8
农业遥感 8
农业用地适宜性评价 8
农业用聚乙烯吹塑薄膜 171
农业用水 9

农业资源 8
农业资源开发 9
农艺;农学 9
农艺类型 9
农艺性状 9,67
农艺修复 9
农用薄膜 8
农用地分等 8,76
农用化学品（包括化肥、农药等） 9
农用石灰 8,121
农用水资源 9
农用塑料 9
农用塑料薄膜 76
农作物;大田作物 79
农作物产量 46
农作物受害 96
农作物指数 45
农作制度（农耕制度） 76
浓（色度） 51
浓的;重的;浓味;厚;大型的 96
浓度;浓缩;蒸浓;提浓 41
浓度梯度 41
浓度梯度测定 41
浓肥;浓缩肥料 41
浓厚香气;沉厚香气 97
浓集的;浓缩的;浓的

41
浓硫酸 41
浓香;香味足 189

浓香气 241
浓香型 28
浓硝酸 41

浓郁 86
(香气)浓郁 70
暖土;早发土 181

O

欧式距离平方 216

沤肥 244

P

pH 标度 163
pH 计 163
pH 计法 163
pH 试纸 163
pH 值(酸碱度值) 165
pH 值;酸碱度 163
耙 95
耙地 95
耙地松土镇压器 47
帕斯卡 160
排根 178
排灌毛沟;排灌犁沟 242
排湿 100
排水 62
排水不良 103,172
排水不良土壤 172
排水不完全 103

排水沟 59,62,152, 157,244
排水技术 62
排水良好 246
排水良好的土地 246
排水尚佳 137,138
排泄物 53,185,241
盘栽(烟)苗 234
判别分析 59
判别函数 59
判别试验 59
判定函数;决策函数 51
判决规则 51
刨茬 58
泡囊丛枝菌根 240
胚;胚胎 70
胚;胚芽;胚根 44
胚根 182

胚茎 70
胚茎;幼芽;胚芽 170
胚乳 70
胚体营养 70
胚芽 70
胚芽;胚芽的 170
胚芽的 170
胚轴 70,170
陪补离子原则 39
培土 99,189
培土;起垄;作垄 189
培土犁;起垄犁 189
培土期;栽培期 47
培土器;起垄机;松土器 189
培养 84,104
培养;培养物;菌种;栽培 48
培养袋 48

· 315 ·

培养法 48,104
培养肥力;提高肥力 27
培养管 48
培养基 93
培养基;培养液 48
培养基成分 131
培养技术 48
培养皿 48,163
培养瓶 48
培养容器 48
培养时间 48
培养条件 48
培养系统 48
培养箱 93,104
培养液 48,150
培养液连续排出设备 42
培育;饲养;提高;栽培 182
培育者 157
配对样本的t检验 159
配方 25
配方比例 25
配方施肥 15,77,84
配方叶组 25
配合肥料 38
配套技术 39
配位反应 40
配叶 117

配叶柜 114
配叶调节器 114
喷灌 158,216
喷灌;人工降雨 158
喷灌;洒灌 216
喷灌系统 216
喷淋设备 216
喷洒次数 229
喷洒间隔 109
喷洒器;喷灌器;人工降雨器 216
喷施 215
喷施;喷雾 216
喷雾 215
喷雾机 137
喷雾剂 215
喷雾器 137
喷雾式苗床 137
喷药伤害;喷雾药害 216
盆地 21
盆栽 173,174
盆栽法 173,234
盆栽培养试验 173
盆栽棚 174
盆栽试验 173
盆栽土 174
盆栽移栽 173
盆栽营养土 174
盆栽植物 173
彭曼公式 161

蓬松结持(度) 82
硼(B) 26
硼肥 26
硼砂 25
硼素营养失调 26
硼酸 25
硼中毒 26
膨压 235
膨胀 27,58,223
膨胀梗丝 74
膨胀水 223
膨胀烟丝 74,178
批量;批号;地段;烟堆 124
偏差 62
偏差系数 37
偏度 202
偏度系数的标准误差 217
偏回归 160
偏磷酸钠 203
偏生产力 160
偏相关系数 160
偏移;偏差 22
偏自相关分析 160
漂白 33
漂白粉 34
漂浮苗床 81
(育苗)漂浮系统 81
漂浮栽培 81
飘逸 158

瓢虫 114
贫瘠的 105
贫瘠的 21,130,152,219
(土壤)贫瘠化 103
贫瘠地;瘠土 105
贫瘠土壤 103,119,172
贫有机质层;矿质层 152
频率 85
频率分布 85
频率曲线 85
品名 75
品位级;质量等级 180
品系;线;轮廓;家系;谱系 122
品系纯度;纯种性 89
品系繁育;系统育种;品系选育 122
品系分离 122
品系间杂交 108,122
品系特异性 47
品系选育 198
品质标准;质量标准 180
品质导向 94
品质等级(质量分等) 180
品质等级(质量分等);质量等级 180
品质分级 180
品质鉴定 15
品质鉴定;品质评价 180
品质评价 180
品质损失 180
品质特征 180
品质统计 18
品质下降;品质劣变 180
品质要求;质量要求 180
品质指数 180
品质指数;质量指数 180
品种 47,239
品种;培育 26
品种纯度田间小区试验 47
品种单一性 46
品种复壮 188,239
品种复壮;品种更新 186
品种改良 104,239
品种混杂 239
品种间杂交 139
品种间杂交;变种间杂交 108
品种间杂交现象 139
品种检索表;品种鉴定说明 112
品种鉴定;品种审定 239
品种特征 239
品种提纯 239
品种退化 239
平板测数 169
平板计数;平皿法 169
平板计数法 169
平板接种试验 106
平板培养;平皿培养物 169
平板培养法 169
平板稀释法 169
平板仪 167
(烟味)平淡 70
平底大玻璃杯 235
平地;平原 81
平地机;压地机 114
平行测定 186
平行试验 160
平衡施肥 20
平衡水分 138
平衡水分;平衡含水率 72
平衡状态 20
平滑的;平展的;(烟叶)平展 73
平均产量 19,130,131
平均抽样个数 19

·317·

平均含水量 19
平均降水量 130
平均链长 19
平均偏差 19,130
平均温差 130
平均验查总数 19
平均值 19,130
平均值(中数) 130
平坡 249
平摊把 81
平原 167
平整土地;填土 114
评分;记分;标记;划痕 195
评价方法 73
评价系统 73
评价原则 73
评价指标体系 73
评吸小组 203
评吸组;专门小组 160
苹果酸 127

苹果酸合成酶 127
苹果酸合酶 127
苹果酸脱氢酶 127
苹果酸盐(或酯) 127
坡地 99,104,202
坡地农田 99
坡度 202
坡度计 36
坡耕地 202
坡积土 36
坡面径流 99,202
坡面治理工程 202
坡向 17,74,202
破烂叶斑病 182
破碎叶 82
(烟叶)破碎指数(分级) 200
破损 106
剖面构型 177
剖面构造类型(剖面构型) 177
脯氨酸 177

葡萄糖 90,91
葡萄糖;右旋糖 56
葡萄糖氧化酶 90
普通过磷酸钙(普钙;过磷酸一钙) 201
普通过磷酸钙;普钙 154
普通黑土 95
普通花叶病 39
普通火山灰土(普通暗色土) 95
普通碱土 95
普通克立格法 154
普通淋溶土 95
普通滤嘴 167
普通潜育土 95
普通试验 43
普通烟草 142,144
普通烟草亚属 220
普通蛀茎夜蛾 217
谱带顶 45

Q

七水硫酸锌 249
栖根真菌 191
畦;垄 188
畦灌 21,32
畦灌;垄灌 189

企业标准 39
起始温度 218
起皱 45,247
气动的;空气的 170
气-固色谱法 88

气候斑点;气候斑点病 245
气候变化 36
气候变化公约 113
气候变化框架公约

· 318 ·

84
气候肥力 36
气候敏感性 36
气候生产潜力 36
气候室;人工气候室 36
气候图 36
气候图集 36
(烟草)气候性病害 59
气候性土壤 36
气候异常 36
气候因子;气候因素 36
气候资源 36
气候资源信息系统 36
气孔 219
气孔的启闭 153
气孔调节 219
气孔运动 219
气流 10
气流复烤 10
气流上升式烤房 238
气流上升式调制室;气流上升式烤房 25
气流下降式烤房 62
气流下降式调制室;气流下降式烤房 25
气溶胶 7
气升式发酵罐 10

气生根 6
气生菌丝 6
气态氮损失 88
气态水 239
气态损失 88
气体采集 87
气体毒害 87
气体肥料 87
气体分析仪 30,87
气体交换 87,187
气体型循环 87
气体熏蒸剂 87
气味浓度 152
气温;常温 18
气雾剂农药 7
气相 10
气相容积 10
气相色谱;气相色谱法;气相色谱仪(GC) 87
气相色谱法 88
气相色谱-红外联用技术 87,88
气相色谱图 87
气相色谱仪;气相色谱分析 87
气相色谱-质谱分析法(气质联用) 87
气相色谱-质谱联用仪 87
气相色谱-质谱仪

88
气象干旱 132
气象因子;气象因素 132
气象站 132
气压 18
气-液色谱法 88
汽化;蒸发作用 239
器官发生;器官形成 156
(烟苗修剪时)掐叶;剪叶 119
掐;摘;摘心;霜害 144
(种子)千粒重 228
迁移 70,135,140,233
迁移机制 234
迁移能力 135
铅(Pb) 117
铅的毒性 117
铅污染 117
前茬作物;前作 176
前作(物) 175
潜伏芽 188
潜伏芽;休眠芽 116
潜性酸度 174
潜育层 90
潜育化土壤 90
潜育期 116
潜育水稻土 90
潜育土 90

319

潜育作用 90
潜在产量 174
潜在的缓冲容量 174
潜在肥力 106,174
潜在饥饿(潜在缺素) 98
潜在缺乏 98,116,174
潜在有效钾 174
潜在有效养分 174
潜在蒸散 174
浅层地下水 199
浅度休眠;非熟休眠 147
浅根 199
浅耕 121
浅耕;浅松土 199
浅耕灭茬;冬浅耕 200
浅耕松土耙 179,223
浅沟灌溉 105
浅褐色 120
(烟叶)浅红色;粉红色;石竹 167
浅黄色(晾烟颜色) 27
浅色烟;淡色烟;烤烟 26
浅色叶背;带灰白色(分级) 159
浅施 199
浅栽 223

浅植 200
浅棕色(分级) 121
嵌合性;花叶病 139
(烟气)呛喉 233
强的(色度);浓(厚)的;(烟味)浓烈 220
强度毒性 220
强度因子 107
强碱 220
强碱性土壤 220
强烈香气;强烈气息 107
强酸 220
强酸化 220
强酸性的 99
强酸性土壤 220
强制通风 83,169
强制性再循环 184
强制循环 83
强壮腋芽 220
羟基 102
羟基丁二酸 127
羟基丁二酸;苹果酸 101
羟基酸 10
切梗丝 218
切丝机 230
茄科 144
茄呢醇 212
茄子 68

侵染剂量 102
侵染前期 175
侵染性病害 105
侵入(过程) 104
侵蚀 72
侵蚀水 109
侵蚀因素 72
侵蚀作用 72
亲水胶体 101
亲水聚合物 101
亲氧的 159
亲子间关系 152
亲子间回归 152
禽肥 175
禽粪肥 84,175
青草气 91
青多黄少 139
青泛白;浮青 92
青痕 92
青黄;黄绿色 92
(烟叶)青黄;青黄色(烟);青黄(烟):代号 GY 93
青黄叶组 92
青灰色的 25,92
青筋黄片 248
青枯症(缺锌) 92
青霉属 161
青色;绿色;青烟 87
青香 92
青香韵 92

青烟 92
青叶子香;鲜叶子香 92
青杂气 92
青杂味 92
青贮饲料 70
轻度毒性 121
轻度感染 124
轻度挂灰 202
轻度花叶病 135
轻度碱性土壤 135
轻度霉变 202
轻度缺乏 120
轻度未熟叶 202
轻度污染的 202
轻粉沙土 121
轻灌(浅灌) 120
轻黏土 120
轻壤土 121
轻霜 120
轻微症状 135
轻盐渍化的 202
轻质壤土 121
轻质土 121
轻组腐殖质 120
轻组碳 120
轻组土壤有机质 136
氢 101
氢化肉桂酸 101
氢化物 101
氢离子浓度 101

氢离子指示剂 101
氢离子指数 101
氢氰酸 101
氢氧化钙 121
氢氧化钾 174
氢氧化镁 126
氢氧化钠 204
氢氧化物 101,159
氢氧离子(羟基离子) 101,102
倾向 234
倾斜度 14
清洁能源 171
清甜香 85
清晰 36
(香气)清新 85
清香型 85
清选出的分级种子 36
清液肥料 36
氰化钾 173
氰化氢 101
氰化物 49
氰化物污染 49
氰化物污染物 49
琼脂 7
琼脂层法 7
琼脂菌落 7
琼脂块法 7
琼脂扩散试验 7
琼脂培养基 7

琼脂平板 7
丘陵地 62,99
丘陵农地 99
丘陵山地 99
秋播 75
秋播型 75
秋地 80
蚯蚓 65
蚯蚓堆肥 65
蚯蚓粪 65,240,247
球菌 37
球状模型 215
区间估计 109
区域保护 123
区域创新能力 185
区域发展;区域开发 185
区域设计 169
区域生态环境问题 185
(定)区域试验 123
区组间 108
驱虫剂 107
趋化物 33
趋化性;趋药性 33
曲霉属真菌 17
曲线拟合 49
取土钻;取土器 89
取消种植 1
取样单元 194
取样工具 194

321

取样密度 194
取样偏差 194
取样器 194
取样设备 194
取样深度 53
取样误差 194
去饱和(作用) 54
去除 186
去除包衣的种子 53
去梗 218
去梗的;抽梗的 218
去梗后叶片 218
去梗烟叶 218
去梗叶片;片烟 220
去离子水 53
去皮种子 51
去湿 184
去污染;除污染 53
权(权重) 245
权重方法 246
全层施肥 246
全氮 232
全钾 233
全量分析 39,232
全硫 233
全面翻耕;全面中耕 39
全面施药;全面施用 158
全苗 86
全苗期 86

全球变暖 90
全球可持续发展 90
全球碳循环 90
全球土壤退化评价 90
全熟种子 86
全碳 232
全叶;整叶 246
全叶卷雪茄 119
全蒸发(过程);蒸发作用 162
全植株 246
全株病;系统性病害 224
全株性症状;系统性症状 224
缺氮;氮饥饿 145
缺氮;氮缺乏 145
缺乏 52
缺乏(油分) 119
缺乏的 52
(营养)缺乏症状 224
缺肥 128
缺钙 28
缺钙绿病 121
缺钾 173
缺钾症 173
缺磷症 15,164
缺硫 222
缺氯 34

缺滤嘴;掉滤嘴 137
缺镁症 126
缺锰褪绿症 127
缺锰症 137
缺钼症 139
缺硼症 26
缺区 137
缺水 243
缺水类型 243
缺素培养液 48
缺素试验 137
缺素营养液 69
缺铁失绿症 109
缺铁褪绿病 109
缺铁症 110
缺铁症;缺铁 109
缺铜 44
缺铜症 44,47
缺锌 249
缺锌症 249
缺穴 137
缺叶绿素症 34
缺株 137
缺株率 136
确认;批准;生效 239
群间选择 108
群落生态学 39
群落生态学;群体生态学 224
群体改良 172

R

燃点;着火点 112
燃烧法 39
燃烧能力 27
燃烧热;热值 96
燃烧速率 28,183
燃烧特性 27
燃烧性 28,38
燃烧质量 28
壤缓冲能力 205
壤黏土 123
壤沙土 123
壤土 123
壤土;沃土 123
壤土组 123
壤香;泥土气味 65
壤质 123
壤质粗沙(土) 123
壤质化 123
壤质黏土 123
壤质土 123
壤质细沙(土) 123
壤质细土 123
扰动层 60
扰动土壤 60
热带农业 235
热导率 227
热法磷肥 96,227
热风调制 96

热风循环式烤房 96
热辐射 96,227
热害 96,100
热害;热逆境 96
热交换 96
热交换器;散热器;换热器 96
热解 179,227
热解温度 179
热浪 96
热量资源 96
热能 96
热能利用率 96
热死点(致死温度) 227
热效率;热能利用率 96
热效应 96
热性肥料 241
人畜粪 100
人粪尿 100,144
人工(造)土壤 128
人工成本;劳务费用 113
人工除草 128
人工醇化 83
人工催熟 17
人工催芽 16

人工打顶 128
人工发酵 83
人工费用 113
人工分类 16
人工腐殖物质 16
人工合成改良剂 224
人工降雨 16,182
人工模拟降雨 128
人工抹杈 95,128
人工培养 16
人工气候 16
人工气候室 43,178
人工气候室;育苗室;环境实验室 166
人工神经网络 16
人工授精 1,107
人工土壤 17
人工团聚体 16
人工移栽 128
人工扎把 95
人工智能 16,107
人力节约;人工节约 113
人力农机具 128
人为表土层 15
人为发生过程 15
人为肥力 15
人为富营养化 48

人为干扰 100
人为排放 15
人为侵蚀 15,17,72,128
人为土壤 15,25
人为因子 15
人员评价 128
人造(雪茄烟)内包皮;人造黏合剂 16
人造堆肥 16
人造厩肥温床 224
人造气候 16
人造烟草薄片 17
忍耐物种 231
任意调查 84
韧度 233
韧皮部 21,163
韧皮部运输 163
韧性结持(度) 226,233
(烟叶)韧性;强度;劲头 220
日变化 50,61
日光温室 97,212,222
日较差 50,60
日平均温度 50
日长敏感性;光(周)期敏感性 164
日照量;太阳辐射 212
日照时间 222

日照时间;日照长度 64
日照小时(数) 222
日照长度 120
日照长度;日长 64
日灼;日烧 222
日灼斑块 222
日最低温度 50
日最高温度 50
茸毛 118,229
茸毛脱落 94
绒毛烟草 142
绒毛叶 44
绒毛状的 240
容积热容量 241
容量 47
容量瓶 241
容重 241
溶胶 38,212
溶解法 60
溶解罐 60
溶解态活性磷 60
溶解态无机磷 60
溶解态有机磷 60
溶解性 60
溶解有机碳 60
溶菌作用 20,86
溶磷真菌 163
溶磷作用 163
溶液肥料 213
溶液培养法 213

溶液培养试验;水培试验 213
溶液态钙 243
溶液态镁 243
溶液中磷 213
溶质扩散 213
溶质势 213
溶质运移 213
熔融法 87
融冻扰动(作用) 47
融化 227
融水灌溉 227
柔和 204
柔和卷烟 135
柔韧性 233
柔软 215
肉豆蔻 149
肉豆蔻酸;十四酸 141
肉桂醛 35
乳酸 114
乳酸盐溶性磷 114
入渗 105
入渗量 105
入渗率 105
入渗强度 105
入渗水(湿润水) 242
入渗速度 105
入渗通量 105
软腐病 204
软化水 53

锐化图像 200
锐形(叶)尖 170
润湿性;亲水性 241

弱分解有机层 202
弱碱 245
弱苗 245

弱酸 245
弱酸化 245

S

撒播 26,86,182
撒播(种)机 26
撒播;撒施;传播;广播 26
撒堆肥机 76
撒施 26
撒施(底)肥 78
撒施肥料 78
塞子 219
三叉地老虎 73
三段式烘烤 228
三角烧瓶;锥形瓶 41
三角试验评吸法 64
三磷酸腺苷 6
三磷酸腺苷酶 5
三氯甲烷 34
三圃制 228
三羧酸循环 113
三通管 228
三要素比例 228
三要素适量试验 228
三元肥料 234
三元复合肥 228
伞形;伞型的 236

散点图 62,195
散点制图 62
散热器;换热器 96
散烟叶 124
散叶(分级) 124
散装掺混肥料(BB肥) 27
扫描电子显微镜 195,198
扫描透射电子显微镜 195
色氨酸 235
色度;色泽 38
(烟叶)色度;饱和(分级) 195
色度淡 120
色度强 220
色很浅的弗吉尼亚雪茄 138
色谱分析;色层分析 219
色谱分析法 34
色谱图 34
色谱图扫描 34

色谱仪;用色谱分析 34
色谱柱 34
色泽;色度 34
涩味 17
森林 83
森林的水源涵养功能 215
森林土壤 83,223
杀虫 60
杀虫(菌)剂;农药 162
杀虫剂 107
<商>杀虫灵 3
杀虫效果 107
杀虫药剂规格(说明书) 214
杀菌 20
杀菌的;抗细菌的 15
杀菌剂 20
杀菌剂的效果 86
杀线虫剂 143
沙砾 91,161
沙砾质土 93

· 325 ·

沙粒 194
沙姆逊(香料烟品种) 194
沙黏土 192,194
沙培 194
沙培试验 194
沙壤土 192,194
沙土 194
沙土率 194
沙质的;多沙的 192,194
沙质粉沙壤土 194
沙质粉沙土 194
沙质黏壤土 194
沙质黏土 194
砂姜黑土 121,199
砂姜土 192
砂岩 194
筛孔 200
筛孔尺寸 202
筛选 197
筛子 195
晒伤;日灼 118
晒烟 222
晒烟架 222
晒烟型 222
晒烟型卷烟 222
晒制 222
晒种 197
山地 140
山地黄棕壤 140

山地农业 98
山地土壤 140,157
山区农业 99,140
伤害;病害;病伤;损坏;受伤 50
商标名称 26
商品肥料 39
商品农药采样方法 194
商品农药验收规则 39
商品有机肥 39
商业化农业 39
上部叶 238
(香料烟)上部叶 238
上部叶(香料烟) 157
上部叶比率 238
上部叶组 20
上层 238
上等农地 176
上等烟 98,222
上二棚 20
上二棚叶(分级) 196
上结下松结构 47
上升水 17
烧杯 21
烧结法 201
烧瓶 81
梢枯病;顶枯的(干顶病) 216
(叶色光泽)稍暗 64

稍薄 120
稍疏松;(丝束)开松 153
(烟叶)稍皱;稍耙松 245
少耕 136,184
少耕播种 136
少耕法 120,184
少耕体系 136
少叶的 152
奢侈供应 125
奢侈吸收 125
设计;布局;布置;陈设 117
设施农业 72,75,178
社会效益 203
摄谱仪;光谱仪 215
伸根期 190
伸长期 161
伸长区 69
身份不足 120
身份厚实;组织紧密;(烟叶分级)身份重 96
身份欠佳的烟叶;马里兰烟叶第二组烟叶;亮色烟叶 227
砷 16
砷;砒霜;三氧化二砷 16
砷酸铅害 117

砷污染 16
深播 52
深层施肥法 221
深层施肥机 78
深橙(橘)黄 52
深耕 51,52
深耕;深松;深翻 52
深耕犁 221
深耕犁;潜叶蝇 135
深耕犁;深耕 234
深耕器 136,221
深耕施肥法 52
深耕细作 52
深耕作业 52
深褐色 50,51
深红色;暗红色 50
深绿色 96
深色 51
深色卷烟 50
深色晾烟 50
深色晒烟 51
深色雪茄;色深味浓的(雪茄) 126
深色烟草 51
深色烟熏烟 50
(烟叶)深棕色 50
深施 27,52
深松耕 34,52
深栽 52
渗出液 161
渗灌 25,80,198

渗漏 119,198
渗水力(水分渗透能力) 243
渗透 161
渗透势 157
渗透试验(穿透试验) 161
渗透水 54
渗透调节 157
渗透性 161
渗透压 157
渗透仪 162
渗透作用 157,162
渗析纸 57
升温降湿 100
升温梯度 189
生产层(土壤) 177
生产工艺 128
生产率;生产力 177
生产能力 157,177
生产潜力 177,248
生产效率 177
生根层 188
生根粉 191
生根激素 191
生根能力 191
生-光降解膜 24
生化抑制 22
生荒地;处女地 144
生活史 120
生境 94

生境分类 94
生境改善 94
生境稳定性 94
生境因素 94
生境综错作用;生境总体 94
生理成熟 166
生理成熟(生理成熟期) 166
生理毒害 166
生理反应 166
生理干旱 166
生理过程 166
生理机理 166
生理机能 166
生理碱性肥料 165
生理生态需水 65
生理生态学 166
生理失调 166
生理酸性肥料 165
生理特征 165,166
生理性病害(生理病) 165
生理有效水 165
生理中性肥料 166
生命必要元素;生物元素 23
生青气 91,183
生青味 91
生石灰 27,125,181,237

· 327 ·

生态安全 66
生态保护 65,66
生态背景 65,66
生态补偿 65
生态场 66
生态承载力 65
生态承载力;生态容量 65
生态脆弱性 66
生态对策 66
生态恶化 66
生态风险评价 66
生态幅 65
生态工程 65,66
生态功能 66
生态功能区 66,67
生态功能区划 66
生态过渡带;生态交错带 67
生态环境 66,67
生态环境恢复 65
生态环境区适应 65
生态环境退化 65
生态环境问题 65
生态环境系统 65
生态环境需水量 65
生态环境质量 180
生态恢复 66
生态监测 66
生态建模 67
生态经济平衡 66

生态可持续发展 67
生态可塑性 66
生态控制;生态防治 65
生态拦截 65
生态模拟 66,201
生态模型 66
生态农业 65
生态平衡 65,66
生态平衡 73
生态评价 65,66
生态破坏 54,65,66,67
生态气候适应 65
生态倾销 65
生态入侵 66
生态社区;生态群落 65
生态生物多样性 65
生态失调 64,66
生态示范区 65
生态适宜性 66
生态适宜性评价 66
生态适应性 65
生态调节 67
生态土壤学 67
生态退化 66,67
生态退化过程 177
生态完整性 66
生态危害 66
生态危机 66

生态威胁 66
生态系统 66,99
生态系统多样性 67
生态系统复杂性 40
生态系统健康 67
生态系统类型 67,236
生态系统敏感性 67
生态系统碳平衡 67
生态系统完整性 67
生态系统要素 67
生态相似性 66
生态小生境 66
生态效益 65
生态效应 66
生态型 66,67
生态需水 67
生态学 24,67,132,152
生态学数学模式 129
生态循环 65,66
生态压力 66
生态异质性 66
生态因子 65,66
生态因子补偿作用 39
生态影响 66
生态影响评价 66
生态用水量 66
生态优势种 66
生态指示植物 66
生态指示种;生态指示

· 328 ·

植物;生态表征 66
生态质量 67
生态治理 67
生态资源 66
生态最适度 66
生土 183
生土;未成熟土 103
生土层 183
生物 123
生物(质)炭 22
生物标志物 24
生物产量 24
生物处理 24
生物刺激 24
生物地球化学循环 23
生物堆制 24
生物多样性 23
生物多样性保护 23
生物多样性的破坏 54
生物惰性碳 24
生物反应器 24
生物防治 23
生物防治措施 23
生物放大效应 23
生物肥力 24
生物肥料 23
生物分解(作用) 23
生物风化 24
生物富集 22,23

生物富集;生物浓缩 23
生物改良 24
生物隔离 24
生物固持(作用) 24
生物固氮 24
生物固定(作用) 24
生物固结 24
生物归还率 24
生物合成 24
生物化学(生化) 22,23
生物化学诊断 22
生物活性 23
生物活性碳 23
生物积累作用 23
生物监测 24
生物碱 142
生物碱测定仪;生物碱分析仪 11
生物碱的 11
生物碱的生物合成 11
生物碱功能 86
生物碱积累 11
生物碱降解 11
生物降解 23
生物降解膜 23
生物降解能力 23
生物降解速度 183
生物界 123

生物净化 23,24
生物矿化 24
生物量 24
生物农药 25
生物农药;生物杀虫剂 24
生物农业 22,23
生物培养诊断 57
生物平衡 23,24
生物气候 23
生物迁移;生物转运 25
生物侵蚀 23,24
生物圈 24
生物圈;生态层 67
生物圈生态学 67
生物群落 23,24
生物燃料 23
生物扰动作用 25
生物人工气候室 25
生物杀虫剂 23,24,123
生物生态学 23
生物试验 24
生物试验法 22,132
生物适应性 23
生物条件 24
生物统计学 24
生物脱甲基作用 23
生物物质生产的能力 29

·329·

生物系统　24
生物相互作用　24
生物信息学　23
生物修复　24
生物序列　24
生物选择性　24
生物学　24
生物学类型　24
生物学耐受性　24
生物学特性;生态学　24
生物学特征　23,31
生物学性质　24
生物循环　23
生物氧化　24
生物因素　24
生物因素;生物制剂　23
生物因子　24
生物有机肥　24
生物有效性　22
生物有效性浓度　22
生物有效性物质　22
生物有效养分　22
生物源农药　23
生物增添作用　22
生物制品　23
生物质能　23
生物转化　23
生物转化作用　24,25
生物资源保护　178

生硬　47
生育期　161
生育期灌水　110
生长迟缓　220
生长促进剂　94
生长袋法　106
生长点　15,93,240
生长发育　93
生长发育迟缓　93
生长发育弱　245
生长发育旺盛　240
生长分析　93
生长规律;生长节律　94
生长环境　94
生长缓慢　172
生长激素　94,226
生长季节　94
生长介质　94
生长茂盛的　125
生长期　93,94
生长期;生长季节　93
生长曲线　93
生长盛期　160
生长室;人工气候箱　93
生长素;生长物质　94
生长素型植物生长调节剂　19
生长速率;增长率　94,183

生长速率减慢　185
生长调节剂　94
生长调节剂与改良剂　94
生长调节物质　94
生长调节因子　94
生长异常　110
生长因素　94
生长诊断　94
生长周期　240
生长周期性　94
生长阻滞;延缓生长;生长减缓　188
生脂固氮螺菌　19
生殖器官　156
生殖器官的发育　56
生殖生长　186
牲畜排放物污染　171
盛花期　86
盛花期打顶　232
盛开花的　177
剩余效应;残余效应　30
失绿病　34
失绿效应　34
失绿症(萎黄症)　92
失调　60,127
施单一肥料　78
施氮肥　145
施底肥　15,21,26
施肥　77,128

施肥;施肥位置 78
施肥次数 85
施肥的经济学 67
施肥方案 78
施肥方法 133
施肥方式;施肥模式 77
施肥过量 73,158
施肥机 128
施肥计划 77
施肥技术 78,225
施肥结构 79
施肥决策 127
施肥量 62,77,183
施肥量不足 236
施肥模式 77
施肥时期 128,229
施肥水平 78
施肥位置 167
施肥信息系统 77
施肥诊断法 57,77
施肥制度 77,78
施肥制度(系统) 224
施基肥 15,21
施钾 173
施磷肥 163
施木克值 195
施药量;使用剂量 15,62
施用方法 132
施用幅宽 15

施用量 15
施用石灰 121
施用时间 15
湿沉降 246
湿地 149,204,246
湿度百分率(含水量百分数) 138
湿度百分数 161
湿度计 102,78
湿法 246
湿法磷肥 246
湿法消化(解) 246
湿害 138
湿球(温度计) 246
湿球温度 246
湿润的 138
湿润灌溉 43
湿润气候 138
湿润水分状况 236
湿润土地 216
湿润指数 138
湿烧法 246
十二醛;月桂醛 61
9,12-十八碳二烯酸 122
十九(烷)酸 146
十字花科 47
十字花科型 46
石膏 94
石灰 112,121
石灰的;含钙的 28

石灰撒施机 121
石灰土 28
石灰效应 121
石灰性褐土 28
石灰性砂姜黑土 28
石灰性水稻土 28
石灰性褪绿病 121
石灰性紫色土 28
石灰需要量 121
石灰需要量测定 121
石灰岩 121
石灰岩土;石灰土 121
石灰诱导的失绿症 121
石块 219
石英坩埚 200
石英砂 181
石油醚 121,163
石油醚提取;石油醚提取物 163
石油醚提取物 161
石质土 122
时间 229
时间尺度 226
时间序列 229
时空特征 226
时域反射法 229
实地调查 152
实际发芽率 175
实际防治效果 5

·331·

实际肥力 5
实际留叶数 148
实际能耗 5
实际显著性 5
实际有效水量 74
实际蒸散 5
实时氮素管理 184
实时养分管理 184
实验结果 74
实验室化学制品 113
实验室模拟 113
实验室认证 113
实验室设备 113
实验室数据库 113
实验样品 113
实验用水 242
食粪性土壤动物 44
食物安全 83
矢量地图 239
矢量数据 239
使成熟;催熟 189
使人不愉快的;讨厌的 237
使用周期;使用年限;生活周期;生活史 120
示范农场 53
示范区;试验区 53
示范试验 53
示范田 53
示范效应;示范作用

53
示意图 2
(同位素)示踪的 113
示踪原子 113
世代 88
世代交替 12,58
世界吸烟与健康会议 247
试点调查 167
试管 226
试管授粉 104
试行标准 226
试剂 184
试验农场 167
试验区配置 169
试验区设计 169
试验田 226
试验项目 74
试验小区 226
试验性研究;探索研究;中间试验 167
试验用地 175
试验站 74
试验值 74
试验植物;试验装置 226
试样 226
试纸 104
适当管理 6
适量 6
适量施肥 6

适时管理 229
适时施用 229
适时移植 229
适熟 80,153
适酸性植物;喜酸性植物 159
适宜产量 154
适宜灌溉 6
适宜含水量 221
适宜温度 175
适宜株行距 178
适应 5
适应不良;适应不全;失调 127
适应能力 5
适应品种 5
适应土壤 5
适应性 5
室(指子房、花药等);子囊腔 123
室内气候 105
室内栽培 105
室外漂浮育苗床 157
释放 186
释放速率 186
嗜碱微生物 11
嗜冷微生物 178
嗜热微生物 227
嗜酸菌 4
嗜酸生物 4
嗜酸微生物 4

嗜压菌 21
嗜压微生物 21
嗜盐微生物 95
嗜盐微生物;喜盐性细
　菌 95
螫刺 5
收割 96
收割;收获 184
收割机 95
收割机(收获机) 46
收购 179
收获方法 95
收获后变化 173
收获量 95
收获面积 95
收获期 95
收获前的 175
收获前期 175
收获指数 95
收烟编号 179
手扶拖拉机 241
手感 95,233
手工采收;分批摘叶
　95
手工打顶 128
手工捏合(雪茄) 128
手工送苗式(栽苗机)
　128
手工移栽器 95
手选 95
受感染的;受侵害的;

受影响的 7
受光发芽 120
受害程度 50
受害的 50
受旱烟草 63
受精作用;施肥;传粉
　77
授粉 171
授粉花数 149
瘦土 149
舒展(烟叶) 153
疏果 86
疏花 82
疏苗;间苗 198
疏水胶体 101
疏松 124
疏松壤土 132
疏松质地 124
熟地 129
熟地;已耕地 176
熟地化 77
熟化作用 132
熟石灰 101
熟石灰;消石灰 202
熟土;松软海绵土
　132
熟土层 132
属 88
属;类 89
属间杂交 89
属性统计 18

束缚水 26,38,41
树脂香 187
数据库管理系统 51
数据库系统 51
数据挖掘 51
数据预处理 51
数量分析;定量分析
　181
数量性状 181
数量因素 181
数学模拟 129
数值转化 233
数字地球 58
数字地形模型 58
数字高程模型 58
数字农业 58
衰变 51
衰老 199
衰老期 152,199
衰退过程 52
衰退期 52
双胺灵 34
双壁开沟犁 122
双侧检验 235
双侧置信区间 235
双管法(测土壤导水
　率) 62
双行播种 62
双行中耕机 235
双降解膜 62
双尾检验 235

· 333 ·

双向电泳 51
双重营养缺陷型 62
双子叶植物 57
霜冻 85
霜冻(叶) 85
霜冻烟叶 85
霜冻致死 85
霜害;霜打 85
霜霉;霜霉病 62
霜霉病 62
霜期 85
水不溶性含氮化合物 242
水层 242
水成土 101
水稻土 159,189
水滴入渗时间 244
水肥管理 127
水肥耦合 241
水分(含水量) 50
水分;水汽;湿气;湿度 138
水分保持 243
水分保持(保水) 242
水分补偿 242
水分大的软叶 204
水分代谢 242
水分供需平衡 244
水分供应 138
水分供应;供水 244
水分过多 73

水分检测 138
水分亏缺 242,243
水分利用 244
水分利用率 68
水分利用系数 244
水分利用效率 244
水分临界点 45
水分临界期 242
水分排除 2
水分平衡 101,242
水分容积百分率 241
水分渗漏 243
水分生产率 243
水分损失(水分流失) 242
水分梯度 138
水分胁迫 244
水分循环 138,242
水分盈亏量 244
水分盈余 244
水分运输途径 244
水分蒸腾 244
水分指数 138
水分状况 101,138,243
水管加热温床 167
水害,淹涝,水淹,涝害 63
水旱轮作 159
水合补偿点 101
水和养分的运输 242

水环境质量评价 242
水窖 242
水窖类型 31
水解 101
水解产物;水解液 101
水解常数 101
水解氮 101
水解反应 101
水解酶 101
水浸叶 58
水浸渍坏叶 242
水涝 244
水涝池;水浸地 244
水涝地 244
水力 241
水流 138,242
水面蒸发 73
水培 16,101
水培(溶液培养) 242
水培法 101,242
水培试验 242
水培养 101
水平传播 99
水平衡 242
水平抗性 99
水平梯田 21,120
水平致病性 99
水圈 101
水溶胶 101
水溶态锰 243

334

水溶态钼 243
水溶态硼 243
水溶态铜 243
水溶态锌 244
水溶性氮 243
水溶性腐殖质 242
水溶性钾 244,245
水溶性磷 245
水溶性磷肥 244,245
水溶性农药 245
水溶性酸 243
水溶性糖的测定 56
水溶性盐类 245
水溶性养分 243
水溶性有机碳 244
水生根 243
水蚀 72,242
水蚀过程 242
水体污染 242
水体重金属离子污染 97
水田耕作制 76
水土保持 41,204
水土保持耕作措施 204,228
水土保持农业 244
水土保持生态效益 65
水土保持生物措施 24
水土保持植物措施 204
水土环境质量 244
水土流失 124,204,242
水-土-植物关系 244
水温 244
水稳性 244
水稳性团聚体 244,245
水污染 242,243
水污染面源 16
水循环 101
水样 243
水样pH测定 56
水样铵离子测定 56
水样采集 243
水样的采集与保存 243
水样钾离子测定 56
水样磷酸根测定 56
水样硫酸根测定 56
水样全氮测定 56
水样全磷测定 56
水样硝酸根测定 56
水浴 244
水质 243
水质标准 243
水质监测 243
水质检验 242
水质控制 243
水质评价 243
水质指标 243
水中农药污染 15
水准仪 120
水资源 243
水资源短缺 243
水资源管理 243
水资源配置 243
水资源信息系统 243
水渍 244
顺丁烯二酸 127
顺丁烯二酰亚胺(MH) 127
顺坡 238
顺坡耕(直耕) 219
顺坡耕作 62
顺式冷杉醇 35
顺序抽样 199
顺序检验 199
顺序统计量 154
顺序下降 54
瞬态流法(测土壤导水率) 233
说明书 214
蒴果成熟期 129,130
丝氨酸 199
丝状真菌 80
死青(烟叶);青暗色 64
死亡浓度曲线 41
四分法 181

四分法(土壤取样) 210
四分位差 181
四氢呋喃 226
松把(分级) 124
松动;松散 124
松结态腐殖质 124
(卷烟)松紧度;紧密度;坚实度;紧实度 39
松软 204
松软肥沃土壤 236
松软结持(度) 132
松软土 138
松软土壤 204
松软土壤;熟土 132
松散 124
松散的(酥性的;脆性的) 85
松散结持(度) 85
松散土壤 85,124
松散性(酥性;脆性) 85
松沙土 124
松土 195
松土播种法 34
松土机;刨土机;种子破皮机 195
松土机;松土犁 124
松土犁;碎土犁 178
松脂气 235

苏氨酸 228
苏打 203
<商>苏云金杆菌 9,20
速测 181,183
速效肥 183
速效肥料 5,181
速效肥料;速效性肥料 183
速效含水量 183
速效钾;速溶性钾 183
速效磷 183
速效性养分 183
塑料薄膜覆盖垛 167,169
塑料大棚 169
塑料废料 169
塑料温室 169
塑性 169
塑性上限 238
塑性下限 125
酸 4
酸不溶性养分 4
酸度 52
酸度(酸性) 4
酸度效应 4
酸害 4
酸化 4
酸化,酸化作用 4
酸化过程 4

酸碱滴定法 4
酸碱电子理论 69
酸碱度计;pH计 163
酸碱反应 4
酸碱平衡 4
酸碱体系 4
酸碱指示剂 4
酸侵蚀 4
酸溶性养分 4
酸式硫酸钾 4
酸式碳酸钙 28
酸式碳酸盐 22
酸洗净的石英砂 4
酸性沉积 4
酸性腐殖化(作用) 4
酸性腐殖质 4,213
酸性粒子 4
酸性硫酸盐土 4
酸性黏土 4
酸性溶液 4
酸性土 4
酸性土壤 4
酸性氧化物(成酸氧化物) 4
酸性重铬酸钾法 4
酸性紫色土 4
酸性棕壤 4
酸雨 4
算数平均 16
随机变量 219
随机变量;随机向量

182
随机变量的变异系数 239
随机变量的标准差 217
随机变量的方差 239
随机分布型式 182
随机函数 219
随机化 182
随机模型 182
随机排列 182
随机区组设计 182
随机取样 182,219
随机误差 182
随机效应 182
随机性 183
随机样本 182
随水浇施 15
碎块结构 125
碎块状结构 36
碎片;烟丝 182
碎土犁 95
碎土器 204
碎叶片 118,195
碎叶组 192
(烟叶)损伤程度 52
损伤叶;伤残叶 50
索晒 222

T

TCA循环;三羧酸循环 234
T-分布 225
t-因子 227
他感作用物质 11
踏勘 184
台地 225
台式发酵罐 21
台田 182
太阳能加热烤房 212
太阳能利用率 68
太阳直接辐射 58
炭 31
炭化 31
炭疽病 14
碳 29
碳捕获和封存 29
碳代谢 29
碳氮比 28,29,30
碳氮分配 29
碳动态 29
碳含量 29
碳化(作用) 30
碳化的 30
碳化作用 30
碳汇 30
碳减排 29,185
碳库 29
碳库管理指数 29
碳硫比 30
碳密度 29
碳排放交易 29
碳平衡 29
碳氢化合物 101
碳失汇 137
碳水化合物;糖类 29,30
碳水化合物代谢途径 29
碳酸 30
碳酸铵 12
碳酸钙 28
碳酸钾 173
碳酸镁 126
碳酸氢铵 12,13
碳酸氢钙;重碳酸钙 28
碳酸氢钾 173
碳酸氢钠 203
碳酸氢钠;小苏打 203

碳酸氢盐;重碳酸盐 22
碳酸体系 30
碳酸盐层 30
碳酸盐结合态 30
碳循环 29
碳氧化物污染 171
碳营养 29
碳源 30
碳源谱 215
碳质氧化 30
弹性;伸缩性 69
羰基键 112
糖 221
糖/烟碱之比;糖/碱之比(率) 221
糖的合成 221
糖分含量 221
糖碱比 183
糖酵解 90
糖类 192
糖类代谢 221
烫片(叶);焦片(叶);糊片 195
桃蚜 141
淘汰品种 59
套袋;装袋 20
套种 237
套作 186
特大的;特长的 112
特级(烟叶;分级) 222
特殊分级 214
特殊气味 214
特征参数 116
特征根 116
特征香气 31
特征向量 69
特征性症状 31
特征值 69
梯田 226
梯田;阶地 226
梯田耕作 226
梯田灌溉 21
梯田旱耕地 226
梯田农地 226
梯田农业 42,226
梯田种植 226
锑 15
提纯 179
提高土壤肥力 104
提高土壤质量 210
提高质量;改善质量 104
提高种子活力 109
提前成熟 189
提前落叶 238
提前收获 175
提水灌溉 179
(根系)提水率 242
(根系)提水作用 101
提早发芽 64

体积 47
体积含水量 241
体外营养的,外生的 67
替代农药 186
替代作用 221
天敌 142
天敌作物 70
天冬氨酸 17
天冬酰胺 17
天冬酰胺合酶 17
天冬酰胺酶 17
天然的;粗的;生烟;生叶 46
添加;添加物 5
添加量 5
田间测定 79
田间持(水)量 79
田间持水量 80
田间持水能力 79
田间定位精确管理 201
田间防治 79
田间观察 79
田间管理;现场管理 79
田间管理法 79
田间技术 80
田间检验 79
田间排列;小区排列 169

田间排水系统 79
田间侵染 79
田间渗透系数 79
田间试验 79,80
田间试验;大田试验;
　肥料试验 80
田间试验计划 80
田间试验解释 79
田间试验设计 79
田间适宜作业季节
　76
田间释放 80
田间水分当量 79
田间水分亏缺 79
田间水分状况 79
田间调查;现场调查
　80
田间调制 79
田间土壤 80
田间萎蔫 80
田间小区试验 80
田间需水量 80
田间烟草长势 174
田间研究;实地研究;
　实地调查 80
田间用水效率 80
田间诊断 79
田间最大持水能力
　79
田间最小持水量 136
田间作业适宜天数
　76
田野;农业;作物 9
甜菜碱 22
甜的;芳香的 223
甜味 223
填充料比例 80
填料 123
填闲作物栽培 30
(卷烟)条 30
条播 63,122,191
条播;沟播 197
条播;条播机;钻头
　63
条播机 191
条播种机 63
条播种子 63
条垛式 247
条件分布 41
条件分析;因素分析
　75
条施 20,78,191
条施(肥) 63
条施;条播 63
条施肥 20
条纹病(缺锰) 220
条纹病斑;条纹;色线;
　层;加条纹 219
条纹病毒 220
条纹状坏死 220
条栽;行栽 168
条栽;列栽 169
条状耕播法 220
条状施肥 20
条作;条播栽培 191
莕子 240
调查范围;研究领域
　79
调亏灌溉 185
调味;醇化;调味品;佐
　料;干燥;变干 195
调温器 227
调香;合香;掺和;配
　叶;配方 25
调制 48
调制不当(烟)叶 136
调制不均 237
调制操作 49
调制管理 49
调制后的(烟叶);调制
　48
调制技术 49
调制架 49
调制期 49
调制设施 49
调制时间 49
调制室;烤房;晾房
　48
调制室风门;烤房 50
调制室面积 81
调制速度 49
调制损失 49
调制条件 48

· 339 ·

调制温度 49
调制系统 49
(烟叶)调制性能 48
调制中烟梗水分过大的烟叶 76
调制组件 49
萜(烯);萜(烃) 226
萜类化合物 226
铁 109
铁、铝氧化物 109
铁蛋白 77
铁肥 110
铁铝土 77
铁锰氧化物结合态 77
铁锹 214
铁缺乏(症) 17
铁素营养失调 110
铁细菌 109
铁氧还蛋白 77
铁质黏土 109
烃 101
通风 240
通风干燥 6,170
通风排湿 100
通风设备 62
通径系数 160
通气降温;通风降温 6
通气搅拌 6
通气孔隙度 6,10

通气良好土壤 246
通气培养 6
通气性;透气性 87
通气性差 103
通气状况 6
通透性土壤 162
通用施肥技术 39
通用施肥模型 88
通用指示剂 237
同分异构体 111
同化;同化作用 17
同化产物 17
同化物 17
同化物质的贮存和运输 234
同化系数 17
同化效率 17
同化性硝酸盐还原作用 17
同晶置换 111
同时成熟 201
同位素 111
同位素标记 111
同位素丰度 111
同位素示踪法 111
同位素示踪技术 111
同位素稀释分析(法) 102
同位素效应 111
同株异花受精 88
铜 44

铜肥 44
铜素杀菌剂 44
铜素营养失调 44
铜污染 44
酮 112
统计的 218
统计分布 218
统计分析系统 218
统计覆盖区间 218
统计假设 218
统计检验 218
统计模型 218
统计图形 32
头状聚伞花序 49
投入产出比率 107
投入产出分析 107
透发性 58
透光率 121
透明薄膜 234
(卷烟纸等的)透气度;透气性 10
透射电子显微镜 225
透水性 243
突变育种 141
图示曲线;图表 103
徒长 7,119,158,215
徒长;徒长的 215
徒长苗 119
涂布培养法 216
涂片 203
土表 211

土层厚度 227
土传病害 205
土地 114
土地报酬 61
土地报酬递减率 116
土地的生产力 115
土地非农化 114
土地分等 114
土地分级图 114
土地覆盖 114
土地改良 16,114
土地干旱化 114
土地耕作 114
土地管理 114
土地规划 114
土地耗竭 114
土地荒漠化 54,114
土地恢复;土壤改良 115
土地开发 114
土地开垦;土地改良 115
土地类型 115
土地利用 114,115
土地利用类型 115
土地利用率 115
土地利用图 115
土地利用现状图 49
土地利用优化配置 153
土地利用专题图 227

土地撂荒 114
土地面积 114
土地平整(整地) 114
土地平整;整地 176
土地评估 115
土地沙化 115
土地生产能力 114
土地使用效率 115
土地适宜性 115
土地适宜性评价 115
土地酸化 4
土地所有权;土地所有制 114
土地调整 16
土地图 114
土地污染 114
土地信息 114
土地信息系统 114
土地休耕 115
土地要素 114
土地整理 16,114
土地质量 115
土地质量变化 31
土地质量等级 115
土地资源 16,115
土地资源开发 115
土地综合生产力评价 40
土耳其烟(香料烟;东方型烟) 235
土纲 154

土块 37,208
土类 92,207
土粒电荷 209
土粒胶结(作用) 209
土粒密度(曾称土壤比重) 209
土粒密度测定 55
土粒凝聚(作用) 209
土粒有效直径 68
土培试验 205
土壤^{14}C 年龄 204
土壤 204
土壤 pH 值 209
土壤 pH 209
土壤 pH 图 209
土壤铵态氮测定 55
土壤薄瘠层 227
土壤饱和水流 194
土壤保持 205
土壤保持覆盖 212
土壤保护 210
土壤报酬 188
土壤本底值(土壤背景值) 204
土壤比表面积 211
土壤比容 211
土壤比重 211
土壤比重(曾用名) 211
土壤表层 128,238
土壤病害 205

土壤波谱特性 215
土壤残留 210
土壤残留性农药 209
土壤测定 55
土壤测试 211
土壤层次 207
土壤层次（土层） 208
土壤超微形态学 211
土壤超显微技术 221
土壤承载力 205
土壤持水率 214
土壤持水性 243
土壤初始含水率 106
土壤穿入阻力 209
土壤传播 89
土壤床发芽试验 89
土壤垂直地带性 204,212
土壤垂直分布 211
土壤大生物 208
土壤大型动物区系 208
土壤单项污染指数 105
土壤氮素表观平衡 208
土壤氮素表观盈亏 15
土壤氮素供应能力 146,208
土壤氮素净矿化量 208
土壤氮素形态 84
土壤氮素转化 146
土壤氮素状况 218
土壤氮素总矿化量 207
土壤导水率 207
土壤的人为熟化作用 15
土壤的熟化度 103
土壤地带性 212
土壤地球化学 161,207
土壤电化学性质 69
土壤动力学性质 64
土壤动态变化 206
土壤动物 204
土壤动物区系 206
土壤动物学 212
土壤多样性 161
土壤发生 161,207
土壤发生层 207
土壤发生分布 207
土壤发生风化过程 161
土壤发生与演变 88
土壤发育 205
土壤反应 210
土壤非饱和水流 237
土壤肥力 206
土壤肥力保持 207
土壤肥力等级 206
土壤肥力等级标准 45
土壤肥力递减定律 116
土壤肥力分级 206
土壤肥力管理 207
土壤肥力监测 207
土壤肥力减退 206
土壤肥力评价 206
土壤肥力特征 31
土壤肥力图 207
土壤肥力学说 227
土壤肥力因子 75,206
土壤肥力指标 206
土壤肥力指标体系 207
土壤肥力质量 207
土壤分布 205
土壤分类 205
土壤分类制 205
土壤分散 205
土壤分散剂 205
土壤分析 204
土壤风化 212
土壤风蚀 205,212
土壤腐殖质 65,207
土壤腐殖质组成测定 55
土壤腐质化(作用)

207
土壤覆盖 208
土壤覆盖;土被 208
土壤改良 184,204,207
土壤改良;复垦 210
土壤改良剂 113,204,207
土壤改良剂;土壤调理剂 205
土壤改良物质 204
土壤钙形态 83
土壤概查 88
土壤杆菌 9
土壤感染 207
土壤干旱 67,161,206
土壤镉测定 55
土壤个体 207
土壤耕性 205,211
土壤汞测定 55
土壤供钾能力 210
土壤固定 207,211
土壤固态水 212
土壤管理 208
土壤还原性物质测定 55
土壤含水量 208,212
土壤含水量测定 55
土壤含盐量 210
土壤耗竭 206
土壤黑炭 205

土壤呼吸 210
土壤呼吸强度 107
土壤呼吸仪 210
土壤化学 205
土壤化学测定 32
土壤化学固定 205
土壤化学诊断法 205
土壤环境 68,206
土壤环境保护 206
土壤环境监测 206
土壤环境容量 71,206
土壤环境因子 206
土壤环境质量 206
土壤环境质量标准 71,181,206
土壤环境质量评价 206
土壤环境质量指数 206
土壤缓冲能力 27
土壤缓冲性 205
土壤缓冲作用 205
土壤缓效钾测定 55
土壤活性酸度 204
土壤基质 208
土壤寄居菌 208
土壤钾素供应能力 174
土壤钾素形态 84
土壤钾素状况 218

土壤坚实度(土壤硬度) 207
土壤坚实度测定 55
土壤剪切 211
土壤碱度 204
土壤碱化作用 204
土壤健康 207
土壤健康质量 207
土壤交换性钙测定 56
土壤交换性钾测定 56
土壤交换性镁测定 56
土壤交换性酸测定 55
土壤交换性盐基总量测定 55
土壤胶体 205
土壤接种 207
土壤结持类型 83
土壤结构 211
土壤结构分类 211
土壤结构类型 211
土壤结构破坏 60
土壤浸出液 206
土壤净化 205,210,211
土壤绝对含水量 204
土壤绝对年龄 2
土壤绝对湿度 2

343

土壤考察 74
土壤颗粒 209
土壤颗粒大小分析 209
土壤颗粒分级 36,209
土壤颗粒组成 209
土壤可溶盐分析 14
土壤空气 204
土壤空气交换 204
土壤空气扩散 204
土壤空气容量 204
土壤空气状况 204
土壤空气组成 204
土壤孔隙 108,210
土壤孔隙度(土壤孔隙率) 210
土壤孔隙度测定 55
土壤库 209
土壤矿物 208
土壤矿物化学 208
土壤类别 211
土壤类型 211
土壤力学 208
土壤利用 211
土壤利用图 211
土壤粒径 202
土壤临界含水量(指植物最低需水量) 45
土壤淋洗 207
土壤磷素供应能力 222
土壤磷素化学固定 32
土壤磷素状况 218
土壤磷吸附 209
土壤流失 208,210
土壤流失量 208
土壤流失总量 93
土壤路线图 191
土壤毛细管作用 205
土壤酶 206
土壤酶活性 206
土壤镁形态 83
土壤密度 205
土壤密度测定 55
土壤描述 54,205
土壤灭菌剂 211
土壤命名 208
土壤母质 209
土壤年龄 204
土壤黏度 212
土壤黏附力(土壤黏着力) 204
土壤黏化 36
土壤黏粒 205
土壤脲酶 211
土壤镍测定 55
土壤排水 205
土壤培养 205
土壤硼形态 83
土壤膨胀 206,211
土壤疲乏(土壤耗竭) 76
土壤评价 204,206
土壤破坏 206
土壤剖面 210
土壤剖面样品 210
土壤普查 88
土壤气态水 212
土壤铅测定 55
土壤潜性酸度 210
土壤潜在生产力 174
土壤潜在蒸发 210
土壤强度 211
土壤切片制备 211
土壤侵染 207
土壤侵蚀(作用) 206
土壤侵蚀程度 206
土壤侵蚀分类 206
土壤侵蚀规律 206
土壤侵蚀过程 206
土壤侵蚀类型 206
土壤侵蚀量 13
土壤侵蚀模数 206
土壤侵蚀强度 206
土壤侵蚀速率 206
土壤侵蚀特征 31
土壤侵蚀退化 206
土壤侵蚀污染 206
土壤侵蚀因素 206
土壤侵蚀营力 206
土壤取样器 210

344

土壤圈 161
土壤全氮 211,232
土壤全氮测定 56
土壤全钾 233
土壤全钾测定 56
土壤全量元素分析 232
土壤全磷测定 56
土壤缺氮区 146
土壤群落 67
土壤扰动作用 161
土壤热通量 207
土壤热状况 207,211
土壤容积 212
土壤容重 205
土壤溶液 211
土壤溶液电导率 41
土壤溶液钾 174,211
土壤溶液硼 211
土壤溶液铜 211
土壤溶液中铵 13
土壤三相结构 228
土壤杀虫剂 208
土壤杀菌;土壤消毒 211
土壤杀菌剂 207
土壤砷测定 55
土壤深层施药 208
土壤深度 205
土壤生产力 205,210
土壤生产力分级 210

土壤生产力恢复 188
土壤生态改良 206
土壤生态环境 206
土壤生态环境效应 206
土壤生态位 68
土壤生态系统 206
土壤生态型 68
土壤生态学 68,206
土壤生物 209
土壤生物动力学 205
土壤生物多样性 205
土壤生物化学 205
土壤生物活性 5,24, 205
土壤生物量 205
土壤生物群落 161, 205
土壤生物数量 181
土壤生物污染 205
土壤生物学 205
土壤施肥 204
土壤湿度 212
土壤湿度;土壤水分; 土壤水 208
土壤石灰位 208
土壤石灰需要量测定 55
土壤适耕性 247
土壤适宜性 211
土壤释水性 243

土壤熟化 212
土壤水(分)常数 212
土壤水(分)蒸发 212
土壤水(分)滞后现象 212
土壤水分 212
土壤水分饱和率 195
土壤－水分－大气－植物模型(SWAP) 212
土壤水分亏缺 208
土壤水分类 212
土壤水分生态分区 67
土壤水分特征曲线 212
土壤水分消耗与补偿 42
土壤水分胁迫 208
土壤水分张力 208
土壤水分张力计 208
土壤水分状况 208, 212
土壤水解性氮测定 55
土壤水流 212
土壤水平地带性 207
土壤水平分布 207
土壤水汽扩散 211
土壤水渗漏 161
土壤水形态 212

土壤水盐运动 212
土壤水资源 212
土壤松散(性) 124
土壤酥度;土壤松散性 85
土壤速测 181
土壤速效钾测定 55
土壤塑性 209
土壤酸度 204
土壤酸度试剂 212
土壤酸度调整 6
土壤酸度中和(酸性土壤改良) 44
土壤酸化 204
土壤酸碱平衡 204
土壤碳 205
土壤碳水化合物 205
土壤碳通量 205
土壤碳循环 205
土壤调查 211
土壤调查规划 211
土壤调查手册 211
土壤通气;土壤通气性 204
土壤通气性 204
土壤通透性 209
土壤铜测定 55
土壤图 161,208
土壤团聚体 7,204
土壤团粒作用 207
土壤退化 205

土壤退化防治 176
土壤微量元素状况 218
土壤微生态 208
土壤微生物 208
土壤微生物(群) 68
土壤微生物(群落) 166
土壤微生物多样性 208
土壤微生物区系 208
土壤微生物群体 208
土壤微生物生物量 208
土壤微生物学 208
土壤温度 211
土壤温度测定 55
土壤温度计 211
土壤温度状况 211
土壤污染 205,209
土壤污染防治 176
土壤污染化学 209
土壤污染控制 209
土壤污染临界值 45,209
土壤污染物 209
土壤污染源 209
土壤污染指数 209
土壤污染综合指数 40
土壤无机氮 208

土壤无机碳 207
土壤无脊椎动物 208
土壤无效水 207
土壤物理学 209
土壤物质 208
土壤吸附 204
土壤硒测定 55
土壤习居菌 207
土壤细菌 204
土壤显微镜学 208
土壤线虫 208
土壤相对湿度 186
土壤详查 54
土壤消毒 205
土壤硝态氮测定 55
土壤小型动物区系 208
土壤锌测定 56
土壤新生体 144
土壤信息 207
土壤信息系统 207
土壤形成 207
土壤形成过程 212
土壤形成条件 212
土壤形成物 161
土壤形成因素(简称成土因素) 207,212
土壤性质 210
土壤性质变异幅度 126
土壤性质空间变异性

214
土壤修复 210
土壤修复技术 185
土壤学 211
土壤熏蒸 207
土壤熏蒸剂 207
土壤压实 205
土壤压实度指标 205
土壤压实作用 161
土壤盐分传感器 210
土壤盐碱化 210
土壤盐渍度 210
土壤盐渍化 193,210
土壤颜色 205
土壤阳离子交换量测定 55
土壤养分 208
土壤养分化学 209
土壤养分流失 124
土壤养分生物有效性 22
土壤养分收支 209
土壤养分形态 83,84
土壤养分有效性 208
土壤养分状况 209
土壤氧化还原电位测定 55
土壤氧化还原反应 209,210
土壤样品 161
土壤样品采集 211
土壤样品储存 210

土壤样品储存容器 210
土壤样品干燥 210
土壤样品筛 210
土壤样品制备 210
土壤遥感 210
土壤液态水 212
土壤因子 68
土壤硬度计 207
土壤硬化 95
土壤有机氮 209
土壤有机磷 155
土壤有机碳 209
土壤有机碳储量 209
土壤有机碳库 209
土壤有机污染 209
土壤有机污染物测定 55
土壤有机-无机复合体 155
土壤有机无机复合体 209
土壤有机质 209
土壤有机质测定 55
土壤有机质的存在形态 84
土壤有机质的平均残留期（MRT） 130
土壤有机质分解率 51
土壤有机质平衡 209

土壤有效肥力 68
土壤有效磷测定 55
土壤有效水 204
土壤有效水分 19
土壤有效养分 204
土壤与地形数字化数据库 204
土壤元素背景值 20
土壤原有机质 142
土壤增温 212
土壤障碍因子 205
土壤蔗糖酶 208,210
土壤真菌 207
土壤蒸汽提取 211
土壤整段标本 208
土壤-植物-大气连续体 212
土壤质地 211
土壤质地分类 36,211
土壤质地剖面 211
土壤质地图 211
土壤质量 210
土壤质量标准 210
土壤质量评价 210
土壤质量指标 210
土壤质量指数 210
土壤中污染物积累 171
土壤重金属污染 209
土壤贮水量 212

347

土壤资源 210
土壤资源信息系统 210
土壤自净化功能 211
土壤自净作用 211
土壤自然结构体 161
土壤自然排水 143
土壤自由水 207
土壤总含水量 99
土壤最佳含水量 154
土属 207
土水势 212
土体层 213
土系 211
土相 209
土腥气 65
土样 210
土杂肥 136
土种 208,211
土柱 205
土柱法 139
土著微生物 104
土族 206
土钻 205
土钻;土壤采样钻 204
团聚(作用) 37
团聚度 52
团聚体 7,37
团聚体稳定性 8,64
团聚体形成(作用) 7

团聚状态 8
团棵期 191
团粒结构 8
团粒结构(粒状结构) 91
团粒状结构 47
推荐施肥 77
推荐施肥;肥料推荐 78
退耕 46
退耕还林 43
退耕还林还草 43,91, 188,235
退耕上限坡度 238
退化 52,54,56,58, 109,188
(品种)退化 52
退化(作用) 52
退化类型 52
退化土壤 52
退化演替 52
褪绿斑点 34
褪绿叶尖 34
褪色 25
褪色(分级) 75
褪色(叶);走色 152
褪色;脱色;变色 59
褪色的;褪色(作用); 脱色 59
褪色烟叶 152
托布津 135,232

脱氨作用 51
脱氮 53
脱氮菌;反硝化菌 53
脱钾作用 53
脱碱(作用) 51
脱碱化土壤 213
脱碱土 213
脱碱作用 213
脱硫作用 54
脱落 63
脱落酸 1,2
脱落物 51
脱氯降解 51
脱氢酶 53
脱水(作用) 51,53
脱水胡敏酸(脱氢胡敏酸) 53
脱硝(作用) 53
脱盐(作用) 53
脱氧核糖核酸甲基化 61
脱氧核糖核酸交联 61
脱氧核糖核酸结合蛋白质 61
脱氧核糖核酸聚合酶 61
脱氧核糖核酸连接酶 61
脱氧核糖核酸体 61
脱氧核糖核酸-脱氧

核糖核酸杂交 61
脱氧核糖核酸限制 61
脱氧核糖核酸重复;脱氧核糖核酸复制 61
脱氧核糖核酸重组 61
拓扑矩阵 232

W

挖掘 73
洼地 53
蛙眼病斑 85
外包叶(雪茄烟) 247
外标法 74
外表面 74
外表面积 74
外部特征 31
外部特征;外观特性 74
外翻耕 30
外共生(现象) 67
外观性状 74
外观质量 15
外寄生(现象);外寄生性 67
外卷叶 188
外生菌根 67
外香型卷烟 74
外源污染物 247
外源性物质 247
丸化种子;丸衣种子;包衣种子 161
(种子)丸粒 161

完全处理 40
完全肥料 40
完全花 40
完全区组设计 39
完全燃烧 161
完全生物降解的 40
完全随机设计 40
完全叶 40
完全营养液 40
完全郁闭 86
完熟 86,132
完熟(叶):代号 H 203
完熟期 86
完熟叶 36
完熟种子 86
晚播 115
晚茬地 116
晚花的 115
晚期采收 115
晚秋耕 115
晚熟 115
晚熟品种 115,116
晚熟性 116

晚霜 115,216
晚烟 115
(马铃薯)晚疫病 115
网斑病 143
网格采样法 93
网格调查 93
网络地理信息系统 245
网室 212
网室研究 195
往复式振荡机 184
旺盛生长 240
旺长期 76
威百亩 201,239
微波 134
微波消解 134
微带青:代号 V 92
微根室法 136
微观结构;显微结构 134
微灌 134
微好氧菌 133
微环境(叶围,根围) 134

349

(烟叶)微红色 167
微碱性 10,202
微碱性;弱碱性 10
微孔玻璃坩埚 201
微孔过滤滤纸 135
微粒 80
微量 133
微量成分 136
微量滴定 134
微量和稀有元素 136
微量营养元素 134
微量元素 134,136
微量元素肥料 134,136,233
微量元素缺乏症 134,136
微量元素叶面肥料 83
微量注射器 133
微喷 134
微喷灌 136
微喷灌溉;微注射灌溉 134
微区试验 134
微弱香气 75
微弱香味 245
微生境 134
微生态环境 134
微生态系统 134
微生态学 134
微生物 133,134

微生物参数 133
微生物除草剂 133
微生物代谢 133
微生物的 133
微生物多样性 133
微生物肥料 133
微生物分解 134
微生物分解(作用) 133
微生物分离 111
微生物分析 134
微生物-根系联合体 134
微生物固持(作用) 133
微生物固定(作用) 133
微生物固定作用 134
微生物过程 133
微生物合成(作用) 133
微生物活性 133,134
微生物鉴定 134
微生物降解 134
微生物降解(作用) 133
微生物接种剂 133
微生物生物量氮 133
微生物生物量碳 133
微生物农药 133
微生物培养 134

微生物培养法 134
微生物区系 134
微生物群 134
微生物杀虫剂 133
<商>微生物杀虫剂;苏云金杆菌 20
微生物生态学 133
微生物生物量 133
微生物属性 134
微生物污染 133
微生物污染物 134
微生物显微镜 134
微生物因素 134
微生物种群 133
微酸的 3
微酸味 3
微酸性 202
微土壤学 134
微团聚体 133
微型动物区系 134
微型发酵罐;微型培养槽 136
维持土壤肥力 127
维生素 241
纬度 116
萎蔫 246
萎蔫;调萎;枯萎 246
萎蔫百分率 246
萎蔫病 246
萎蔫点含水量 242
萎蔫含水量 246

萎蔫含水率 246
萎蔫湿度 246
萎蔫现象 246
萎缩;收缩;皱缩 200
卫星摄影 194
未成熟 103
未成熟;欠熟 237
未成熟胚 103
未成熟种子;未成熟籽 103
未发酵的 237
未发酵烟叶 237
未腐解有机物质 237
未腐熟堆肥 236
未干(烟叶);未烘的 237
未耕的;未开垦的 236
未耕地 238
未耕耘土壤 191
未加工的烟草原料;叶用烟草 119
未开垦地 237
未垦地;成土母质 157
未萌发的种子 175
未熟 103
未熟;未成熟;成熟度不足 103
未熟果 47
未探明的汇 137

未污染土;未感染土 236
未扎把烟叶 237
未知态 237
未知态氮 237
喂丝 49
温(度)差 226
温床播种 84
温床栽培 84,100
温带气候 225
温度 225
温度补偿 225
温度补偿点 225
温度分布 226
温度计 227
温度控制 225
温度控制室 226
温度上限 238
温度梯度 226
温度调节 18
温度雨量图;温湿图 102
温和香气 135
温湿度 225
温湿度记录仪 225
温湿度昼夜变化 60
温湿度自动记录仪;温湿计 102
温室 41,90,92,100,241
温室管理 92

温室类型 92
温室农业 9
温室气体(GHG) 92
温室试验 92
温室通风 92
温室效应 90,92,100
温室研究 92
温室育苗 92
温室栽培 90,92
温室植物 219
吻合度 90
稳产栽培 216
稳定腐殖质 216
稳定入渗率 216
稳定碳 236
稳定性 216
稳定性团聚体 216
稳定性有机氮 216
稳结态腐殖质 216
稳态流法(测土壤导水率) 218
蜗牛 203
沃地;肥地 77
乌敏酸 236
污染"避难所" 171
污染标度 195
污染标准 171
污染当量 171
污染的土壤 84
污染毒性 171
污染防治技术 171

· 351 ·

污染控制作用 171
污染类型 171
污染浓度 171
污染土壤 42,171
污染土壤的化学修复 32
污染土壤的物理修复 165
污染土壤的植物修复 166
污染土壤改良 104
污染土壤修复技术 42
污染危害 42,106,171
污染物的降解 52
污染物的净化 36
污染物的容纳 3
污染物的生物降解(作用) 23
污染物的物理吸附与物理沉淀 165
污染物环境标准 71
污染物降解 171
污染物控制 171
污染物浓度 171
污染物迁移 234
污染物阈值 171
污染物指数 171
污染源 42,156,171,213
污染源控制 43

污染指示物 104
污染指数 104,171
污水 21,84,171,199,241
污水灌溉 199,241
(烟叶)污脏;不洁净 59
无柄叶 199
无病毒种子 240
无残留性农药 147
无定形有机物质 172
无覆盖播种 142
无根的 191
无公害农产品 92
无公害农业 171
无公害杀虫剂 24
无害的 95,106,147
无灰滤纸 17
无机氮 106
无机氮;矿质氮 135
无机氮肥 106
无机氮化合物 106
无机肥料 106
无机化学品污染 106
无机焦磷酸酶 106
无机磷 106
无机农业 106
无机酸 106,135
无机态硫 107
无机污染 106
无机污染物 106

无机物质 107
无机盐 135
无机盐类 107
无机养分 106
无机阴离子 106
无机营养 106,122
无机营养;矿质营养 135
无机元素 106
无机质;无机物质;矿物质 135
无结构土壤 147,220
无菌操作 17,219
无菌技术 17
无菌培养 17,19,20
无菌培养(单一种纯净培养) 219
无菌培养试验 219
无菌栽培植物 90
无氯肥料 34
无偏估计 236
无偏估计值 236
无沙土(烟叶);健全(烟叶);健康叶 213
无霜季;无霜期 85
无霜期 85
无霜日 51,85
无霜生长期 85
无水(乙)醚 2
无水酒精 2

· 352 ·

无土农业　212
无土栽培　212
无土栽培基质　212
无污染农药　146,171
无效降水量　237
无效孔隙度　104
无效水　104,236
无效性　236
无效养分　146,236
无性的;中性的;中和的　144
无序分布型式　182
无烟气(烟草)制品　203
无氧培养　13

无叶的　119
五氧化二磷　164
戊二醛　90
戊酸　238
物候图　163
物理肥力　165
物理风化　165
物理干旱　165
物理降解　165
物理吸附　165
物理吸附水　165
物理信号　165
物理性黏粒　165
物理性沙粒　165

物理修复　165
物体;身体;身份;(香气)丰满度　25
物质流　129
物质迁移　129
物质循环　49
物种多样性　214
物种丰富度　214
物种平衡　214
物种生态学　88
误差;错误　72
误差分析　72
误差控制　72
误差条形图　72

X

X射线衍射　248
X射线衍射图谱　248
西柏三烯二醇　31
吸肥力　2
吸附(作用)　6
吸附　6
吸附动力学　6
吸附剂　6
吸附力;吸附能力　6
吸附容量　6
吸附水　6
吸附速率　183

吸附态硼　6
吸附态铜　6
吸附态锌　6
吸附性;吸附能力　6
吸附引力　6
吸附作用;吸持作用　213
吸管　131,167
吸管法　167
吸光度　2
吸光度单位　2
吸光系数　2

吸滤瓶　221
吸气式播种机　170
吸入水　103
吸湿　138
吸湿的　101,102
吸湿曲线　138
吸湿性;吸湿度　101
吸收度　2
吸收度单位　2
吸收量　2
吸收速率　183
吸收系数　2

·353·

吸收效率 238
(根系)吸水和失水 244
吸水深度 242
吸水纸发芽试验法 89
吸烟机 203
吸烟室 203
吸烟习惯 203
吸胀种子 103
吸阻 62,178
吸阻测定仪 62
硒 198
硒污染 198
烯丙醇 1,11
烯醇 70
稀释率 58
稀释平板分离法 58
稀释平板计数 58
稀释效应 58
稀土元素;稀土 183
稀植 227
熄火 74
(烟支)熄火 74
习惯施肥法 49
洗盐 193
喜氮植物 146
喜高温的 179
喜光的 164
喜光性种子 97
喜旱植物 248

喜旱植物;适旱植物 248
喜酸性的;适酸性的 159
喜盐植物 193
喜阳植物;适阳植物 97
喜雨植物 152
喜雨植物;适雨植物 152
系谱育种;系统育种 161
系数 37
系统变异性 224
系统分析 224
系统取样;系统选择;系统抽样 224
系统误差 224
系统性入侵 224
细胞(呼吸)色素氧化酶 49
细胞壁 31
细胞分裂素;植物胞裂素;植物胞激素 166
细胞构造 49
细胞减生;抑生(现象) 102
细胞解剖学 49
细胞膜 50
细胞溶酶体 49

细胞融合 50
细胞色素 49
细胞调节作用 31
细胞质 50
细粉 80
细菌 20
细菌(性)的 20
细菌的溶解作用 20
细菌肥料 20
细菌肥料(菌剂) 20
细菌分泌物;细菌渗出物 20
细菌酒精发酵 20
细菌菌落计数器 20
细菌密度 20
细菌培养 20
细菌区系 20
细菌性坏死 20
细菌营养 20,195
细菌总数 232
细粒 80
(烟气)细腻 203
(烟味)细腻;(烟叶)平滑;(烟气)和顺;柔和 203
细平耙 203
细沙 201
细沙(细沙土) 80
细沙的 80
细沙土 80
细筛 80

354

细土　80
细叶烟　14
狭叶的　142
下部(叶):代号 X
　125
下部叶　125
下部叶组;下二棚;脚
　叶(白肋烟)　247
下等烟　172
下二棚　125
下二棚(分级);中下部
　叶　125
下降水　54
下脚料　152
夏旱地区　222
夏旱水分状况　247
夏收作物　222
夏烟　222
先进采样法　177
纤维;叶脉;筋脉　79
纤维根;须根　79
纤维素分解(作用)
　31
纤维素分解细菌　31
纤维素降解菌　31
纤维素类生物质　31
纤维素酶　31
纤维有机土壤物质
　79
酰胺　12
酰胺酶　12

酰胺态氮　12
酰胺态氮肥　12
鲜(叶)重　85
鲜明　200
鲜烟(叶)　85
鲜烟叶;青烟叶;原烟
　(初烤烟叶)　92
鲜叶挑选　85
鲜叶重　85
鲜重;复烤前重量　92
闲地;闲田　238
嫌光性种子　120
嫌气分解　13
嫌气固氮细菌　13
嫌氧培养皿　13
显微成像系统　134
显微分析法　134
显微结构诊断　57
显著差异　200
显著水平　200
显著性检验　200,226
显著性水平　120
现代化农业　138
现代集约持续农业
　138
现代农业　138
现蕾期　27,82
现蕾期打顶　232
限定变量　122
限制因子　122
限制因子律　116

线虫　143
线虫;蠕虫　68
线虫病害　143
线性回归　122
线性回归方程　122
线性回归分析　122
线性燃烧速度　122
线性统计分析　122
腺毛　89
腺毛密度　172
腺毛体;腺毛状体　89
相伴阳离子　3
相对饱和度　186
相对产量　186
相对防治效果　185
相对含水量　186
相对空气湿度　185
相对密度　185
相对密度;粗密度　46
相对频数　185
相对湿度　181,185
相对温度　186
相对误差　185
相对蒸腾　186
相关　44
相关分析　14,44
相关函数　44
相关图　44
相关系数　37,44
相关性状　44
相关与回归　44

相互平均法　184
相克生物　15
相克现象　14
相似性　201
香的;芳香的　84
香豆醛　44
香豆素　44
香豆素内酯　44
香豆酸　44
香精　72
香料　84,215
香料;有气味的　152
香料型卷烟　156
香料烟　235
香料烟(东方型烟;土耳其烟)　16
香料烟调制　156
香料烟香型　156
香料烟叶　156
香气;芳香　16
香气;气味;香水;嗅;闻　195
香气不纯　104
香气成分　152
香气成分;香气组分　16
香气持久性　152
香气纯正　179
香气分类　152
香气量　181,241
香气浓馥　96

香气品质;香气质量　180
香气前体物质　81
香气强度　152
香气清雅　85
香气弱　245
香气特征　152
香气主要成分　152
香气状态　139
香气足　81,189
香味浓馥　25
香味浓郁　245
香味特性　16
香味特征;香韵　81
香味物质;食用香料;调料　81
香味型烟草　81
香味质量　180
香味组分　81
香味组分;(食用)香精　81
香型　81,152
香烟　35
香叶基丙酮　89
香韵;香型　148
箱;盒;套;　30
箱线图　26
详细的勘测　54
响应时间　188
向顶性坏死　4
向气性;向氧性　7

向上卷曲　49
向下卷曲　49
向氧性　159
向雨性　152
象甲;象鼻虫　203
橡胶塞　191
橡皮管　192
消毒　59
消毒土壤　160
消毒作用　60
消毒作用;消毒　219
消化;蒸煮;浸提　58
消石灰　101
消石灰;熟石灰;氢氧化钙　121
硝胺　144,145
硝化的　145
硝化细菌　144,145
硝化抑制剂　106,145
硝化作用　145
硝基腐殖酸制品　177
硝石;硝酸钾　144,145
硝酸　19,145
硝酸铵　13
硝酸钙　28,144
硝酸还原酶　144,148
硝酸钾　144,174,193
硝酸磷肥　145,146
硝酸钠　204
硝酸盐　19,144

硝酸盐的还原(作用)
144
硝酸盐毒害 145
硝酸盐供应 144
硝酸盐淋失 144
硝酸盐污染 144
硝态氮;硝酸盐氮
144
硝态氮肥 144
小比例尺调查(小比例
尺测图) 203
小地老虎 9
小而僵直的叶片 203
小规模试验 167
小环境;微域环境;微
环境(叶围,根围)
134
小流域治理 203
小农经济 75
小气候 134
小气候差异 134
小区 169
小区划分;土壤小区划
分 169
小区径流 169
小区试验 169
小山;丘陵 99
小十字期 235
小型根箱 136
小型拖拉机 39
小雪茄 35,122

小循环 136
小样本 203
小叶 83,119
小雨 121
小长蝽象 151
(生理)小种;亚种;族
181
小组评吸 160
校正 184
校正保留时间 44
效益分析 22
效用递减规律 116
协方差 44
协方差分析 14
协方差函数 44
协方差设计 44
协合作用 224
协合作用;增效作用;
协生性 224
协调发展 44,95
协同进化 37
协同效应 224
协同运移 75
协同转运 44
胁迫 220
斜长石 167
谐调 246
(香气)谐调 95
缬氨酸 239
心土层 221
心土破碎机 221

芯叶;填充料烟叶 80
辛辣;辛辣味 4
辛辣味 179,215
辛辣味;刺激性;苦涩
味 25
辛辣味;尖锐度;清晰
度;尖刺 200
辛辣味;辛辣调料
200
辛香;辛辣的 215
锌 249
锌肥 249
<商>辛硫磷 165
锌素营养失调 249
锌污染 249
新采收的(烟叶) 85
新陈代谢 132
新成土 70
新翻地 144
新黄质;新叶黄素
143
新积土 143
新技术推广 74
新开垦的地 26
新垦荒地 144
新品种 144
新闻报道;印刷;镇压
(指土壤) 176
新鲜厩肥 85
新鲜有机物质 85
新鲜种子 85

· 357 ·

新烟草碱 14
新植二烯 143
信息 105
信息传递 106
信息激素;性引诱激素;味诱激素;性外激素 163
信息理论 106
信息农业 105
信息社会 106
信息时代 105
信息系统 106
信息载体 105
形成(生成) 83
形成层 29
形成过程 83
形态变化 139
形态诊断 139
形态诊断法 139
1:1型层状硅酸盐 116
1:1型层状结构的铝硅酸盐 116
1:1型矿物 236
2:1型层状硅酸盐 116
2:1型层状结构硅酸盐 117
2:1型矿物 236
性引诱法 199
雄蕊 217
休(闲)耕 75

休耕 146
休眠 181
休眠阶段 62
休眠期 62,161
休眠芽 188
休眠种子 116
休闲地 75,103
休闲地耕作 75
休闲栽培 75
修复技术 186
修复目标 186
修复植物的筛选 189
修剪机 178
修建;抑制 178
锈病 192
锈褐色 192
锈褐色细长条斑 192
溴百里酚蓝 27
溴代甲烷;溴甲烷 26
溴化氢 101
溴甲酚绿 26
溴甲烷 133
须根 190
须根;毛壮根 94
须根系 79
需肥规律 185
需肥量 78
需光量 121
需光种子 121
需水关键期 112
需水量 13,242,243

需水临界期 45
需氧堆肥 7
需氧呼吸 7
需氧微生物 7
需氧性细菌 6
序列图 199
蓄水减沙 243
蓄水坑 244
蓄水坑灌 244
悬浮 158
悬浮(悬浮液;悬浊液) 223
悬浮肥料 223
悬浮苗床 81
悬浮体系(悬浊体系) 223
悬挂式农具 140
悬着水 95,162,223
悬浊态 223
旋耕;旋涡式耕作 191
旋花科 44
旋花属 44
旋转犁;单向犁 153
旋转设计 191
旋转式切丝机 191
旋转式振荡机 191
选苗移植 176
选叶 118,213
选叶台 195
选择取样 198

选择性吸收 198
选种 197
削减;削弱 246
穴(坑)植 167
穴播 99,214
穴播;点播 27
穴播机 99
穴播机;穴植机 169
穴灌(点浇) 99
穴距 99
穴施 57,99,215
穴移植;穴移栽 167
雪茄;雪茄烟 35
雪茄内包皮 35
雪茄内包皮;胶粘剂;粘合剂;捆扎机 22
雪茄内包皮烟叶 22
雪茄内包叶 35
雪茄外包皮烟叶类型 35
雪茄外包烟叶 35
雪茄香型 35
雪茄芯烟 35
雪茄芯烟类型 35
雪茄芯烟烟叶 35
雪茄型卷烟 35
雪茄烟草 35
熏蒸(消毒)法 86
熏蒸步骤 86
熏蒸操作 86
熏蒸法 86
熏蒸技术 86
熏蒸剂 86
熏蒸条件 86
熏蒸药害 86
熏蒸应用装置 15
熏制烟 80
循环 35,49,124
(再)循环 184
循环经济 35
循环利用 35
循环系数 49

Y

压力势 176
压青;施绿肥 92
压实 39
压实的烟草 176
压实度 52
压实性 39
芽孢杆菌培养基 20
芽孢杆菌属 20
(人工)芽接 106
蚜虫 15,27,168
蚜虫寄生蜂 160
蚜茧蜂科 15
亚表层(心土层) 221
亚表土 221
亚地表径流 221
亚耕作层 220
亚甲蓝 133
亚类 220
亚麻酸 122
亚目 221
亚硝化(作用) 146
亚硝基去甲基烟碱(NNN) 146
亚硝酸 146
亚硝酸盐 145
亚硝酸盐还原酶 145
亚硝酸盐氧化酶 145
亚硝态氮 145
亚油酸;9,12-十八碳二烯酸 122
亚油酸;亚油酸酯(或盐) 122
亚种 221
烟把 95,230
(香料烟)烟把 160
烟把排列整齐 230
烟把破损率 183
烟包内的损耗率;包内废物率 241

烟饼　170
烟草(叶)调制　230
烟草;烟叶;烟丝　229
烟草白粉病　231
烟草斑点病　231
烟草斑点病毒　235
烟草胞囊线虫　230
烟草薄片　230
烟草病虫害;烟草农业害虫　231
烟草病毒病害　240
烟草病害　230
烟草病害分级及调查方法　91
烟草病害药效试验方法　133
烟草仓储害虫　163
烟草赤星病　230
烟草醇化　229
烟草丛枝病　231,247
烟草大田病害　59
烟草低头黑病　229
烟草番茄斑萎病毒病;烟草番茄斑萎病毒　235
烟草分级　230
烟草分类;烟草分级　230
烟草根腐线虫病;烟草根褐腐线虫病　231
烟草根黑腐病　229

烟草根黑腐病菌　227
烟草根结线虫病　231
烟草国外引种技术规程　192
烟草行业烟草栽培重点实验室　112
烟草褐斑病　27
烟草黑斑病　230
烟草黑胫病　229
烟草黑霉病　229
烟草化学　230
烟草坏死病毒病;烟草坏死病毒　229
烟草环斑病毒病;烟草环斑病毒　235
烟草灰斑病　230
烟草蓟马　231
烟草甲虫　115
烟草剑叶病　230
烟草角斑病　229
烟草金针虫　41
烟草金针虫;烟草叩头虫　231
烟草茎点病　230
烟草卷叶病　118,230
烟草卷叶病毒;烟草曲叶病毒病　229
烟草菌核病　231
烟草砍茎收割机　217
烟草科学家研究会议　235

烟草科学研究合作中心　44
烟草空茎病　230
烟草枯萎病　230
烟草类型　231
烟草立枯病;烟草猝倒病　230
烟草镰刀菌根腐病　230
烟草马铃薯X病毒病;马铃薯X(型)病毒　174
烟草马铃薯Y病毒病　231
烟草马铃薯Y病毒病;马铃薯Y(型)病毒　174
烟草脉斑驳病毒病(病)　235
烟草脉带花叶病毒病;烟草脉带花叶病毒　235
烟草脉曲病毒病;烟草脉曲病毒　231
烟草盲蝽象　230
烟草煤污病　213
烟草品种　231
烟草品种改良　138
烟草品种抗病性鉴定　102
烟草品种命名原则

192
烟草破烂叶斑病 231
烟草破烂叶斑病菌 17
烟草普通花叶病毒 144
烟草普通花叶病毒;烟草普通花叶病 229
烟草气候性病害 231
烟草潜叶蛾 230
烟草青枯病 229
烟草生产发展趋势 234
烟草生物碱 144,229
烟草蚀纹病 230
烟草蚀纹病毒(病) 226
烟草属 144
烟草属植物起源 88
烟草霜霉病 230
烟草税收 230
烟草碎叶病 230
烟草炭疽病 229
烟草特有亚硝胺 235
烟草提取物 230
烟草添加剂 229
烟草褪绿斑驳病毒病 225
烟草蛙眼病 85,230
烟草吸味;烟草吃味 231

烟草细菌病害 20
烟草细菌性黑腐病 229
烟草细菌性叶斑病 229
烟草线虫病害 143
烟草香气 229
烟草香味物;烟草香料 230
烟草斜纹夜蛾 230
烟草野火病 231
烟草叶片(不含梗) 231
烟草夜蛾;烟青虫 230
烟草夜蛾姬蜂 102
烟草引种 230
烟草营养失调症 127
烟草原种、良种生产技术规程 192
烟草真菌病害 86
烟草制品 231
烟草种植者;烟农 230
烟草蛀茎蛾 195
烟草专卖法 116
烟草专卖管理 6
烟草专卖零售许可证 120
烟草专卖生产企业许可证 120

烟草综合利用 233
烟草组分 230
烟草作物 230
烟囱;堆;垛 216
(烤房)烟道 82
烟斗;斗烟;管(子) 167
烟竿(杆) 231
烟竿;烟杆 219
烟梗 231
烟梗比例 135
烟梗干燥;干梗期;干筋期;干筋 218
烟褐花蓟马;烟褐蓟马 84
烟灰 229
烟架层数 116
烟碱;尼古丁 144
烟碱分解细菌 144
烟碱过滤效率 144
烟碱含量;烟碱量(烟气) 144
烟碱含量范围 183
烟碱合成与降解 144
烟碱调整 185
烟筋;烟梗 135
烟量充足;烟气饱满 86
烟霉 50
烟苗 231
烟苗生根期 191

·361·

烟农互助组 141
烟气捕集器 203
烟气成分 203
烟气刺喉 63
烟气分析 203
烟气分析仪;气体分析仪 87
烟气焦油 203
烟气焦油;烟气中的焦油 225
烟气浓度 53
烟气柔和 135
烟气特征 203
烟气组分 231
烟潜叶蛾 165
烟潜夜蛾 90
烟青虫(烟草夜蛾) 97
烟青虫;烟草夜蛾 156
烟青枯病 91
烟丝 48,49,200,231
烟丝填充能力 80
烟丝填充性(能力) 230
烟味丰富 86
烟酰胺 142
烟蚜 141,229
烟蚜茧蜂 15
烟叶 230
烟叶;上部烟叶(代号 B);上二棚(分级);叶片;上中部叶 117
烟叶变化 117
烟叶标样 217
烟叶表面光滑程度;柔和性;平滑度 203
烟叶部位;叶位 118,217
烟叶部位组 118
烟叶仓库 230,231
烟叶仓库熏蒸 231
烟叶测湿仪;烟叶含水率测定仪 230
烟叶陈储期 202
烟叶成分与烟气释送 117
烟叶成熟度 118
烟叶储存期 119
烟叶打滑 118
烟叶的背面 119
烟叶等级 91
烟叶等级控制因素 118
烟叶发酵 230
烟叶发育 118
烟叶分级要素 230
烟叶分拣机 166
烟叶分组要素 230
烟叶加工 118
烟叶净重 143
烟叶品质因素 231
烟叶品质因素等级 91
烟叶品质与可用性 118
烟叶破损度 52
烟叶燃烧性 39
烟叶色素;烟叶色泽;烟叶颜色 230
烟叶沙土检验 194
烟叶收购 231
烟叶收获;采收烟叶 230
烟叶衰老 118
烟叶损耗 118
烟叶特性与品质 117
烟叶调制 118
烟叶调制方法 49
烟叶脱落 118
烟叶香气和香味 117
烟叶颜色组 117
烟叶阴筋 138
烟叶长度;叶长 118
烟叶脂肪类 118
烟叶质量;烟叶品质 118
烟叶中拟除虫菊酯杀虫剂残留量的测定方法 1
烟叶装框 117
烟叶状态 118

烟叶组织;烟叶质地 119
烟用醋酸纤维丝束 35
烟折 159
烟株 231
烟株;烟茎;拐头 217
烟株顶部 15
烟株定苗 199
烟株挂起调制法 214
烟株下面第一片烟叶; 脚叶 176
烟株长势 168
烟蛀茎蛾 165
烟蛀茎蛾;烟草麦蛾 231
烟嘴 35,167
烟嘴;切丝机刀门 140
淹水土壤 220
延迟发芽 173
延迟授粉 53
延迟移栽 116
延期发芽 173
延期萌发 53
严重挂灰 199
严重缺乏 199
岩石圈 122
研钵 139
研磨(作用) 235
研磨;研磨作用 93

研磨机 18
研磨器 235
盐 193
盐成风化 193
盐度适应 193
盐分和水淹胁迫 193
盐分胁迫 193
盐害 193
盐化 193
盐基饱和度 21,52, 161
盐基交换 21
盐基交换量 21
盐基淋溶 21
盐基脱饱和(作用) 21
盐碱地 193
盐碱地烟草 193
盐碱土 193
盐碱土改良 104,184
盐浓度;含盐量 193
盐生植物 95
盐水;含盐的 193
盐酸 101
盐土 213
盐析 193
盐胁迫 193
盐性环境 193
盐性土 193
盐渍化 192,193
盐渍化程度 52

盐渍土 193
颜色 38
颜色;色度 38
颜色不合格 152
颜色代号 38
颜色强度;(叶片)色度 38
厌水物质 101
厌氧堆肥 13
厌氧发酵 13
厌氧降解 13
厌氧菌;嫌气微生物 13
厌氧生物处理 13
厌氧条件 13
厌氧微生物(厌氧菌) 13
厌氧微生物;厌氧细菌 13
验级;再鉴定 185
验级室 91
验收 3
验证 240
羊粪 200
阳离子 173
阳离子;碱离子 21
阳离子;正离子 30
阳离子固定 30
阳离子交换 30
阳离子交换(位)点 30

阳离子交换常数 30
阳离子交换量 30
阳离子交换色谱(法) 31
阳坡地 213
阳生植物 222
养地作物 46,207
养分;养料;营养品;营养的 149
养分补偿 150
养分的临界值范围 45
养分的有效性 19
养分动态 149
养分分层现象 150
养分富集 150
养分高效基因型 151
养分供应量 13
养分供应能力 150,151
养分固定 150
养分管理 150
养分归还 150
养分归还学说 227
养分过量 151,157
养分过剩 150
养分含量 42,149,181
养分耗尽 149
养分互作 107
养分活化 150
养分积累 3

养分交互作用 150
养分解吸 149
养分库 150
养分矿化 150
养分亏缺区 149
养分利用率 150
养分利用率;养分的回收 184
养分利用效率 150
养分临界值 45
养分淋失 117,150
养分流 150
养分浓度 41,149
养分浓度梯度 149
养分平衡 149,150
养分平衡施肥 20
养分平衡指数 151
养分迁移 150
养分缺乏;营养不足 149
养分缺乏的诊断措施 57
养分缺乏症状 149
养分三要素 228
养分生物有效性 22,149
养分失调;营养失调 151
养分释放 150
养分释放峰值 150
养分释放曲线 150

养分释放特性 150
养分收支 149,150
养分输出;养分流失 149
养分损耗区 149
养分损失 150
养分吸附 149
养分吸收 2,150,151
养分吸收拮抗现象 14
养分吸收量 2,13
养分吸收效率 150
养分限制因子 150
养分向根表面的迁移 150
养分胁迫 150
养分形态 83
养分循环 149
养分循环特征 149
养分循环再利用 238
养分阳离子 149
养分有效性 149
养分元素循环 149
养分再利用 150
养分再循环 150
养分周转率 150
养分转化 150
养分转移 150
养分资源综合管理 107
养分总量 233

氧 159
4-氧代-α-紫罗兰酮 112
氧化电位 158
氧化钙 27
氧化钙;生石灰 28
氧化硅 200
氧化还原(作用) 158,184
氧化还原滴定 159
氧化还原滴定法 158
氧化还原电位 158
氧化还原反应 158,159,184
氧化还原过程 158,184
氧化还原酶 159
氧化还原酶系 158
氧化还原平衡 184
氧化还原体系 159,184
氧化还原指示剂 158
氧化还原转化 233
氧化还原状况 184
氧化淋溶土 158
氧化磷酸化作用 159
氧化铝 12
氧化镁 126
氧化镁药害 126
氧化锰 128
氧化铁 110

氧化物 159
氧化亚氮(N_2O) 146
氧化紫罗兰酮 159
氧化作用 158,159
氧自由基 159
样本 193
样本;样品 214
样本变异系数 194
样本标准差 193
样本方差 194
样本均值 193
样本量;样本大小 193
样本数 193
样本相关系数 193
样本协方差 193
样本中位数 193
样点法 170
样点观察法 170
样品保存 193
样品袋;样本袋 193
样品干燥 64
样品柜 193
样品前处理;样品预处理 193
样品室 193,214
样品收集 193
样品烟 203
样品制备 193,214
样品贮藏 193
腰叶 28

遥测 186
遥测法 225
遥感 186,191
遥感成像(遥感图像) 186
遥感估产 248
遥感技术 186
遥感监测 186
遥感诊断 57
遥感诊断法 186
药草气 97
药草香 97
药害 162
药剂拌种 197
药效 162
药效持续法 64
药液罐 33
药用烟;药物型卷烟 131
野火病 246
叶(尖)端 117
叶把 117
叶斑(病) 119
叶背发青 246
叶变色 118
叶柄 119
叶柄 163
叶部斑点病 195
叶部病原菌 83
叶部施氮 83
叶部水分吸收 83

· 365 ·

叶部吸收 83
叶层厚度 53
叶耳 18
叶耳带绿 92
叶发黄 248
叶分泌物 118
叶腐烂病 118
叶梗比很低的窄薄烟叶 142
叶梗分离 199
叶痕 118
叶黄素 166
叶黄素;胡萝卜醇 247
叶基 83,117
叶基;叶肩 118
叶尖 119
叶尖发黄 119
叶尖坏死 119
叶尖角 229
叶尖烧伤 28
叶尖下卷 62
叶焦病;灼糊叶 117
叶节;断叶片 118
叶宽 119
叶龄 7,117
叶绿醇 166
叶绿素 34
叶绿素含量 34
叶绿素降解 34
叶绿素缺乏 34

叶绿素缺乏型 34
叶绿素荧光 34
叶绿酸 34
叶绿体 34
叶绿体分解 38
叶脉 119
叶脉病 231
叶脉角度 14
叶脉较细 227
叶脉褪绿 240
叶脉显露;很窄 220
叶脉状态 119
叶面;表面光滑程度(叶片) 119
叶面病害 83
叶面肥料 83
叶面积 117
叶面积比 117
叶面积系数 117
叶面积指数(LAI) 117
叶面颗粒 118
叶面喷洒 83
叶面喷洒;茎叶喷洒;叶面施肥;叶面施药;根外追肥 83
叶面平坦 119
叶面稍皱缩 119
叶面生物 119
叶面施肥 83
叶面施用(药) 83

叶面受光强度 120
叶面蒸腾作用 83
叶面脂质 119
叶面皱缩 119
叶面状态 119
叶片;片烟 118
叶片;片烟;去梗后叶片;薄片;薄板;层状体 114
叶片变厚 227
叶片表面化学 119
叶片尺寸 114,118
叶片打包机 114
叶片大小的测定 56
叶片分析 83,117
叶片分析诊断法 57
叶片分选筛 114
叶片复烤机 114
叶片厚度 119
叶片呼吸 118
叶片加工线 114
叶片加料机 114
叶片较小 203
叶片结构;叶片组织 119
叶片卷缩 49
叶片老化 118
叶片破裂 85
叶片起皱 178
叶片翘曲 48
叶片稍小的 237

叶片损伤 114
叶片特征 118
叶片营养 83
叶片营养分析 118
叶片预热机 114
叶片早衰 64
叶片长度 118
叶片长度;叶长 120
叶片长宽比 120
叶片诊断 83,118
叶鞘 118
叶肉 117,118,132
叶色 117
叶色变化 117
叶色诊断法 57,117
叶上着生的 165
叶伸展 119
叶身;叶片 117
叶数 118
叶态;叶状;(类)型;形态;形状;方式 83
叶围 165
叶萎蔫 247
叶位 26
叶温 119
叶形 118
叶形和大小 118
叶形指数 118
叶性器官;叶原体 165
叶序 165

叶序(列) 117
叶芽 88,117
叶腋 117
叶缘 118,128
叶缘烧伤 27
叶缘失绿 128
叶质 118
叶中带梗率 218
叶中含梗 218
叶重 119
叶状的 118
叶子边缘卷曲 49
叶子脱落 2
叶子中脉;主脉 135
叶组 118
页岩 199
夜蛾 146
液氨 122
液氨施肥机 122
液氨保藏法 122
液溶胶 125
液施 122
液态氮 122
液态发酵 82
液态水 122
液体氮肥 122
液体地膜 122
液体肥料 122
液体肥料;流体肥料 82
液体复合肥料 122

液体过滤器;流体过滤器 82
液体厩肥;粪水 122
液体培养基;流体介质 82
液体悬浮培养 82
液体熏蒸剂 122
液相 122
液相色谱(法) 117
液相色谱法 122
液相色谱仪 122
腋;叶腋 19
腋芽 19,196
腋芽生长处 200
腋芽生长势 174
腋芽抑制剂 221
一般线性模型 88
一次抽样检验 201
一次单株选择 152
一次性施肥 153,201
一次性收获 152
一代;世代;时代 88
一年二作 235
一年两熟(两茬复种) 62
一年三熟制(三茬复种) 234
一年生植物 14,227,248
一年生作物 14
一年四作(四茬复种)

367

180
一年一熟制 152
一年一熟制;单作 201
一水硫酸铜 48
一水硫酸锌 249
一氧化氮 145
一氧化碳 29
伊利石 103
仪器方法 107
仪器分析 107
移动极差 140
移入 103
移液管 233
移栽 234
移栽;移栽的;移植的;假植的 234
移栽后 173
移栽后施用 173
移栽后天数 51
移栽机;移栽者 234
移栽机具;移栽器 234
移栽密度 234
移栽苗 197
移栽期 234
移栽前施用 176
移栽前土壤熏蒸剂 176
移栽损伤 234
移栽穴 234

移栽穴施底肥 128
移植 234
移植;植入 103
遗传标记 89
遗传多样性 88
遗传抗性 106
遗传漂变;遗传漂移 89
遗传潜力 89
遗传生态学 88
遗传信息 89
遗传型 - 环境互作 89
遗传型 - 环境相关 89
遗传因子 89
乙醇 72,73
乙醇胺 72
乙醇发酵 10
乙二胺四乙酸 73
乙二醛 90
乙二酸 158
乙醚萃取物 73
乙醛 3,72,73
乙醛酸循环 90
乙酸 3
乙酸钾 173
乙酸盐 3
乙酸乙酯 3
乙缩醛;缩醛 3
乙烯 72

乙烯基塑料育苗盘 240
乙烯利 34,73
乙酰胺 3
2 - 乙酰吡咯 3
乙酰呋喃 3
2 - 乙酰呋喃 3
乙酰基去甲基烟碱 3
乙酰基新烟碱 1
乙酰降烟碱 3
乙酰乙酸 3
乙酰乙酸乙酯 3
乙酰乙酸酯 3
已改良土地 103
已改良土壤 184
已耕地 228
异常成熟 238
异常代谢 1
异常低温 238
异常高温 238
异常气象灾害 50
异常现蕾 238
异常早熟 1
异丁酸 110
异佛尔酮 111
异构酶 111
异花传粉栽培作物 46
异花传粉植物 46
异花受精的;异体受精的 11

368

异化性硝酸盐还原作用 60
异亮氨酸 111
异苗长素 102
异色(与叶脉颜色有差别) 43
异味 152
异味;杂气 152
异戊二烯 111
异养生物 97
异养硝化细菌 97
异养型微生物 19
异养硝化作用 97
抑菌剂 20
抑菌作用 20
抑芽 27
抑芽剂 54,221
抑真菌作用 86
抑制 106
抑制(菌)土壤 107
抑制发芽 106,216
抑制剂 53,106
抑制效应 106
抑制性土壤 14
抑制腋芽 43,106
抑制腋芽;腋芽控制 221
抑制作用 188
易被固定养分 184
易倒伏 223
易分解腐殖质 113

易分解有机氮 113
易烘烤性 65
易氧化碳 184
易氧化有机质 113
溢流管 81
因果分析图 31
因素;因子 75
因子方差 75
因子分析 75
因子克立格分析法 75
阴地植物;耐阴植物 199
阴离子(负离子) 14,143
阴离子交换 14
阴离子交换容量 14
阴坡地 148
阴燃 203
阴燃持火力 27
阴燃区 203
阴天 37
荫生烟叶(高档雪茄烟叶) 199
荫栽烟草 199
荫植烟草 199
泅筋 223
<俗>泅片 119
银 201
银量法 16
引诱剂 18

引诱剂;诱虫剂 18
引诱作用 18
引种烟草 152
引种者 109
吲哚 22,105
吲哚乙酸 102
吲哚乙酸(IAA) 105
吲哚乙酸氧化酶 102
隐翅虫 191
隐形症状 129
茚三酮反应 144
英国有机及合成肥料协会 1
英式纯烤烟 70
英式卷烟;烤烟型卷烟 70
荧光分光光度仪 82
荧光分析 82
荧光特性 82
营养 151
营养(溶)液 150
营养(液)栽培;水培 149
营养比例 151
营养钵 48,151
营养不良(营养不足) 53,64,102,104,151,160,220,221,237
营养不良;营养不足;营养失调 127
营养不足症 52

营养袋 83,151
营养根 151
营养恢复学说 150
营养价值 151
营养阶段 151
营养块;泥炭(腐殖质)块 160
营养临界期 45
营养临界期和最大效应期 45
营养培养基 150
营养配比 149,150
营养器官 151,156,240
营养缺乏;营养不良 106
营养缺乏;营养不足 151
营养缺乏病 149
营养缺素诊断(缺素诊断) 149
营养缺陷型 19
营养生长 240
营养生长期 240
营养生长期缩短 200
营养失调 151
营养失调症 127
营养衰竭 150
营养水平 151
营养土块 235

营养物质 150
营养物质污染 150
营养限制 150
营养相互关系 235
营养效率 149
营养胁迫 151
营养信号 235
营养性病害 151
营养性疾病 57
营养叶 235
营养液 48
营养元素 149
营养诊断 149,151
营养诊断施肥 149
营养周期 240
营养状况 151
营养最大效率期 130
应(施)用次数 229
应用生态学 15
硬块 28
硬磐 157
硬实 95
硬脂酸;十八酸 218
永冻层 47,73,162
永久电荷 162
永久电荷表面 162
永久凋萎;永久萎蔫 236
永久萎蔫 162
永久萎蔫点 162

永久萎蔫含水率 162
永久萎蔫系数 162
永久性腐殖质 162
蛹化;成蛹 151
用量 62
用水率;灌溉率 242
优化配置 153
优化施肥 153
优级纯试剂 94
优良品种 73
优美香气 80
优势度指数 62
优势腋芽 220
优雅香气 32
优质 98
优质肥料;优质厩肥 88
(美俗)优质烟叶里夹藏劣质烟叶 143
优质种子 180
油分 152
油分尚有 152
油分稍有 120
油锅;油浴 152
油酸 152
油渣饼;豆饼 152
游离氨 84
游离氨基酸 84
游离胺 84
游离碱 84

· 370 ·

游离酸 84
游离态颗粒有机物 84
游离氧化物 84
友好的土壤环境质量 71
有斑;花片 239
有斑点的 50
有偿使用 39
有翅成虫的 72
有毒物质 148,170,33
有毒元素 233
有毒元素污染 170
有害成分 151
有害物质 95,96
有害物质数量 42
有害因素 6
有害元素 53,95
有害杂草 106
有害作用 56
有机(粪)肥 154
有机残留物 155
有机残体混入 104
有机层 154
有机除草剂 154
有机氮 155
有机氮;有机态氮 155
有机氮肥 155
有机氮化合物 155

有机氮库 155
有机氮农药 156
有机氮农药污染 171
有机氮杀虫剂 156
有机氮转化 235
有机的;有机质的;有机物的;组织的 154
有机毒物 155
有机肥料 154
有机肥料养分分析 149
有机废弃物肥料 155
有机氟杀虫剂 156
有机覆盖物 154
有机汞化合物 156
有机汞农药 156
有机汞农药中毒 156
有机还原性物质 155
有机合成农药 224
有机化合物的合成 224
有机胶体 154
有机结合态 154
有机颗粒 155
有机矿质胶体复合体 156
有机矿质颗粒 155,156
有机矿质土 155

有机垃圾好氧生物反应器 7
有机磷 155
有机磷残留 156
有机磷除草剂 156
有机磷的矿化 136
有机磷毒物 156
有机磷化合物 155,156
有机磷农药 155,156
有机磷农药污染 171
有机磷农药中毒 155
有机磷杀虫剂 155,156
有机磷杀真菌剂 156
有机硫化合物 155
有机硫农药 156
有机卤代物 155
有机络合态锌 155
有机氯化物 154
有机氯农药残留量 156
有机氯农药污染 171
有机氯农药中毒 154
有机氯杀虫剂 155
有机农药 154,155
有机农业 154
有机溶剂 155
有机杀虫剂 154
有机砷残留 155

371

有机砷化合物 156
有机砷农药 155
有机砷农药中毒 155
有机食品 154
有机酸 154
有机酸新陈代谢 154
有机态硫 155
有机碳 154
有机碳密度 154
有机土 99
有机土层次 155
有机土壤物质 155
有机污染 155
有机污染物 155
有机污染物的植物修复 166
有机污染物降解 155
有机-无机复混肥料 155
有机物 155
有机物(质)降解 154
有机阳离子 154,155
有机叶面杀(真)菌剂 154
有机营养 155,156
有机营养肥料 155
有机营养学说 227
有机质 154
有机质积累 3
有机质年矿化率 136

有机质土(有机土) 155
有结构土壤 220
有胚植物 70
有身份的烟叶 25
有霜期 85
有限的资源 80
有限花序 52,54
有香味的;有滋味的;味浓的 81
有效成分 5,19,68
有效持水量 19
有效氮 19
有效肥力 68
有效肥料 19
有效辐射 68
有效根长 68
有效积温 13,68
有效钾 19
有效钾库 19
有效降水 19,68
有效降水量;有效降雨量 68
有效降雨量 68
有效孔径 68
有效孔隙 68
有效孔隙度 5,68
有效磷 19
有效硫 19
有效面积 68

有效浓度 68
有效期限 238
有效容积 5,68
有效容量 5,19
有效湿度;有效水分 19
有效数字 200
有效水 19
有效水容量 19
有效碳 19
有效田间持水量 19
有效透气孔隙度 5
有效土层 19
有效温度 68
有效系数 19
有效性 19
有效性指标 19
有效阳离子交换量 68
有效养分 19
有效扎根深度 68
有效植物养分 19
有氧呼吸 7
有益共生物 22
有益元素 22,97
有益作用 22
幼虫 72
幼虫死亡率;幼虫致死率 115
幼虫养育 184

幼苗 197
幼苗;发芽 216
幼苗活力 198
幼苗抗病性;苗期抗性 198
幼苗评价 197
幼苗生长速率(活力) 197
幼苗试验 197,198
幼苗质量 180
幼年土(沉积物上发育的);原生土壤 111
幼年土壤 248
幼期;幼龄期 111
诱捕法 18
诱虫灯 121
诱虫饵 107
诱导开花 82
诱导酶 105
诱导缺乏 105
诱导效应 105
诱导休眠因素 62
诱发感病性 105
淤积(作用) 201
淤积土 241
淤积物 157,241
淤泥土 136,140
余味 7
余味干净;(烟味)干净;无残留味 147

愉快香韵 169
雨季 182
雨量 182
雨量分布 182
雨量计;雨量器 170
雨蚀(作用) 182
雨水 182
雨水冲刷 182
雨水利用 182
雨水排水道 219
雨水资源 182
雨养农业 182
玉米;黄色(的) 127
育苗 182
育苗室 166
育苗移栽 198
育种;繁育 26
育种材料 26
育种过程;育种程序 26
育种目标 26
育种学;养育学 228
预测 175
预测;预报 83
预测产量 175
预测模型 175
预处理 176
预处理方法 176
预防为主原则 175
阈值 228

元素氮 69
元素的临界值浓度 45
元素地球化学 69
元素毒害 69
元素分析;最后分析 236
元素分析仪 69
元素迁移 69
元素协作用 69
元素组成 69
原产地 167
原核生物 177
原核微生物 177
原生矿物 156,176
原生盐渍化 157
原始肥力 106
原始记载;原始记录 156
原始结构 70
原始数据 156
原始数据;原始资料 183
原始土壤 106,177
原始土壤;未开垦的土壤 240
原位采样方法 104
原位化学修复 104
原位生物修复 104
原位土壤淋洗法 104

373

原烟 183
原烟(初烤烟叶) 92
原烟打叶 90,92
原烟加工厂 92
原叶;青烟;鲜(烟)叶 92
原叶打叶复烤 92
原种 157,219
原种;良种 69
原种圃 69
原状结构(未扰动结构) 237
原状土;未扰动土 237
原子发射光谱法 18
原子吸收分光光度法 18
原子吸收分光光度计 18
原子吸收分析 18
原子吸收光谱测定法 1
原子吸收光谱分析 18
原子吸收光谱仪 18
原子吸收火焰分光计 1
原子荧光光谱仪 18
圆和香韵 203
圆和香韵;柔和香气 203
圆盘犁 60
圆盘耙 60
圆盘耙耕作 59,169
圆盘式移栽机 60
圆锥形株型 41
源 213
源库假说 213
远红外 75
远缘(有性)杂交 60
远缘杂交 60
远缘杂种 60

远紫外 75
月桂醛 116
月桂酸;十二酸 116
月平均温度 139
月最低温度 139
月最高温度 139
越冬 158,247
越冬;越年份 158
越冬场所;越冬巢 97
越冬防治 158
越冬卵 157
越冬世代 158
越冬作物 158
越冬作物;冬作 247
云母 133
匀地试验 25
允许(剂)量 231
允许土壤流失量 208
运输效率 234
运算 153
运转周期 192

Z

杂草 245
杂草丛生 245
杂草为害 245
杂交;混种 137
杂交;杂种;混种 137

杂交不育的 46
杂交繁育;杂交育种 46
杂交后代植株 97
杂交品种 101

杂交亲和性 46
杂交育种 26
杂交组合 46
杂气;杂味 237
杂色(或各种)类型

136

杂色(烟叶);青块 239

杂色(烟叶分级);灰色 111

杂色:代号K 239

杂色;斑点;花斑;斑驳 139

杂色斑 239

杂色成分 136

杂色橙色 112

杂色品系 239

杂色叶 239

杂种 46

杂种优势 97,101

甾醇 219

栽培 182

栽培措施 47,48

栽培管理 48

栽培技术 48

栽培季节 46,48

栽培类型 47

栽培模式 48

栽培品种 47

栽培品种;栽培品系 47

栽培特性 93

栽培体系;栽培制度 48

栽培性状 47

栽培休闲 48

栽培植被型 47

栽培种 47

栽前施肥 78

栽植带 168

栽植穴 169

栽种时期 168

载玻片 90

载气 30

再度感染 196

再耕 157

再利用 188

再生生长 185

再生作用(更新作用) 185

再现性 186

再循环 184

再运转 188

再造烟叶 184

再造烟叶生产 231

暂行标准 226

暂时萎蔫 226

遭受虫害 107

早播 65

早果 64

早花 64,175

早期采收 64

早期低打顶 64

早期高打顶 64

早期抗病性 65

早期抗性;幼龄抗病性 111

早期生长发育 94

早熟 64,159

早熟的 175,183

早熟品种 64

早衰 65,175

早烟 64

早栽烟 65

造林 7

造碎 195

增浓香气 103

(烟叶)增湿房;回潮室 154

增效剂 224

扎根深的 234

栅格数据 183

栅栏组织 159

摘心 144

摘心;打尖 232

窄行栽培 142

窄垄(行)播种 159

窄垄(行)栽培 159

窄谱性农药(专效性农药) 142

窄叶型 142

窄叶烟 14

沾污烟叶组(分级) 216

展叶 118

· 375 ·

辗平;镇压 189
蘸根 190
张力计 226
张力计法 226
沼气 23,80
沼气发酵肥 132
沼气肥 23,80
沼泽;沼地 129
沼泽地 129,140
沼泽灰壤 129
沼泽土 25,129
照度 103
(光)照度计 125
照明质量 121
照射 110
照准仪 10
遮蔽 158
遮光;(雪茄烟支)分色;遮阴的 199
遮光发芽 50
遮光膜 199
遮阳网 222
遮阴烟叶 25
遮阴栽培 199
折晒 222
折线图 122
蔗糖 192,221
蔗糖酶 109,192,221
针刺 5
珍珠岩 162

真比重 184
真菌 86
真菌病害 86
真菌毒性 86
真菌区系;真菌群落 141
真菌修复 86
真菌学 141
真空泵 238
真空干燥器 238
真空干燥箱 238
真空回潮 238
真空滤器 238
真空蒸馏 238
真叶期 235
真值 235
诊断 56
诊断技术 57
诊断特性 57
诊断推荐(施肥)综合法 56
诊断指标 57
振荡器 199
振动耕作 240
振动式碎底土机;振动式翻底土机 240
镇压 39
镇压(指土壤) 176
镇压作业 176
蒸发(作用) 73

蒸发;发散 234
蒸发耗损 73
蒸发量 73,183
蒸发皿 73
蒸发曲线 73
蒸发速率 73
蒸发指数 73
蒸发作用 162
蒸馏 60
蒸馏(作用) 60
蒸馏罐 10
蒸馏水 60
蒸片 218
蒸汽处理 218
蒸汽灭菌器 218
蒸汽消毒;蒸汽灭菌 218
蒸汽蒸馏 218
蒸渗系统 125
蒸腾(作用) 234
蒸腾流 234
蒸腾率 183
蒸腾强度 234
蒸腾速率 234
蒸腾系数 234
蒸腾效率 234
蒸腾作用失调 60
整地 80,93,210
整地机具;耕作机具 228

整理;种子拌药(消毒) 62
整群抽样 37
整株砍收 246
整株晾制 217
整株调制法 217
正常 148
正常持水量 148
正常光照;正常照明 148
正常苗;健苗 148
正常土 148
正常土壤剖面 148
正常叶 148,157
正电荷(阳电荷) 173
正黄 52
正激发效应 173
正交函数 157
正交回归 157
正交拉丁方设计 157
正交设计 157
正交试验设计 157
(烟叶的)正面 189
正态分布 148
正态离差 148
正甜香 179
正长石 157
症状 224
支(卷烟) 166
支链淀粉 13

支脉;叶脉 240
只施基肥 21
芝麻饼 199
知识库 113
脂肪 76
脂肪合成 76
脂肪酸 76
脂肪酸成分 76
脂类代谢 122
脂类的合成 122
脂溶性农药 76
直播 58
直播栽培 47
直方图 38,99
直方图断点 99
直方图匹配 99
直根;主根 225
直接播种 58
直接计数法 58
直立型白肋烟 218
直链淀粉 13
直射阳光 58
植(物)醇;叶绿醇 166
植保素;植物抗毒素 166
植被(植物覆盖层) 239
植被层 239
植被覆盖 240

植被破坏 54,56
植被碳 239
植被资源 239
植入(法);移植 103
植酸 166
植酸盐(植素;肌醇六磷酸) 166
植物(水分)压力势 168
植物螯合 166
植物必需养分 72
植物必需养分的供给能力 222
植物必需营养元素 72
植物必需元素 72
植物病毒 168
植物病害 167
植物病害防治 43
植物病理学 168
植物残体分解 122
植物粗灰分测定 54
植物防御 167
植物非必需元素 147
植物分泌物修复法 168
植物根际修复法 166
植物化学生态学 167
植物化学诊断法 57
植物挥发 166

377

植物激素 168

植物激素;植物活菌素 166

植物监测 168

植物碱 167

植物碱;生物碱 11

植物降解 166

植物矿物质营养律 116

植物矿质营养 168

植物老化 8

植物内源激素 167

植物全氮测定 54

植物全钾测定 54

植物全磷测定 54

植物缺磷响应机制 131

植物缺素症 149

植物色素 164

植物生理学 166

植物生态学 166

植物生物修复 26

植物生长激素 168

植物生长激素;植物生长素 19

植物生长模型 137

植物生长素 19

植物生长调节剂 168

植物生长物质 168

植物生长抑制剂 168

植物水分和干物质测定 54

植物水分亏缺(PMS) 168

植物提取 166

植物体内养分运输 151

植物-土壤界面 169

植物稳定化 166

植物吸收 167

植物纤维 239

植物形态学 168

植物性农药 166

植物性杀虫剂 26

植物修复 166

植物养分 168

植物养分比例 168

植物样品的采集与制备 168

植物液汁分析 168

植物营养平衡 20

植物营养缺乏症 168

植物营养生理学 168

植物营养性状 151

植物营养遗传学 168

植物营养诊断 57

植物营养状况 168

植物有机营养 168

植物有效性 166

植物与害虫间相互影响 169

植物栽培;植物栽培学 167

植物栽植 168

植物蒸腾(植物蒸发) 73

植物资源 168

植物组织发育不正常 1

植物组织培养 168

植株;工厂 167

植株部位 168

植株残体 168

植株丛生 28

植株发育 167

植株分析 167

植株根系 168

植株化学诊断法 167

植株枯萎 57

植株密度 167

植株群体;种植密度 168

植株新陈代谢 168

植株有机体 168

植株与叶片总数(密度) 167

植株长势 240

纸色谱法 160

纸质烟嘴 30

指示剂 104

指示生物 104
指示植物 168
指数方程 74
指数分布 74
指数函数 74
指数模型 74
指数曲线 74
指数增长 74
指纹;指纹结构;指纹技术 80
指纹法;指纹分析 80
指纹图 80
指组织疏松、稍有身份、填充力好的烟叶 82
制丝车间;烟草加工车间 231
制丝设备;烟草加工设备 231
制造工艺 128
质地等级(指土壤) 226
质地分级 227
质地分类 226
质地名称 227
质量 129
质量保证体系 180
质量标准 180
质量参数 180
质量分析;质量性状 180
质量改进 180
质量管理;质量控制;品质监督 180
质量管理体系 180
质量含水量 129
质量监督 181
质量监督检验 181
质量检测 180,181
质量目标 180
质量认证制度 180
质量手册 180
质量体系 181
质量因素 180
质量综合判定 180
质流 129
质膜渗透性 132
质谱(图) 129
质谱法 129
质谱法;质谱分析法 129
质谱学;质谱分析(法) 129
质外体运输 15
致病细菌 160
致病性分化 239
致密 203
致密层 39
致密质地 37
致死低温 76
致死干旱度 76
致死高温 76
致死剂量 120
致死剂量;致死药量 117
致死浓度 117
致死湿度 76
致死温度 51,76,112,225
致癌药物 5
蛭石 240
滞后现象(平衡阻碍) 102
滞水土壤水分状况 216
置信区间 41
置信区域 41
置信水平(置信概率) 41
中(色度) 137
中部(叶):代号 C 49
中部叶 135
中部叶组,中部叶 28
中部叶组;腰叶 49
中成分(浓度)肥料 131
中档烟 135
中等规模调查 131
中等碱度(中等碱化)

137
中等碱性的 137
中等－偏厚 131
中等养分水平 131
中等质地 131
中低产田 131
中毒农药 132
中毒症状 170,224
中度感病;中感 137
中度抗病;中抗 137
中度缺乏 137
中度胁迫 135
中度有效性 137
中耕 108
中耕;耕作 47
中耕;栽培;培土 47
中耕除草 47
中耕机 48
中耕培土 47,189
中耕培土犁 200
中耕拖拉机 191
中耕休闲 48
中耕作物 108,191
中国卷烟销售公司 33
中国生态研究网络 33
《中国烟草》(杂志) 33
《中国烟草科学》(杂志) 33
中国烟草标准化研究中心 34,47
中国烟草博物馆 33
中国烟草机械(集团)公司 33
中国烟草进出口(集团)公司 33,37
中国烟草科技信息中心 33
中国烟草物资公司 33
中国烟草学报 5
中国烟草学会 33,47
中国烟草总公司 33,37
中国烟叶公司 33
中和滴定法 144
中和值 144
中和作用 144
中间价;平均价;中间汇率;平均汇率 135
中间试验 167
中间香型 179
中焦油 131
中焦油卷烟 135
中量元素 196
中量元素肥料 196
中量元素养分 196

中脉 135
中脉;主脉 176
中黏土 131
中壤土 131
中途掉温(烘烤) 125
中晚熟的 135
中纬度 134
中位数 131
中位数图 131
中温期 132
中下部(叶):代号CX 49
中下部黄烟组 135
中下部叶比例 135
中心距 31
中心趋势 31
中型动物区系 132
中性成分 144
中性化合物 144
中性溶液 144
中性土壤 144
中性香味卷烟;中度加香的卷烟 144
中性紫色土 144
中盐渍化的 137
中雨 137
中早熟的 135
中长生育期 132
中值 135
中子测水仪 144

中子法 144

中子仪 144

终点 70

终霜 115

种 214

种肥 197,214

种间关系 108

种间基因转移 108

种间杂交 108

种内关系 109

种内竞争 109

种皮 72,196

种皮处理 196

种皮厚度 37

种群 172

种群密度 172

种群数量 172

种植 46

种植管理 93

种植机 168

种植机;播种机 168

种植集约化 107

种植计划;种植设计 169

种植季节 46

种植季节;种植期 169

种植结构 46

种植距离 169

种植密度 53,168,217

种植面积 93

种植苗床 168

种植苗圃 169

种植模式 46

种植区 168

种植深度 169,199

种植顺序 46

种植业用地 168

种植指数 46

种植制度 46

种质 89

种质资源 89

种子;籽;结籽;播种 196

种子拌药(消毒);拌种;施肥;追肥 62

种子包衣 196,197

种子包衣机 196

种子包装 197

种子标准化 197

种子饼;饼肥 196

种子产量 197

种子成熟 197

种子处理 197

种子处理(拌种) 197

种子纯度 179

种子催芽 176

种子袋 196

种子等级 197

种子淀粉含量;种子淀粉率 112

种子发芽;种子萌发 197

种子发芽力 197

种子发芽势 174

种子发育 56

种子分级 197

种子覆土;埋种;盖种 196

种子改良 196

种子呼吸 197

种子混杂度 104

种子活力 197

种子检疫检验 181

种子鉴定;种子证书 196

种子搅动器;种子搅拌器 196

种子结构;种子质地 112

种子库 196

种子老化 8

种子萌芽率低 172

种子年龄 7

种子培养 196

种子品质差 197

种子破皮(处理) 197

种子破皮机 195

种子清选 36

种子生产者;制种者

381

197
种子寿命 197
种子习性;种子特性 196
种子消毒 197
种子携带的病菌 197
种子休眠 62
种子准备 176
仲胺 196
众数 137
重茬危害 106
重点施肥 198
重度胁迫 98
重粉沙土 97
重复 186
重复测定 186
重复次数 149,186
重复耕作 186
重复试验 186
重复小区 186
重复性 186
重铬酸钾 173
重铬酸钾氧化法 173
重铬酸盐 57
重过磷酸钙 41,62, 141,153,234
重金属 96
重金属复合污染 97, 137
重金属富集 96

重金属生物有效性 97
重金属污染土壤的植物修复 166
重金属污染物 97
重金属污染指标 97
重金属胁迫 97
重力沉降 92
重力计 91
重力侵蚀 91
重力势 91,245
重力水 92
重量的;重量分析的 91
重黏土 88,96,161
重沙质黏土 97
重沙质壤土 97
重施肥法 96
重盐渍化的 199,220
重组腐殖质 96
重组土壤有机质 96
周期率 116
周期系统抽样 162
周转期 235
轴;轴线;坐标轴;坐标系;中心线;晶轴;主根;茎轴 19
昼风;日风 51
昼夜温差 226
昼长;日长 51

皱缩花叶 45
(烟叶)皱缩 247
(烟叶)皱缩(分级);因水分和造碎引起的烟叶重量损失 200
皱褶;皱叶病 45
株高 168
株冠;冠层 29
株间施药 46
株距 60,168,214,247
株系隔离 122
猪粪 99,166,223
猪粪水 166
猪圈废水 166
潴育层 101,244
潴育水稻土 101
逐步的(分段的) 218
逐步回归 218
逐步回归方法 219
逐步判别分析 218
逐叶采摘法 176
主(直)根系 225
主产量;主收获 177
主成分;主要元素 176
主成分分析 177
主处理 127
主动吸收 5
主动吸烟 5
主动运输 5

主分级员 33
主根 19,21,177
主根;直根 126
主茎 127
主茎;主轴 126
主料烟 81
主流烟气 127
主脉 126,127,177
主脉;叶脉 143
主脉;中脉;烟筋;烟梗 135
主排水沟;总排水管 126
主区;整区;主(试验)小区 126
主香 126
主芽 126
主要病害 117
主要产烟国 127
主要成分 33,72,177
主要成分;大量成分 127
主要的污染源 126
主要害虫 127
主要基质 127
主要黏土矿物 126
主要品种 117
主要特征;主要性状 127
主要土层 127

主要土类 126
主要烟草类型 127
主要养分 72
主要营养元素 127
主要元素 177
主要致病因素 176
主要作物 117,126,177
主叶 177
主因子分析 177
主轴;总轴;主传动轴;主茎 126
贮仓;烟支库 126
贮藏病害;收获后病害 173
贮叶 119
注册农药 185
注册品种 185
柱状结构 38
柱状图表 21
专家经验 74
专家系统 74
专性好氧菌 151
专性互利 151
专性嗜热菌 151
专性吸附 214
专性吸附钾 214
专性厌氧菌 151
专性营养 103
专业烟农 86

专一浸染 201
专用肥料 214
砖红壤 116
砖红壤性土 116
转氨酶 12
转化酶 109
转基因农业 89
转基因烟草 233
转基因作物 233
转移酶 233
转置矩阵 234
装肥机 128
装炕 123
装炕方法 21
装盘机 234
装盘机;(育苗盘)基质装填机 234
装盘装置;装盘机 234
装烟门;装料门 123
壮苗 213,240
壮株 240
追肥 232
追肥;表施(指追肥) 232
追肥;根外追肥 232
追肥 5
锥形烧瓶 72
灼叶;叶焦;叶腐烂病 118

383

资源　187

资源安全;资源问题　177

资源保护　187

资源多样性　187

资源分布　60

资源高效利用　98

资源高效利用评价　68

资源管理　187

资源化;资源回收　187

资源环境　187

资源环境信息系统　187

资源节约型经济　187

资源节约型农业　187

资源开发　187

资源可用性　187

资源枯竭　187

资源利用　187

资源利用效率　187

资源平衡　20

资源评价　187

资源缺乏　187

资源生态系统　187

资源退化　187

资源胁迫　187

资源循环　187

资源遥感　187

资源要素　187

资源与环境综合承载力　224

资源最优化利用　22

子实体　86

子样本　221

子叶　44,197,198

子叶柄　44

紫褐色叶尖　179

紫花苜蓿　125,131,179

紫罗兰醇　109

紫罗兰酮　109

紫色土　179

紫色土壤　179

紫外(线)吸收　2

紫外分光光度法　236

紫外光　236

紫外－可见分光光度计　236,238

紫云英　17,34,135

自动滴定仪　18

自动供水控制　18

自动灌溉系统　18

自动化移栽　18

自动记录仪　198

自动喷灌机　18

自动吸烟机　18

自动烟气采样器　18

自动烟叶采收机;自动采叶装置　18

自动移液器　18

<商>自动移液器　167

自动移栽机　18

自回归模型　18

自净作用　18,198

自控温室　198

自流灌溉　92,110

自流灌溉农田　39

自满滴定管　249

自然醇化;自然陈化　142

自然肥力;天然肥力　142

自然干扰　142

自然含水量　142

自然降雨　143

自然净化特性　142

自然净化作用　142

自然平衡　142

自然侵蚀　142

(卷烟的)自由燃烧　84

自然沙土率　143

自然授粉　182

自然衰减　142

自然条件下发酵　142

自然条件下调制　142

自然通风　143

自然土壤 143
自然团聚体 142
自然循环 142
自然烟叶把［非平坦状烟叶把］ 47
自然烟叶扎把 47
自然杂交 142
自然状况容重 142
自然资源生态系统 143
自然资源信息系统 143
自生固氮菌 84
自生固氮细菌 84
自生固氮作用 84
自我调控 198
自相关 18
自相关分析 18
自养生物 18
自养硝化(作用) 18
自养型微生物 19
自由采样 84
自由度 52
自由扩散 84
自由燃烧速度 84
自由水 84
自由阴燃速度 143
自走犁；动力犁 139
渍水 244
综合改良 40

综合考察 107
综合利用 40
综合利用的 128
综合试验 40
综合土壤肥力指数 107
综合效益 40,224
综合养分管理 107
综合养分管理规划 40
棕(色)化；棕化；挂灰 27
棕钙土 27
棕红壤 27
棕化(作用) 27
棕化反应；棕色化反应 27
棕榈醛；十六醛 159
棕榈酸 97
棕榈酸；十六酸；软脂酸 159
棕榈油酸 159
棕壤 27
棕色坏死斑点 27
棕色森林土 27
总残留量 233
总产量 233
总氮 229,232
总氮的测定 56

总肥力 88
总耗水量 233
总灰分 232
总挥发碱 233
总挥发碱的测定 56
总挥发酸类 233
总碱度 232
总孔隙 233
总孔隙度 233
总粒相物；总微粒物 233
总磷 233
总面积 232
总生物碱 232
总湿度(总含水量) 232
总糖(量) 233
总体标准差 172
总体参数 172
总体方差 172
总体分布 172
总体平均值 172
(根系)总吸收面积 233
总悬浮颗粒物 233
总养分 233
总叶数 232
总有机碳 231,233
总有机碳分析仪 233

·385·

总植物碱的测定 54
足够养分 6
组 35
组氨酸 99
组胺 72
组成分析 40
组蛋白 99
组分;成分 42
组分分析 40
组距 36
组限 35
组织;(烟叶)组织 229
组织测定 229
(烟叶)组织紧密 228
组织培养 229
组织疏松、吸湿性好的烟叶 91
组织疏松烟草 153
组织速测诊断法 57
(根)钻法 18
最大持水量 130
最大持水能力 130
最大固定作用 130
最大降水量;最大沉淀量 130
最大净固定作用 143
最大毛管持水量 130
最大密度 130

最大湿度 130
最大田间持水量 130
最大吸附量 130
最大吸湿度 130
最大吸收率 130
最大吸收期 130
最大需要量 130
最大叶 115
最大允许值 130
最大蒸散量 130
最大值 130
最低绝对致死温度 2
最低气温;最低温度 136
最高残留限制 130
最高产量施肥量 130
最高经济产量 130
最高绝对致死温度 2
最高生长温度 130
最高收益用量 139
最高温度 130,160
最好的上二棚(烟叶) 22
最佳 153
最佳管理 22
最佳含水量 153
最佳留叶数 153
最佳生长温度 153
最佳施氮量 153

最佳施肥量 153
最佳水平(适宜量) 153
最佳压实度 153
最佳质地 154
最适 N/K 比值 153
最适产量 153
最适成熟度 153
最适耕作 154
最适供应 154
最适经济产量 153
最适量 153
最适密度 153
最适温度 153
最适叶面积 153
最适宜温度;最佳温度 154
最适移栽期 154
最适因子律 116
最适作物产量 153
最小持水量 119
最小二乘法 133
最小二乘法拟合 119
最小检出量 136
最小密度 136
最小田间持水量 136
最小显著差 119
最小养分律 116
最小因子定律 116

· 386 ·

最小值 136
最优回归 154
最优设计 153
作垄 21,189
作物/肥料价格比 46
作物保护 46
作物布局 45
作物残体;作物残茬 46
作物脆弱性 46
作物大幅度持续增产 115
作物单一性 46
作物肥料效应函数 86

作物分布 60
作物根系活力诊断法 57
作物灌水率 45
作物耗水量 242
作物回收率 46
作物监测 45
作物轮作顺序(茬口顺序) 46
作物轮作制 46
作物气候生态型 45
作物气候适应型 45
作物缺素症防治 176
作物生态生理学 45

作物生长率 45
作物水分胁迫指数 46
作物需水量 46,64
作物移走率 46
作物营养需求 45
作物栽培 45
作物长势 45
作物指标 45
作穴 31
坐标 44
坐标;参考文献 185
坐果习性 86

β分布 22

β-胡萝卜素 30

γ分布 87